PLANNING IN THE USA

This new book by Barr scope, purpose and practice of
American planning. T nd the issues that arise in the
implementation of pol the USA and the growth of
planning powers, land anning and zoning discussed.
This discussion also ex rols and historic preservation.
Growth management y an analysis of problems of
transportation, housin ies and their limitations are
treated at length. The he book, including the extraordinary difficulty of f rights issue, the inadequacy
of local controls over l nd the role of government in
modern America.

Planning in the USA is a lematic nature of the planning
process, the fallibility nning decisions. It will be of
interest not only to pla re of contemporary urban and
environmental problem

Barry Cullingworth i , University of Delaware; and
a Senior Research Fello ambridge.

PLANNING IN THE USA

Policies, Issues, and Processes

Barry Cullingworth

London and New York

First published in 1997
by Routledge
11 New Fetter Lane, London EC4P 4EE

Simultaneously published in the USA and Canada
by Routledge
29 West 35th Street, New York, NY 10001

©1997 Barry Cullingworth

Typeset in Garamond 3 by Keystroke, Jacaranda Lodge, Wolverhampton

Printed and bound in Great Britain by Butler & Tanner Ltd, Frome and London

British Library Cataloguing in Publication Data
A catalogue record for this book is available from the British Library

Library of Congress Cataloging in Publication Data
Cullingworth, J. B.
Planning in the USA : policies, issues, and processes / Barry Cullingworth
p.   cm
Includes bibliographical references and index.
1. City planning—United States. 2. City planning—Environmental
aspects—United States. 3. Land use, Urban—United States.
4. Urban policy—United States. I. Title.
HT167.C85 1996
307.1′2′0973—dc20   96–33158

ISBN 0–415–15011–6
0–415–15012–4 (pbk)

CONTENTS

PLATES

FIGURES

BOXES

ACKNOWLEDGEMENTS

The author would like to thank the following for permission to reproduce photographs in this publication:

Chicago Historical Society for plate 10; Cincinnati Historical Society for plates 31 and 32; Cornell University Library for plate 3; Fairfax County Public Library Photographic Archive for plates 7 and 8; Jay Hicks for plates 13 and 14; Alex MacLean/Landslides for plates 2, 11, 12, 17, 20, 21, 22, 23, 29, 30, 34, 35, 36, 43; Library of Congress for plate 4; I. N. Phelps Stokes Collection, Miriam and Ira D. Wallach Division of Arts, Prints and Photographs, The New York Public Library, Astor, Lennox and Tilden Foundations for plates 6 and 9; Seaver Center for Western History Research, Natural History Museum of Los Angeles County for plate 5; Camilo Vergara for plates 39, 40, 41, 42; Viewfinder for plates 1, 16, 18, 19, 24, 25, 26, 27, 28, 33, 37, 38; and Wright Runstad & Co. for plate 15.

The author and publishers would also like to thank Curtis Brown Ltd for permission to reproduce 'Song of the Open Road' by Ogden Nash.

PREFACE

This book has two objectives. First, it is intended to give an outline of policies relating to land use, urban planning and environmental protection. Second, it aims to provide an introduction to the policy-making process in these fields. The central concern is with the way in which policy issues are identified, defined, and approached.

The coverage of the book is wide: it includes the nature and limitations of planning and governance, land use regulation, the quality of the environment, growth management, transportation, housing, and community development, as well as an extensive discussion of current environmental issues. The focus is on the problems facing policy-makers in their search for solutions (though the term 'resolutions' is preferred). It also discusses the difficulties of separating facts and values. This is particularly clear with environmental issues where even 'experts' are protesting that their expertise is limited, and that questions of 'risk' have no scientific answers. It is now widely accepted among scientists that determining acceptable degrees of risk is a matter for public policy, not for science. Such professional modesty is increasingly apparent in the professions dealing with environmental hazards, but it is also growing in the professions concerned with urban and land use planning issues. It is against this background, together with an associated mistrust of government, that public involvement in the planning process takes on a new meaning.

This book has its origin in my earlier *The Political Culture of Planning: American Land Use Planning in Comparative Perspective* (Routledge 1993). That book was more narrowly concerned with land use planning and the ways in which its character in the USA differs from that in other countries. This comparative analysis has been replaced in the present volume by a more extensive treatment of the generic problems of planning and by a major section on environmental policy. However, much of the material on various aspect of US land use planning has been reproduced, updated where appropriate.

I wrote this book while at the Department of Land Economy at the University of Cambridge, England. This deprived me of the direct help from numerous American colleagues who contributed so much to the previous volume, but their influence is still very apparent. Some helped me overcome the problems of distance by sending me contemporary materials and even by commenting on draft chapters. Federal and state officials were particularly helpful in responding to my constant barrage of transatlantic letters. All this direct assistance has been supplemented by many who have unknowingly helped me through their writings: the wide coverage of this book implies a huge debt of gratitude.

I am grateful for comments on a draft of the housing chapter made by Douglas E. Peterson and Laurence S. Newman, Housing Development Specialists in the Department of Economic Development, Arlington County, Virginia. On the transport chapter, I had the benefit of advice from Robert L. Moore, Chief of the Transportation Planning Division of the Fairfax County Office of Transportation.

Jay Howenstine helped me in the seemingly endless task of keeping track of changes in housing policy. The concluding discussion in Chapter 9 owes much to Caroline Torma, who introduced me to the widening world of 'historic preservation'. Jim Hecimovich, Assistant Director of Research, American Planning Association, was helpful to me in the laborious task of tracking down photographs. I owe a particular debt to Jay Hicks, formerly International Real Estate Specialist, US Department of State, and now Director of Development, Manor Care Inc., Silver Spring, Maryland, who spent an inordinate time in searching for elusive material on the operation of planning at the local level, obtaining materials, and contributing photographs: he greatly eased my problems of transatlantic communication.

Special thanks go to Wendy Thurley, Librarian of the Department of Land Economy at the University of Cambridge who demonstrated how invaluable a good librarian is. Sally Jones drew the figures, thereby adding considerably to the clarity of the exposition at important points. The diagrams of the urban government models, reproduced in Chapter 3, are based on drawings originally published in Robin Hambleton and Huw Thomas, *Urban Policy Evaluation: Challenge and Change*, London: Paul Chapman Publishing, 1995. Thanks are also due to Curtis Brown Ltd for permission to reprint 'Song of the Open Road' by Ogden Nash.

Barry Cullingworth

ACRONYMS AND ABBREVIATIONS

Though abbreviations abound in the planning and (even more so) environmental fields, they have been largely avoided in this book: they are confusing and frequently difficult to remember. Nevertheless, they are useful on occasion and, in any case, readers who follow up the references will soon find themselves immersed in them. Hence the following list may be helpful.

ACHP	Advisory Council on Historic Preservation		EPA	Environmental Protection Agency
ACIR	Advisory Commission on Intergovernmental Relations		EZ	Enterprise zone
ACSC	Area of critical state concern (Florida)		FHA	Federal Housing Administration
APA	American Planning Association		FHWA	Federal Highway Administration
			FONSI	Finding of no significant impact (NEPA)
BART	Bay Area Rapid Transit		FTA	Federal Transit Administration
CAA	Clean Air Act		GAO	General Accounting Office
CAC	Citizen advisory committee (Oregon)		GPO	Government Printing Office
CDBG	Community development block grant			
CDC	Community development corporation		HCFCs	Hydrochlorofluorocarbons
CEQ	Council on Environmental Quality		HOV	High occupancy vehicles
CERCLA	Comprehensive Environmental Response Compensation and Liability Act 1980 (Superfund)		HUD	Department of Housing and Urban Development
CFCs	Chlorofluorocarbons		ISTEA	Intermodal Surface Transportation Efficiency Act
COG	Council of government			
DDT	Dichloro-diphenyl-trichloroethane		LOS	Level of service standards (Florida)
DOT	Department of Transportation		LULUs	Locally unwanted land uses
DRI	Developments of regional impact (Florida)		MPO	Metropolitan planning organization
EA	Environmental assessment		NEPA	National Environmental Policy Act
EC	Enterprise community		NHPA	National Historic Preservation Act
EIS	Environmental impact statement		NIMBY	'Not in my back yard'

NIMTO	'Not in my term of office'	SHPO	State Historic Preservation Officer
NPL	National priority list (hazardous waste sites)	SIP	State implementation plan (clean air)
		SMSA	Standard metropolitan statistical area
NRC	National Research Council	SOV	Single occupancy vehicle
NTHP	National Trust for Historic Preservation	SSZEA	Standard State Zoning Enabling Act
		TDC	Transfer of development credits
OTA	Office of Technology Assessment	TDM	Transport demand management
		TDR	Transfer of development rights
PCB	Polychlorinated biphenyls	TIP	Transportation improvement program
pH	measure of acidity	TMA	Transport management area
PRP	Potentially responsible party (hazardous waste)	UDAG	Urban development action grant
PUD	Planned unit development	UGB	Urban growth boundary (Oregon)
SAUS	*Statistical Abstract of the United States*	VOC	Volatile organic compound

INTRODUCTION

Standing on its own, the term 'planning' often relates to land use planning but, of course, it can apply to many other areas of public or private activity – economic development, health, housing, social security, defense, energy, and so on. This book focuses on land use, urban and environmental planning. However, as will be very apparent, problems have a habit of becoming 'interconnected': they refuse to be neatly parceled into separate areas which can be conveniently dealt with by individual government agencies, policies, programs or budgets. This poses difficulties for policy-makers and implementors; indeed, it often seems that it is this interrelationship of problems which is the central problem of government. The issue is neatly highlighted in Donna Shalala's lecture on urban policy:

> Every time Treasury changes the Tax Code, every time Congress alters a welfare program, every time the Defense Department awards a military contract, urban policy is being made.

Yet such measures are often not even recognized as constituting 'urban policy'.

Debates on urban policy seem endless. Attempts to tackle 'the problem of the cities' have proved to be of extraordinary complexity because there is not a single problem: there is a host of interconnected problems. These include housing and community development; employment, training, and economic development; poverty and social security; city finance and local, state, and federal taxation – a complete list would be very long. To deal with any one of these problems is not easy: taken together they are extremely difficult to

comprehend, let alone to attempt to solve. Moreover, political and ethical issues constantly arise: is the problem of teenage pregnancy one of education, neighborhood, culture, poverty, or moral turpitude? Should such households be provided with training, employment, accommodation, income support, or incarceration?

As with a policy package, this book attempts to overcome these difficulties and present a clear and reasonably succinct account of the web of issues which constitute planning and public policy. But, again like a policy package, it must fall short of comprehensive goals. In a world where everything is related to everything else, it is impossible to deal with all things at once, and therefore problems have to be broken down into manageable issues. Yet, once this is done, important relationships are separated, and the 'manageable issue' also necessitates a limited approach. The boundaries of this book mirror these wider matters. It focuses on land use planning: land use, the connection between land uses (transport), the interaction between land uses and the natural elements (environment), and the ways in which land uses develop (urbanization), and are controlled ('land use planning' in its narrow legal sense). It also discusses the ways in which controls over land uses interact with systems of government (public participation, support, and prejudice), and with specific aspects of urbanization (economic development, the provision of infrastructure, the demand and supply of housing). This is a formidable list, but it is by no means complete. For instance, it omits natural resources, public finance, architecture and civic design, information systems, demography and income distribution. The list could easily be lengthened, but the point does not need laboring any further.

This is an important factor to bear in mind constantly when each individual policy issue is under consideration. Contemporary problems are severe in their number, extent, and complexity. Yet resources are available on an unprecedented scale to deal with them; and public opinion presses for quick solutions. Why do so many solutions prove to be inadequate and, seemingly, give rise to additional problems? So great is the disillusionment that strong arguments are canvassed that it is better to leave at least some of the problems to 'free market forces', and that 'government is the problem, not the solution'.

In this book it is suggested that the answer does not lie in that direction (though a less regulated approach can sometimes help). There is no such thing as 'free' market forces: their freedom depends on a legal and political framework which protects them from contrary forces, and provides a framework of security for action. Without that framework, there would be anarchy, not a market. Even more tendentiously, it can be argued that there is no point in pursuing a 'minimum government' approach: the electorate will not allow it. Politicians frequently face the imperative to 'do something', even if they are unclear what it is they should do.

Many of these are matters which are not to be settled by dictat, appeals to reason, or argument. They involve differing values and beliefs. It would be helpful if the questions of value could be neatly separated from those of fact but, unfortunately, they cannot. (Try separating them out in relation to the previously quoted policy conundrum of what to do about the 'teenage pregnancy problem.')

The reader will already have detected some of the author's values. Though this book is intended as an academic text, it is inevitably influenced by personal views. Hopefully, these have been made explicit but, such is the interaction between facts and values, this cannot always be so. Perhaps it can be suggested that there is educational value in being asked to sort out what is fact and what is value? This is not a flippant comment: one of the themes of this book is that

policy debate inevitably involves both, and that much difficulty is created by their confusion.

Questions of value are important not only because they get mixed up with those of fact, but also because there are so many issues where the 'facts' are limited. When this is admitted, the decks are clear for a confrontation of values. These may be modified by greater understanding but, in the final analysis, they are matters of individual beliefs. Democratic systems of government allow these to have full play.

If this argument is accepted, it is easier to understand the essentially political nature of the policy process. It needs to be constantly borne in mind that the words 'policy' and 'politics' have the same root. Interestingly, as scientific knowledge increases, uncertainty grows: scientific matters are increasingly becoming concerned with 'chance' and 'risk'. The implications of this are particularly clear in the case of environmental policy; but (so it is argued in the last chapter) they go much further, and demand some rethinking about the formulation and implementation of public policy.

FORMAT OF THE BOOK

The book is divided into six main parts. Part I contains three chapters on planning and government. Chapter 1 expands on the discussion started in this Introduction: what is the character of planning? Is it based (as some theories contend) on rationality and, if so, what kind of rationality? Can it be comprehensive or must it be essentially incremental? What are the roles of the different levels of government and of the courts? How important are interest groups? What are underlying cultural attitudes? How do private and public planning processes differ?

Chapter 2 summarizes urbanization trends. The United States is a uniquely mobile society, and always has been. There is a restless search for improvement in the quality of life which is seen in many different fields. In environmental terms it has led to a high degree of urbanization and, later, sub-urbanization. The chapter describes this, and analyzes the role which government policies have played.

Plate 1 New York
Courtesy David Williams, Viewfinder Colour Photo Library

Chapter 3 discusses a number of issues relating to the government and planning of urban areas. The first colonial settlements quickly saw the need for a simple system of local government which had sufficient powers to deal with the problems of a relatively simple society. As economic development and urbanization gathered speed, governmental systems had to develop, though the forces of privatism were (and remain) strong. The reform movement had several dimensions, ranging from the battle with corruption to the promotion of the City Beautiful. Planning had a difficult birth, and its present uncertainties have historical roots.

Part II has three chapters on land use regulation. Chapter 4 gives a brief history of the emergence of zoning (as distinct from planning) as a method of controlling land use which served the dominant interests of the time. The institutional and legal framework, discussed in Chapter 5, was fashioned by the federal government (with the preparation and dissemination of model 'standard state enabling acts'), and by the support of the courts. Zoning has developed a profusion of techniques which go far beyond the simple districting of the early ordinances. The multiplicity of techniques is illustrated in Chapter 6, while Chapter 7 focuses on the schemes which have been devised for apportioning the infrastructure costs of development. Interwoven with issues of land use control are wider ones such as those of economic growth and the exclusion of unwanted social groups.

Land use planning has been predominantly concerned with quantity rather than quality, but in Part III two concerns about quality are discussed.

Chapter 8 deals with the long history of the legitimation of aesthetic controls, while Chapter 9 chronicles the evolution of policies relating to historic preservation. Aesthetic controls started with billboards, at the end of the nineteenth century. These gave rise to a long battle between planners and the powerful billboard lobby which still rages. The control of good design may seem to some to be an oxymoron: certainly good design is an elusive quality which is difficult to define, let alone to control. Historic preservation has been a legal battleground, in which New York's Grand Central Terminal looms large. The scales have been further tipped in favor of this special aspect of land use control with the realization that historic areas have economic benefits for tourism.

Though zoning might be regarded essentially as a form of 'growth management', its use for this purpose is relatively recent. Part IV deals with this in two chapters: Chapter 10 discusses its use by local governments; Chapter 11 deals with the role assumed by a number of states. Growth control policies can take zoning nearer to the concept of comprehensive planning, though the attainment of comprehensiveness has seldom been sought and even less frequently attained.

Part V is devoted to three 'development issues': transportation (Chapter 12), housing (13), and community and economic development (14). Transportation is the circulatory system of the economy and, in the metropolitan areas at least, it is suffering from sclerosis. To continue the medical metaphor, a wide range of remedies are being prescribed, including controls, incentives, charges, and comprehensive planning. Unfortunately, neither the diagnosis nor the treatment is proving simple, and it seems to be clear that there are no easy answers. Housing is problematic in a different way: the objectives have always been clear. They were set out in the Housing Act of 1949: to provide 'a decent home and a suitable living environment for every American family'. However, reaching that goal presents formidable problems of finance, politics, and planning. The outlook is bleak for many. Community and economic development is put forward as one solution, and it presents some promise if adequate resources are made available.

Environmental issues have grown in political significance (and in reality) over the last quarter of a century. Part VI is devoted to a selection of issues. Chapter 15 deals with the growth of environmental concerns, the National Environmental Protection Act, and three major areas of environmental policy: clean air, clean water, and waste. Chapter 16 is a more discursive discussion of the limits of environmental policy. This takes up and develops points made earlier about expertise, scientific uncertainty, and risk. Policy involves the calculation of the degree of risk which is unacceptable. Since this is essentially a value issue, not a scientific one, it has to be settled in a democratic manner.

This theme is developed in the final chapter which attempts to bring together significant points made in the body of the book. Since it is not the author's intention to enter the lists of those who attempt to provide programs of reform, the discussion is focused on the framing of questions and the manner in which they might profitably be debated.

PART 1

PLANNING AND GOVERNMENT

Planning is a purposive process in which goals are set, and policies elaborated to implement them. Such is the theory; but the theory is affected in many ways. There are problems with the concept of planning, with the forces of urbanization which it seeks to regulate, and with the very nature of government. This first part of the book introduces this complex of issues which make planning such a difficult, frustrating, and fascinating subject.

Chapter 1 discusses the nature of planning. How far is it a rational activity akin to mathematics? The question appears absurd at first sight: surely a planner would not proceed in an irrational way? The discussion shows that the issues are much more complicated than this suggests. Rational goals are elusive, and apparently sensible methodologies are strangely difficult to implement. Moreover, though it may seem intelligent to attempt to plan comprehensively, experience shows that this is an ideal which faces formidable obstacles. Rational planning is a theoretical idea. Actual planning is practical exercise of political choice that involves beliefs and values. It is a laborious process in which many public and private agencies are concerned. These comprise a wide range of conflicting interests. Planning is a means by which attempts are made to resolve these conflicts. This is particularly difficult in land use planning because of the cultural, legal, and constitutional aspects of property rights.

Since this book is focused on urban planning issues, it is appropriate to examine the nature of urbanization and its trends in the US. Chapter 2 does this in a summary way. The original settlers formed a rural society, and towns were very small.

Urbanization accelerated in the second half of the nineteenth century, and by 1920 a half of the population was urban. The proportion increased to almost 70 per cent in 1960, and 75 per cent in 1990. Urbanization was followed by suburbanization, largely as a result of developments in transport, highways, and innovations in the finance of home ownership. In this the federal government played a major role. Suburbanization eventually led to inner-city decline as people and (later) shops and jobs moved out. By 1990, virtually a half of the American population lived in the suburbs. Over a sixth of the population moves every year. Though most of these moves are short-distance, huge regional movements have taken place. These have resulted in enormous growth in states such as California and Florida. The statistical population center of the US is now in Indiana, south-west of Indianapolis; in 1850, it was in West Virginia. Migration is still taking place on a large scale, but its character is now complex and volatile.

Having discussed the nature of planning and the history of urbanization, Chapter 3 extends the historical account and examines the development of urban government and planning. This is of intrinsic interest, but it also helps in an understanding of the historical and intellectual heritage of urban planning. Previous generations have battled with questions of how to make planning effective in a democratic society: their experience is of relevance to the contemporary scene. Issues highlighted in this account include the pervasiveness of privatism, the reform movement, the City Beautiful movement, and the growth of planning.

1

THE NATURE OF PLANNING

If we can land a man on the moon, why can't we solve the problems of the ghetto?

Nelson 1977: 13

We must first exorcize the ghost of rationality, which haunts the house of public policy.

Wildavsky 1987: 25

THE CHARACTER OF PLANNING

Planning is a process of formulating goals and agreeing the manner in which these are to be met. It is a process by which agreement is reached on the ways in which problems are to be debated and resolved.

Definitions of planning abound: there is a large literature devoted to exploring the meaning of the term. One generally common element in these definitions is that planning is forward-looking; it seeks to determine future action. At the simplest, one may plan to go to the library tomorrow. Such a 'plan' involves a choice between alternatives – not to go to the library tomorrow, to go elsewhere, or to stay at home. The plan may also be based on explicit assumptions: for example, the decision to visit the library may be dependent upon finishing the books that have already been taken out of the library, or on the weather being fine. On the other hand, if the books will be overdue and subject to a fine if they are not returned tomorrow, the plan may override other considerations.

This is a trite example, but it does contain important elements which are present in more sophisticated forms of planning: forethought, choice between alternatives, consideration of constraints, and the possibility of alternative courses of action dependent upon differing conditions. Of course, when a plan involves other people (which it usually does), it must incorporate an acceptable way of reconciling differences among the participants: this is a major feature of any type of planning; and the more numerous and diverse the participants, the greater the difficulties of planning. At the extreme, fundamental clashes in outlook, beliefs, or objectives may make planning impossible. At the worst, there is a resort to violence – of which there are, tragically, all too many examples around the world.

This underlines another important aspect of planning: there has to be a sufficiently sound basis of agreement for planning to be possible. In democratic societies, large numbers of diverse interests not only have to be considered but also have to be involved in the planning process. Much of 'planning' then becomes a process of reaching agreement on objectives. But, as will be shown repeatedly through this book, objectives and ways of reaching the objectives are not easily separable. Many may agree that a comprehensive system of health care is needed, but it may prove impossible to fashion an acceptable method of providing this – a point dramatically illustrated by the collapse of President Clinton's health proposals. In the debate, differences appear in both means (such as methods of financing) and ends (such as the extent to which health care is to be 'comprehensive' in terms both of the people to be included and the health conditions to be covered). Very speedily, ends and means become confused.

By contrast, where there is full agreement on a planning objective (putting a person on the moon, for example), the debate focuses on methods. When the nation agreed that it was a national priority to devote the necessary resources for this incredible feat, there was no problem with defining the problem, or of obtaining the necessary funding. Though the objective was incredibly difficult, it was simplified by the agreement that supported it. There was, for example, no argument on whether it might be better to build a transoceanic tunnel, or to build a ten-mile high city, or to attempt any other seemingly impossible enterprise. More realistically, there was little serious debate as to whether the resources could not be put to better use in, for example, eliminating poverty.

The planning of wars contains many lessons on the problems of planning, but consider how much more difficult is the planning of a 'war on poverty'. In his first State of the Union message in 1964, President Lyndon Johnson declared such a war: 'This Administration today, here and now, declares unconditional war on poverty in America.' Sad though it is, no such 'war' was possible: the single aim of destroying the enemy of poverty inevitably broke up into a myriad problems concerning a proliferation of programs aimed at constituent parts of 'poverty', their financing, their administration, their adequacy, and their effects. Poverty proved to be a hydra-headed monster, encompassing an incredible number of issues – from food stamps to regional development, from model cities to education, from health to income maintenance: 370 new programs of assistance to states and local governments were introduced between 1962 and 1970.

The issues are discussed further in Chapter 14. Here it is important to note the types of questions that the plans raised. Is poverty an economic issue? (in which case the answer would lie somewhere in the policy area of maintaining incomes); or is it a matter of personal inadequacy? (in which case, what scope is there for remedying this?); or is it a market failure? (which might be dealt with by market incentives). Then there were questions as to how far the state could – and should – interfere with the market. Do public programs destroy individual initiative? Where should resources be concentrated: on individuals, communities, urban redevelopment, or job creation? And so the questions multiplied. Distinctions between ends and means proved baffling. Poverty became seen as an umbrella term for a wide range of problems of modern post-industrial society. The problems were difficult to define, let alone to resolve.

Similar problems arise with any form of planning where there is not a single, clear, and accepted objective – which is usually the case. To take a further example, which is a major focus of this book: land use planning. How does one plan urban development? The first question is why it should be planned at all. The answers to this are legion, and they are usually expressed in very general terms: to achieve 'orderly' development, to minimize the loss of agricultural land, to reduce transport needs to the minimum, to encourage economic development, or to facilitate private investment in property.

Typically, there are several objectives. There may be a general desire to provide a spacious environment while, at the same time, maximizing the use of public transport and safeguarding rural land adjacent to the built-up area. These objectives involve conflicts: spacious environments consume a greater amount of land (often previously in agricultural use); and low densities present problems for public transit which operates most effectively in high density corridors.

Planning necessarily involves restraint on the actions of individual land owners and residents. Such restraints arouse opposition and claims that property rights are being infringed. This is an important limitation on the scope of planning, more so in the US than in those countries where a high degree of public command over land development is politically acceptable. In fact, there is a considerable amount of control over the use of land in the US and, though it is a source of continual controversy, the principle of some degree of regulation is generally accepted. (To use the customary example, no one wants a glue factory to be located in their neighborhood.) The issue then is not whether there should or should not be planning, but how much of it there should be and how it should operate. There is a huge literature on this, replete with a wide variety of concepts. Immediately apparent is the divorce between planning theory and planning practice.

PLANNING THEORY AND RATIONALITY

Central to much planning theory is the concept of rationality. Since rationality requires all relevant matters to be taken into account, the use of the concept readily leads to a comprehensive conception of planning. This stems from the simple (and valid) idea that, in the real world, everything is related to everything else, and the planning of one sector cannot properly proceed without coordinated planning of others. Rationality also requires the determination of objectives (and therefore – though not always explicitly – of values), the definition of the problems to be solved, the formulation of alternative solutions to these problems, the evaluation of these alternatives, and the choice of the optimum policy. Much of the difficulty of this approach (quite apart from matters of implementation, discussed below) is that it can mask the essentially political nature of the process. The overriding consideration easily becomes procedural efficiency, which places planning on a 'scientific' level 'above politics'.

The persuasiveness of the concept of comprehensive rational planning is persistent: it can be seen in a succession of federal governmental initiatives based on concepts of coordination and systematic targeting of resources as with the Model Cities Program for instance (see Box 1.1). Of course, the planning process produces 'objectives', definitions of problems, and proposed 'solutions'. But all this is done in the context of the politics of the place and the time, and against the background of public opinion and the acceptability or otherwise of governmental action. Some important issues may be regarded not as problems capable of solution but as powerful economic trends which cannot be reversed. Others may be of a nature for which possible solutions are conceivable but untried, too costly, too administratively difficult, too uncertain, or even dangerous to the long-term future of the area. And, as will be apparent from later chapters, these acutely difficult problems (of urbanization, congestion, inner-city decay, for example) have continually proved beyond the powers of governments to solve, at least in the short run; and the long run is unpredictable. Major differences of opinion exist among experts, politicians, and electors on these matters. As a result there are severe constraints operating on the planning process, and there is little resembling a logical, calm set of procedures informed by intellectual debate.

Perhaps the most misleading concept of planning is the theatrical analogy. Planning is likened to the production of a play – involving the coordination of many roles: the actors, the backstage hands, the management, the marketing, and so on. But the analogy is false in that with the theater there is a common objective to which all are committed: the production of a play – 'the show must go on'. Of course, there may be cross-currents and disagreements, but these are subservient to the overall objective (if not, there are resignations, replacements, or – at the very worst – abandonment of the play). The same holds true with the more complex productions of opera and movies. But it is not complexity that is the crucial factor. It is difficult to think of anything more complex than putting a person on the moon, but agreement on that single objective, coupled with the provision of the necessary resources, enabled problems of great complexity to be resolved.

THE PRACTICE OF PLANNING

Practitioners are quick to point out that planning involves deciding between opposing interests and

BOX 1.1   RATIONAL COORDINATION – THE MODEL CITIES ATTEMPT

In the 'war on poverty', a model cities program was proposed which would involve (in the words of the legislation) 'concentration and coordination of federal, state and local public and private efforts'. As originally conceived, the program was to be of a 'demonstration' character restricted to the poverty areas of a very small number of central cities. This implied more than the coordination of programs: it explicitly envisaged the redistribution of resources.

The 'demonstration' concept did not survive the political process: the need to obtain political support for the legislation led to an increased number of cities – eventually to 150. More than this, Congress was not willing to see funds diverted from other programs into model cities; and so the congressional commitment became largely to another categorical program rather than to a coordinative mechanism for reforming other grants-in-aid.

While there was support for the idea of better coordination among urban programs, there was also a fear about 'a concentration of power within any single executive agency'. Bureaucracies can be viewed not only as machines for the efficient implementation of policy but also as a dangerous concentration of power in 'monolithic organizations where a few powerful men at the top concentrate control over a vast range of activities. The implication of this view is that efforts to strengthen coordination among agencies are potentially dangerous, because they may upset the existing balance of power that permits considerable freedom of action for many interest groups.'

In case this is thought to be an extreme view, the reader is cautioned that 'after Watergate and after Vietnam, the dangers of excessive White House power are all too obvious'. It is concluded that 'if the designers of future urban policies take away any single lesson from model cities, it should be to avoid grand schemes for massive, concerted federal action'.

Sources: Frieden and Kaplan 1977; Downs 1967

objectives: personal gain versus sectional advantage or public benefit, short-term profit versus long-term gain, efficiency versus cheapness, to name but a few. It entails mediation among different groups and compromise among the conflicting desires of individual interests. Above all, it necessitates the balancing of a range of individual and community concerns, costs, and rights. It is essentially a political as distinct from a technical or legal process, though it embraces important elements of both.

To illustrate, a small town may wish to preserve its character and to 'protect' it from further development, but individual local businesses may look to the advantages of increased trade, and land owners may see the profits to be made from additional development. Complicating matters further, the school board may welcome growth because additional students will provide the rationale for improving the range of educational provision, while utilities may oppose growth since they are already stretched to the limit and incapable of expanding services at reasonable cost. Additionally, there could be local issues relating to road capacity, wetlands, scenic beauty, waste disposal, or parking. A full list would be very long and, though not all issues will arise with every development proposal, it is not at all uncommon for there to be many conflicts.

SECTORAL AND COMPREHENSIVE PLANNING

The concept of comprehensive planning in theory may be contrasted with the narrowly focused planning which takes place in practice. Each administrative agency takes its decisions within its particular sphere of interest, understanding, resources, and competence. How can it be otherwise? The task of any agency is to undertake the task for which it is established, not to take on the complicating and possibly conflicting responsibilities of others (which, in any case would be resistant to a takeover). Thus, a conservation agency will take decisions of a very different character from an economic development agency: they have separate and potentially conflicting goals. The idea that there is some level of planning (presumably to be administered by superhuman planners?) which can rise above the narrow sectionalism of individual agencies is not only inconceivable in terms of implementation: it also assumes that an overriding objective can be identified and articulated. This is typically expressed in terms of the public interest; yet there are very many 'publics'. They have conflicting interests which are represented by, or reflected in, different agencies of government.

The example of metropolitan government is a case in point. Such a tier of government could rise above local interests and take decisions for the benefit of the region as a whole. There are some good theoretical arguments in favor of this, but the practical point is that people live locally, not regionally. They view any regional policy in terms of its local impact. Thus, it is in the interests of the metropolitan area that adequate provision of affordable housing is made. It is part of the metropolitan planning process to ensure that sufficient sites are identified for this purpose. However, even if the electorate agree in principle to the provision of affordable housing, they may well – and typically do – object to it being located in their particular neighborhood. The same issues arise with a wide range of provisions which, while necessary for the metropolitan area as a whole, are unpopular locally. It is for this reason that proposals for metropolitan government are usually defeated, and why there are so few metropolitan governments in the US. The various constituencies in a metropolitan area have such a wide range of conflicting interests that any agreement is difficult to achieve.

It might also be noted at this point that a number of factors are leading to an increasing privatization of space in urban areas: this trend, seen with the growth of so-called 'common interest communities' and,

more generally, of 'private governments', is explicitly aimed at safeguarding and promoting very local interests (Barton and Silverman 1994; McKenzie 1994). Indeed, the result – if not the aim – may be isolation from the troublesome problems of the adjacent areas.

INTEREST GROUPS

The resolution of differences of interest (and the establishment of acceptable means of dealing with them) is a central problem of planning. Obvious differences of interest arise along lines of economic position, age, race, occupation, and a host of others. Their variety is illustrated by the enormous number of organized interest groups. These groups are the organizations of a democratic society. Individuals separately can exert little influence (unless they are of great wealth or extraordinary charisma). Influence is gained by combining with other like-minded people to pursue shared goals. Indeed, interest groups form the core of political activity in general, and of planning activity in particular.

Interest groups are of extraordinary variety. At the national level are organizations of the professional bodies (planners, architects, engineers, and experts in water systems, environmental pollution, soil science, and so on). Developers, builders, suppliers of building materials have their own organizations, as do various forms of local government. Then there are 'lobbies' based in Washington who keep a close eye on the legislative process and on any proposal that may affect their interests. Think-tanks may be allied to these, or they may have varying degrees of independence, though generally they will exhibit some political or philosophical leaning. Similar bodies exist at the state level and, to a much lesser extent, at the local level. Not surprisingly, local organizations tend to be preoccupied with issues affecting their locality, and take a broader geographical interest only when co-operation with bodies in other areas promises more effective action.

This short selection illustrates the range of interest groups which operate in the field of land use planning.

Plate 2 Boston
Courtesy Alex MacLean/Landslides

An important part of the planning process consists of negotiating with such organizations. This is not only because wide participation is a hallmark of a democratic society: it is also efficient to bring into the process those who are to be affected by it and those whose cooperation is needed if it is to be effective. Indeed, without the support of the more powerful groups, planning will not work. (As will be shown later, this is a situation that is frequently encountered.) It does, however, present a difficulty for those who are weakly organized and who may include the poorest members of society; without effective organization and power, the interests of these groups are ignored or overridden.

A selection of interests represented organizationally at the national level is given in Box 1.2. The list is only illustrative, and the divisions are not as clear-cut as the headings may suggest since many organizations fulfil a range of functions, from research to lobbying, and from professional concerns to political action. Nevertheless, it gives some idea of the huge range of organizations whose influence is brought to bear, with varying degrees of effectiveness, on the planning process at the national level.

LOCAL INTEREST GROUPS

At the local level matters are much more complicated. Each area will have its complement of permanent organizations, many of them related to national bodies. In addition, there will be myriad local groups of varying degrees of permanence established to influence particular neighborhood issues, or to campaign for the provision of some local amenity, or to organize opposition to some unwanted development.

By definition, interest groups share a concern about a specific issue or range of issues, but they may have very differing ideas in other directions. Thus, a local group organized to protect the character of a suburb may contain a diversity of attitudes concerning taxes, car-parking, aesthetics, street lighting, schools, and recreation. Moreover, since any one individual has many interests, some of these may present internal personal conflicts. An individual member of our suburban protection organization is not only a suburbanite, but can be also a motorist, a parent, a shopper, a business person, a golfer, a commuter, a gardener, a member of a political party, and a property-tax and income-tax payer.

Individuals are quite capable of living with internal conflicts – often rationalizing them in such a way that they do not appear to be in conflict. Thus a suburban resident may be a strong supporter of a local growth management policy which severely restricts new development in the area; but, if he is prevented from developing a piece of land which he owns, he may argue that this is a case where the policy is being imposed far too stringently and insensitively. Human beings have a remarkable ability to reconcile conflicting views when their own interests are at stake.

Since land use planning is essentially a local matter, local interest groups naturally exert a considerable influence on planning policy. Above all, given the high proportion of home owners, their interests tend to predominate, particularly in suburban areas. Indeed, pressures to introduce or to extend planning controls often come from home owners concerned about the effect of change on their property values. The widespread use of exclusionary zoning policies is the result. Homogeneous communities, with no low-income housing, are seen as the guarantee of stable property values. The morality of this is, of course, questionable, but morality is often not a primary consideration in a process as political as land use planning. This political system responds to the views of the powerful constituents. Home owners and development interests are frequently the most powerful. Where developers go against the interests of existing home owners (for example, in proposing to build low-income housing), the home owners are likely to win. On the other hand, where development and environmental issues clash, the development interests tend to win, particularly when jobs are at stake. (This is still generally true, even though there is now a heightened concern for the environment: real though this is, it typically takes second place to the need for local employment.)

BOX 1.2   SOME NATIONAL INTEREST GROUPS IN LAND USE PLANNING

Governmental
Advisory Council on Intergovernmental Relations; Council of State Governments; National Association of Counties; National Association of Regional Councils; National League of Cities; National Association of Towns and Townships; National Governors Association; United States Conference of Mayors

Professional
American Planning Association; American Society for Public Administration; American Institute of Architects; American Society of Landscape Architects; American Society of Civil Engineers; Council of American Building Officials; National Association of Housing and Redevelopment Officials; International City Management Association; American Public Health Association; International Association of Chiefs of Police; International Association of Fire Chiefs; Institution of Transportation Engineers; American Park and Recreation Society; American Bar Association

Developmental
National Association of Home Builders; Urban Land Institute; National Association of Real Estate Brokers; National Association of Realtors; Manufactured Housing Institute; National Council for Urban Economic Development; Partners for Livable Places; American Road and Transportation Builders Association; Waterfront Center

Public Works
American Public Works Association; American Water Resources Association; National Solid Wastes Management Association; Airport Association Council International; American Association of Port Authorities; Association of Metropolitan Sewerage Agencies;

Research
Urban Institute; Brookings Institution; Regional Science Association International; Transport Research Board; Environmental Design Research Association; National Center for Preservation Law; Environmental Law Institute; Housing and Development Law Institute; Resources for the Future; Rand Corporation

Disadvantaged Groups
National Association for the Advancement of Colored People; National Urban League; National Urban Coalition; National Association for State Community Service Programs; Center for Community Change; National American Indian Housing Council; Community Transportation Association of America

---

ADVOCACY PLANNING

Interest groups which have influence are typically well organised, well funded, and highly articulate. So who speaks for the unorganized, inadequately funded, and powerless? Who represents minorities, the poor, the disadvantaged? The answer, of course, varies from place to place. Nationally, there are bodies such as the National Association for the Advancement of Colored People and the National Urban League (see Box 1.2). Many national organizations have local representation. But there are also countless organizations which have been formed locally to represent local interests. Sometimes these are short-life bodies set up to deal with some specific local issue (opposition to a road scheme, for instance). Others keep a watching brief on some aspect of local conditions or politics. Some are active in promoting the development or improvement of local opportunities or living conditions through neighborhood organizations such as community development corporations. (A few examples are given in Chapter 14.)

The role of the planning profession in promoting the interests of the disadvantaged is a problematic one on which there is continual debate. The professional code of ethics, however, seems quite clear:

> A planner shall seek to expand choice and opportunity for all persons, recognizing a special responsibility to plan for the needs of disadvantaged groups and persons, and shall urge the alteration of policies, institutions, and decisions which militate against such objectives.

This clause was added to the ethical code of the American Institute of Planners (now the American Planning Association) as a result of strong and eloquent pressure from Paul Davidoff. He was a planner and lawyer who founded the Suburban Action Institute as a pressure group to increase access to suburban jobs. He set a pattern which many followed of combining research and action. Research establishes the needs of the disadvantaged and the ways in which they are being denied, while action is promoted by making these issues known, by organizing communities, and by legal and political initiatives.

Davidoff stressed the need openly to invite debate on the political and social values that underlie plans. The planner should not, so he argued, be a mere technician: the planner must act as an advocate. Davidoff certainly aroused the conscience of the planning profession, though there have been more words than actions. Planners have employers who pay their salaries and define what their jobs are. Employers are often unsympathetic to 'alternatives' for the disadvantaged. Developers are concerned with profitability, politicians with majority votes. Though they may have social concerns, these are unlikely to be predominant and, in any case, may well be interpreted somewhat differently from the people affected by their actions. Planners may be squeezed between conflicting groups. The few planners who have nevertheless been able to follow an active 'equity agenda' such as Norman Krumholz (quoted below, p. 17) are the exception.

Experience has also shown that open advocacy can be self-defeating. It may raise overwhelming opposition that could possibly have been avoided by more subtle methods. A classic case is integration, where explicit initiatives have so often failed. By contrast, schemes such as the Gautreaux experiment in Chicago and the current *Moving to Opportunity* program (noted in Chapter 17), which have been termed 'stealth programs', have had positive results. The essential feature of these programs is that they operate on a small scale (with only a small number of black families moving into a white area): they thus avoid raising resistance.

Such an argument may raise moral doubts, but the intense opposition to open programs has created increasing pessimism about the viability of traditional approaches. Another alternative has been proposed (and implemented) by Chester Hartman. This involves the stimulation and utilization of research which is focused on issues that enable activists to be effective in carrying out an advocacy agenda. His own organization, the Poverty and Race Research Action Council, has had some success with such an approach (Hartman 1994: 159).

PLANNING VS. IMPLEMENTATION

It might reasonably be assumed that plans are prepared in order to be implemented. Though this may often be the intention (even if a vain one), it is not always the case. Some plans are basically pieces of propaganda intended to boost the attraction of an area (usually for development), or to promote one type of future over another (such as one with greater leisure provision, or one which is more ecologically sustainable), or to press for some particular character of development (as with the classical architecture of the City Beautiful movement). Plans can serve many functions: inspiration may be more important than implementability. Or the preparation of a plan may be the short-term answer to a particular political pressure 'to do something' about the future of an area: the plan is thus seen as the first step; but by the time the plan is completed, the enthusiasm for change may have dissipated, or the plan may be seen as impracticable or too costly. Again, a plan may be required as part of a submission to a higher level of government for grant-aid: once the grant is obtained, the plan has served its purpose. Not infrequently, plan-makers indulge in a dream: they know that they cannot forecast what influences will exert themselves in the future, but they feel compelled to try; they thus 'resolve the conflict by making plans and storing them away where they will be forgotten' (Banfield 1959).

The rational model of planning embraces the simplistic view that there is a logical progression through successive stages of 'planning', culminating in implementation. The beguiling logic does not translate into reality. On the contrary, it is highly misleading – and dangerous – to separate policy and implementation matters. In fact, sometimes policy emanates from ideas about implementation rather than the other way round. Thus, a policy of 'slum clearance' or 'redevelopment' focuses on the clearly indicated types of action. The implementation becomes the policy, and the underlying purpose is left in doubt. If the objective is to improve the living conditions of those living in slum areas, there might be better ways of doing this, such as rehabilitation or area improvement through local citizen action. With such an approach, demolition might be merely an incidental element in the local program. With clearance as a policy, however, there is a danger that different objectives might be served (such as central city commercial interests). Demolition might even be detrimental if it reduces the quantity of affordable housing. With hindsight, it is not surprising that this is what happened with the urban renewal policy. Clearance and redevelopment (later expanded to urban renewal – 'the renewal of cities') became a policy of economic development of central city areas.

Even when policy is not framed with a particular form of implementation in mind, it is frequently modified by implementation; it may even be transformed. The crucial difficulty is the void between the purpose and hopes of a paper plan and the realities on the ground.

In Banfield's words, policy 'is an outcome which no one has planned as a "solution" to a "problem": it is a resultant rather than a solution' (Banfield 1959). The interrelationship between policy-making and implementation arises from the necessity of collaboration among a multiplicity of public, private, and voluntary agencies. But since each of the agencies has its own agenda, and even its own way of looking at problems, such collaboration necessitates compromise; and compromise means that the policy is changed (Pressman and Wildavsky 1984).

INCREMENTALISM

The obvious failure of comprehensive planning to attain the goals that are theoretically possible has led to a number of alternative theories. Many of these revolve around the problem of making planning effective in a world where market and political forces predominate. Meyerson (1956) proposed a 'middle-range bridge' (between *ad hoc* decision-making on minor issues and long-range comprehensive planning) which would monitor and interpret market and

---

BOX 1.3   PROBLEMS OF IMPLEMENTATION

1 Many policies represent compromises between conflicting values.
2 Many policies involve compromises with key interests within the implementation structure.
3 Many policies involve compromises with key interests upon whom implementation will have an impact.
4 Many policies are framed without attention being given to the way in which underlying forces (particularly economic ones) will undermine them.

Source: Barrett and Hill 1993: 105

community trends. Lindblom (1959) went further, and dismissed rational-comprehensive planning as an impractical ideal. In his view, it is necessary to accept the realities of the processes by which planning decisions are taken: for this he outlined a 'science of muddling through'. Essentially, this incrementalist approach replaces grand plans by a modest step-by-step approach which aims at realizable improvements to an existing situation. This is a method of 'successive limited comparisons' of circumscribed problems and actions to deal with them. Lindblom argues that this is what happens in the real world: rather than attempt major change to achieve lofty ends, planners are compelled by reality to limit themselves to acceptable modifications of the status quo. On this argument, it is impossible to take all relevant factors into account or to separate means from ends. Rather than attempt to reform the world, the planner should be concerned with incremental practicable improvements.

An alternative is provided by Etzioni's 'mixed-scanning' model: this incorporates elements from both comprehensive and incremental planning theories. It holds that decisions on 'fundamental' issues – such as primary goals – are followed by detailed examination of alternative programs of implementation (Etzioni 1967). This is an attractive theory, though skeptics are not convinced! Many argue that political forces are stronger that those of rationality. Altshuler, for example, categorically states that 'the city planner like almost everyone in American politics controls so little of his environment that unquestioning acceptance of its major features is a condition of its own success' (Altshuler 1965).

BOX 1.4   MODELS OF DECISION-MAKING

Rational-comprehensive

1(a) Clarification of values or objectives distinct from and usually prerequisite to empirical analysis of alternative policies

2(a) Policy-formulation is therefore approached through means–end analysis: first the ends are isolated, then the means to achieve them are sought

3(a) The test of a 'good' policy is that it can be shown to be the most appropriate means to desired ends

4(a) Analysis is comprehensive; every important relevant factor is taken into account

5(a) Theory is often heavily relied upon

Successive limited comparisons

1(b) Selection of value goals and empirical analysis of the needed action are not distinct from one another but are closely intertwined

2(b) Since means and ends are not distinct, means–end analysis is often inappropriate or limited

3(b) The test of a 'good' policy is typically that various analysts find themselves directly agreeing on a policy (without their agreeing that it is the most appropriate means to an agreed objective)

4(b) Analysis is drastically limited: (i) important possible outcomes are neglected; (ii) important alternative potential policies are neglected; (iii) important affected values are neglected

5(b) A succession of comparisons greatly reduces or eliminates reliance on theory

Source: Lindblom 1959: 154–5

The few practicing planners who have written about planning theory point to the validity of this statement. Even the exceptional Norman Krumholz, who explicitly placed 'equity planning' at the top of his personal agenda, makes it clear that his high ideals had to be mediated through the rapids of blatantly political forces. After describing the Cleveland political scene in the 1970s, he asks what the implications were for planning:

> First, there was probably little interest in city planning above the level of project planning. Any deals that had been cut between developers and politicians would be difficult or impossible for planners, speaking the language of 'consistency with the general plan', 'long-range significance' or 'the public interest', to modify. Second, Council took a great interest in zoning because it might be marketable . . . Third, there was little interest in general medium- to long-range planning, since its implications and marketing opportunities were unclear . . . Fourth, appeals to 'rationality' had little capacity to stir action or support. Who cared how rational a policy was if it didn't produce patronage? Finally, new physical developments of all kinds were welcomed, and if the city had some subsidies to offer, they would be made available on generous terms.
>
> (Krumholz and Forester 1990: 14–15)

Though this may read like an indictment, Krumholz's intention is to show the framework within which he and his planning staff had to operate.

LOCAL VS. CENTRAL CONTROL

Land use planning in the US is largely a local matter. Though there are important exceptions (the state and federal governments have specific planning functions in relation, for example, to environmental and coastal concerns), the scope and character of land use planning is mainly determined locally. This means that there is a great variety in the ways in which planning is carried out. At one extreme are areas where there is virtually no planning at all: land owners are free to build where and what they wish. At the other extreme are areas where there is a highly sophisticated planning machine which controls the location, character, quality, and design of all development.

In some other countries there is much more uniformity. For instance, each of the Canadian provinces has planning laws which operate in a generally consistent way, and are subject to certain controls operated by the provincial governments. In Britain, there is a planning code which operates over the whole country, subject to mandatory central government requirements, and coordinated by an elaborate governmental apparatus.

No such central control exists in the US. On the contrary, the governmental system was explicitly designed to prevent centralization. It is characterized by 'checks and balances' which other countries find baffling since it clearly reduces the efficiency of government – which is precisely what was intended. In the words of Richard Hofstadter, what has emerged is 'a harmonious system of mutual frustration' (Hofstadter 1948: 9). This deep concern about the dangers of government, together with the high esteem accorded to the Supreme Court (and the legal system generally) is distinctively American, and it profoundly affects the character of US planning.

There are other features of US planning which are distinctive. One is the limited amount of discretion which the constitutional framework allows to local governments. Discretion implies differential treatment of similar cases, and therefore runs foul of the equal protection clause of the constitution. The Bill of Rights guarantees that individuals are to be free from arbitrary government decisions. This is a major constraint on planning in the US. By contrast, the British planning system provides for a great deal of discretion. This is further enlarged by the fact that the preparation of a local plan is carried out by the same local government that implements it. In the US, the 'plan' is typically prepared by the legislative body – the local government – but administered by a separate board. The British system has the advantage of relating policy and administration, but to American eyes 'this institutional framework blurs the distinction between policy making and policy applying, and so enlarges the role of the administrator who has to decide a specific case' (Mandelker 1962: 4).

Another striking characteristic of US land use planning is its domination by lawyers and the law. In this, it is different only in degree from other areas of American public policy. All government is assumed to be 'an intrinsically dangerous and even an evil thing, to be tolerated only so long as its disadvantages are not outweighed by its defects' (Nicholas 1986: 11). By contrast, the law is a thing of great reverence. The particularly strong presence of law in land use planning derives, of course, from the strong attachment to property – an attachment that is enshrined in, and protected by, the Constitution. Land use planning is thus inherently a matter of law.

UNDERLYING ATTITUDES TO LAND AND PROPERTY

Perhaps the most tangible illustration of the American attitude to property is the constitutional safeguard. The Fifth Amendment provides that:

> No person shall be ... deprived of life, liberty or property without due process of law; nor shall private property be taken without just compensation.

Many countries have nothing equivalent to this; and they have very varying attitudes to property. The Netherlands and Britain, for example, have a positive attitude to government controls over land. There is a popular support for the preservation of the countryside and the containment of urban sprawl. Without these attitudes, the systems of land use planning that operate in these countries would be impossible. In the US, land has historically been a replaceable commodity that could and should be parceled out for individual control and development; and if one person saw fit to destroy the environment of his valley in pursuit of profit, well, why not? There was always another valley over the next hill. Thus the seller's concept of property rights in land came to include the right of the owner to earn a profit from his land, and indeed to change the very essence of the land, if necessary to obtain that profit. 'Cheap land has as one of its consequences that of stimulating and universalizing acquisitive instincts and respect for property rights' (Philbrick 1938: 723). However, the time came when the ever-receding frontier ceased to be so; it was overtaken, and land became more valuable.

One might have expected the growth of a conservationist ethic, as is prevalent in western Europe. However, though this happened to a limited extent, particularly with environmentally valuable resources, the main effect was in the opposite direction: to increase the attractiveness of land as a source of profit. Speculation has never been frowned upon in the US: on the contrary, it has been a notable feature of the economic landscape. In many countries, land is regarded as something special, to be preserved and husbanded. In the US, the dominant ethic regards land as a commodity, no different from any other. This, of course, is related to the sheer abundance of land in the US. Indeed, until recently, it seemed limitless; and even now there are many parts where this still seems true. In some areas, however, the rate of urbanization has given rise to the emergence of 'growth management' policies which seek to channel growth into areas judged to be acceptable. That these have failed to prevent urban growth is unsurprising: the forces at work are strong, and the governmental powers of control are weak (as is discussed at length in Chapters 10 and 11).

PRIVATE AND PUBLIC PLANNING PROCESSES

On a simple view of the development process, the private sector is responsible for development proposals, while the public sector is responsible for regulating them. This ignores the important role which the public sector plays in the provision of infrastructure and in rendering essential services; but it approximates the reality, even if in a somewhat distorted way. Certainly, the two sectors have different functions and methods of operation; it is worth examining these briefly in this concluding section of the first chapter.

The private development process is a series of stages by which a proposed development is brought to fruition. At its simplest, the process is conceived and followed through by the person or company which undertakes the construction: he or she decides that there is a market demand for a particular development;

a site is selected; finance is arranged; permits are obtained from the appropriate regulatory agencies; the project is constructed; and the finished product is marketed either for sale or for rent.

Of course, developments vary greatly. The initiator may be the client (a home buyer, for example) who has the necessary finance but requires a site, an architect, and a builder; or the client may already own a site and require a developer to coordinate all the planning and building operations. The prime mover could be an insurance company seeking an investment, a farmer wanting to convert the farm into a subdivision (the 'last cash crop'), or a municipality attempting to expand its tax base.

Another variation arises when the body providing the finance wishes to be a joint partner in the venture and share in the profits (and the risks). Here the finance company would be looking at a balance between risk and high profits – a very different situation from that of a long-term equity investor who seeks security and plays a passive role in the development process.

The private planning process typically involves a degree of risk. There therefore needs to be good judgment about the state of the market, the likely demand for development, the prospects for specific locations, the trend in interest rates, the political outlook, and a host of similar uncertainties. Good judgment can pay off handsomely; so can good luck. Bad judgment or bad luck can be disastrous. The success of the private developer is measured simply: does the development return a profit?

The public planning process is very different. It is essentially reactive to private initiatives. (Even the provision of infrastructure often follows, rather than leads, private investment.) It is concerned to ensure that development accords with the standards set out in legislative instruments. Its regulatory character means that it is on the lookout for deviations, misinterpretations, and errors. Its chronic shortage of resources, and its perpetual concern about future costs of maintaining the public estate, lead it to try to secure the maximum amount of public benefit from developers.

The public process also involves a multiplicity of agencies each with its own objectives, plans, finances, and concerns. These numerous agencies operate in their specific areas of competence and responsibility (water, roads, schools, parks, libraries, waste disposal, clean air, fire services, and so on). A major problem for a developer is finding a way through the maze of agencies that have to be satisfied, or at least consulted, about development proposals. (The developer may not know that the agencies may be equally confused about the overall process.)

The public sector has no equivalent to the developer's profit (though dissatisfaction with local and state services can be bluntly registered through the ballot box). Indeed, all too often, it is by no means clear where blame – or praise – lies. Even if the responsible agency can be identified, it may be difficult to identify the responsible person.

There are thus two different worlds, with different objectives. One is characterized by willingness (and necessity) to take risk; the other, being publicly accountable, is averse to taking risks. One is opportunistic; the other is bureaucratic. One seeks financial reward; the other good husbandry and probity.

Given these very different frameworks, there is a serious communication gap between the two sectors (Peiser 1990). Bridging this gap is a major part of both the private development and the public planning processes. How it works in practice is discussed at length in the following chapters. Here and in Box 1.6 an overview is given of the formal steps in the planning process. (The adjective 'formal' is used to indicate that there may be informal ways of dealing with problems as they arise.)

THE ELEMENTS OF THE PLANNING PROCESS

The planning process encompasses the preparation of a plan and its implementation. In theory, a local government has a comprehensive plan which forms a framework for its zoning and subdivision ordinances. Approval of the plan and the ordinances are legislative acts which are the responsibility of the elected legislative body (the council), though they may be prepared by an advisory planning commission. The

---

BOX 1.5   DIVISION OF PLANNING RESPONSIBILITIES

POLICY: Legislative body (Council) approves plans and ordinances

ADVICE: Planning Commission holds hearings and makes recommendations to the legislative body concerning plans and policy matters

IMPLEMENTATION: Planning Department

APPEALS: Zoning Board of Adjustment

---

implementation of the ordinance rests with the planning department, but applications from owners for changes in zoning are dealt with by a separate zoning board of appeals (sometimes termed a board of adjustment). This rather complicated system is a product of history, which is discussed later (Chapter 5). (State enabling acts differ, and there are variations on this model.)

In addition to these planning and zoning controls, there are many other areas of control which may fall to the responsibility of other departments of the local authority or to other agencies. These include public utility connections and building codes. A large development might also involve negotiations with the departments or agencies responsible for transportation, education, and environmental protection. A successful developer will have considerable skills in maneuvering a route through this network.

Not surprisingly, 'regulatory barriers' are a common target of criticism, and measures are intermittently taken to reduce them. However, they perform an important role in modern society, and there are strong public pressures which may lead to an increase in regulation rather than a decrease.

FURTHER READING

A good discussion of concepts of planning, together with an excellent bibliography is Alexander (1992) *Approaches to Planning: Introducing Current Planning Theories, Concepts, and Issues*, particularly chapter 4. (Another version of this chapter, published in 1991, is 'If planning isn't everything, maybe it's something'.) Useful collections of papers on planning theory are Faludi (1973) *A Reader in Planning Theory*; Burchell and

---

BOX 1.6   STEPS IN THE PLANNING PROCESS

Step 1: Identify issues and options
Step 2: State goals, objectives, priorities
Step 3: Collect and interpret data
Step 4: Prepare plans
Step 5: Draft programs for implementing the plan
Step 6: Evaluate potential impacts of plans and implementing programs
Step 7: Review and adopt plans
Step 8: Review and adopt plan-implementing programs
Step 9: Administer implementing programs; monitor their impacts

Source Anderson 1995

Sternlieb (1978) *Planning Theory in the 1980s*; and Mandelbaum *et al.* (1996) *Explorations in Planning Theory*. Other important books are Friedmann (1987) *Planning in the Public Domain: From Knowledge to Action*, Wildavsky (1987) *Speaking Truth to Power: The Art and Craft of Policy Analysis*.

The classic studies on implementation include: Meyerson and Banfield (1955) *Politics, Planning and the Public Interest: The Case of Public Housing in Chicago*; Altshuler (1965) *The City Planning Process*; and Pressman and Wildavsky (3rd edn, 1984) *Implementation*; Levy *et al.* (1973) *Urban Outcomes*. A most interesting case study is Derthick (1972) *New Towns In-Town: Why a Federal Program Failed*. On the problems of defining problems, see Rittel and Weber (1973) 'Dilemmas in a general theory of planning', and Nelson (1977) *The Moon and the Ghetto*.

The classic paper on advocacy planning is Davidoff's 'Advocacy and pluralism in planning' (1965). See also his 'Working toward redistributive justice' (1975). An interesting and useful set of essays on 'Paul Davidoff and advocacy planning in retrospect' is edited by Checkoway (1994). The major text on 'equity planning' is Krumholz and Forester (1990) *Making Equity Planning Work: Leadership in the Public Sector*. (There is an extract from this in Stein, *Classic Readings in Urban Planning*.) Questions of equity also lead into questions of ethics, on which see Hendler (1995) *Planning Ethics: A Reader in Planning Theory, Practice and Education*; Howe (1994) *Acting on Ethics in City Planning*; and Wachs (1985) *Ethics in Planning*. Varady and Raffel (1995) *Selling Cities: Attracting Homebuyers through Schools and Housing Programs* argue that cities need to achieve a balance between greater equity and maintaining their social and economic ability through educational and housing programs to attract and hold middle-income families.

On 'private governments', see Barton and Silverman (1994) *Common Interest Communities: Private Governments and the Public Interest*; and McKenzie (1994) *Privatopia: Homeowner Associations and the Rise of Residential Private Government*.

For a discussion of the different outlooks and perspectives of planners and developers, see Peiser (1990) 'Who plans America? planners or developers?'. A standard text on real estate is Miles *et al.* (1996) *Real Estate Development: Principles and Process*. On the methodology of plan preparation, see Anderson (1995) *Guidelines for Preparing Urban Plans* (Box 1.6 is taken from this most useful book). The classic text on this subject is Kent (1964) *The Urban General Plan*.

QUESTIONS TO DISCUSS

1 **Much has been written about the importance of factors other than rationality in planning. Discuss whether this amounts to an argument that planning is *irrational*.**

2 **Do you think that US land use planning would be different if the country contained the same number of people on 5 per cent of the land area?**

3 **Discuss the merits of an incrementalist approach to problems of public policy. Is such an approach compatible with long-term planning?**

4 **What role do interest groups play in the planning process?**

5 **Discuss the extent to which planners are able to advocate planning approaches which favor the disadvantaged.**

6 **'Issues of implementation are crucial in the policy-making process.' Discuss.**

7 **In what ways do the private and public planning processes differ?**

2

URBANIZATION

In the heart of the continent arose a new *homo Americanus* more easily identified by his mobility than by his habitat. He began to dominate the scene in the years between the American Revolution and the Civil War, and he was now shaping the new nation into a New World.

Boorstin 1965: 49

A CULTURE OF MOBILITY

If one word were to be chosen to describe the character of the US, it might well be 'mobility'. The land was settled by migrants from other continents — first Europe, later Asia and South America. The growth of the US is characterized by movement: movement from the coastal settlements to the interior, from the East to the mid-West and then to the far West. And the 'non-migrants' did not stay still: they moved constantly in search of new opportunities or better employment or improved housing conditions (and later for improved environments for leisure or retirement). The same restlessness or, more accurately, the same keenness to discover, to initiate, and to experiment, the same desire for advancement, the love of the new, is deeply ingrained in the American psyche. The US was born of the search for better ways of living, and the search continues.

In this chapter some of the main features of this continual mobility will be summarized. It is important for the subject of this book, since it is the underlying motive power of land development and urbanization.

THREE CENTURIES OF URBAN GROWTH

In the colonial era the US was a rural country: apart from native habitations, settlements were small and scattered. This was an agrarian subsistence economy. Though the first settlements were in the nature of towns, they were very small and, as farming developed, the growth of the rural population soon outstripped that of the towns. In 1690, the urban population made up about a tenth of the total; a century later the urban proportion had halved, with only twenty-four places being 'urban' (defined generously as having populations in excess of 2,500). It was not until 1830 that the urban population attained the 1690 level — whereafter the growth was phenomenal. By 1860, there were a hundred cities with populations in excess of 10,000, of which eight exceeded 100,000. The second half of the nineteenth century saw an acceleration of urban growth, and by 1910 the number of 100,000+ cities had increased to fifty. Boston increased in population from 43,000 in 1820 to 251,000 in 1870, and to 748,000 in 1920. The comparable figures for Philadelphia were 64,000, 674,000, and 1.8 million, while New York topped the league with 137,000 in 1820, 1.5 million in 1870, and 5.6 million in 1920.

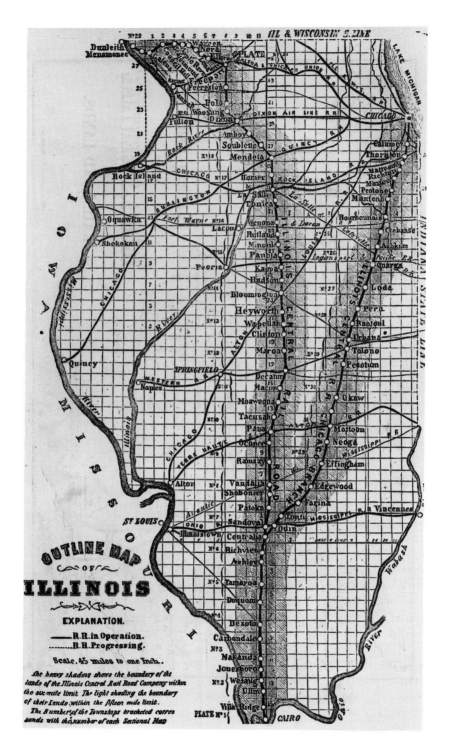

Plate 3 Lines and Stations of the Illinois Central Railroad, 1860
Courtesy of Cornell University Library

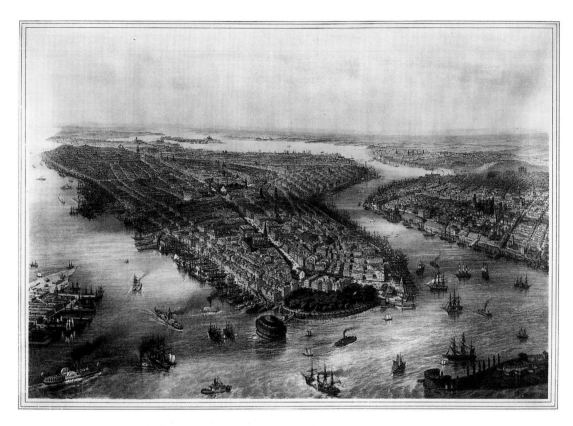

Plate 4 New York and Brooklyn, 1855
Courtesy of the Library of Congress

The maritime cities of the colonial period looked more to their colonial masters across the Atlantic than to the hinterland. They were outposts of empire and had more association with England than with each other. This was to change, of course, particularly after Independence. The rate of change was phenomenal, and was caused by a number of interacting forces. The rapid development of commerce and international trade, the development of manufacturing and transport networks, and the massive immigration from Europe worked together to transform a scattering of maritime centers into a complex, industrialized nation.

Urbanization was a major feature of this metamorphosis. The driving force was a dynamism of enterprise, mobility, experimentation, and exploration which

played itself out in many ways. On the geographical dimension it was seen in the growth of existing towns and the settlement of new places which, it was always hoped, would expand and prosper. Town promotion became highly competitive but the winners were those who by chance of geography, luck, or success in securing new transport links, founded a base which could attract further investment. The competition for transport investments was intense. Particularly successful were those places that, by one means or another, secured the terminus of a canal or railroad, though these new routes also became 'life-promoting arteries all along their way' (Boorstin 1965: 168).

Some of the initiatives were huge in scale, and many had strong federal or state governmental backing. Plans for the National Road were started in 1806, and

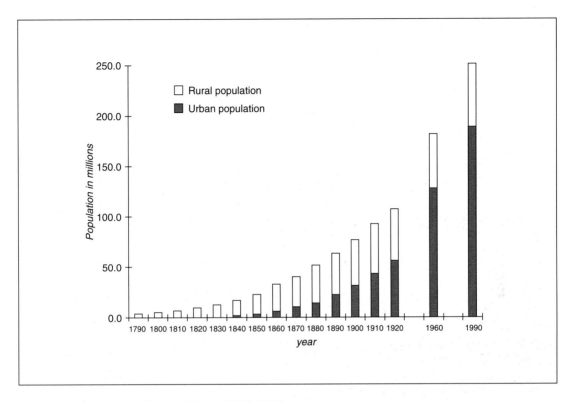

Figure 2.1 Urbanization in the United States, 1790–1990
Source: Judd and Swanstrom 1994: 22; *Statistical Abstract of the United States 1995*, Table 44

it was later constructed by the federal government roughly along what became US Highway 40. 'In 1808, Albert Gallatin, Jefferson's imaginative Secretary of the Treasury, gave the Senate his remarkable *Report on Roads and Canals*, a comprehensive scheme he had worked out with Jefferson for a federally-aided transportation system to cover the nation, connecting the eastern rivers with the Mississippi basin' (Boorstin 1965: 252). Though delayed by the war with Britain, most of Gallatin's projects were built over the next sixty years. The 363-mile-long Erie Canal – 'the wonder of the age' – was opened in 1825. Philadelphia tried to repeat the success with the creation of the Pennsylvania canal system, while Baltimore opted for the Baltimore and Ohio Railroad. These and many other schemes opened up the continent and provided avenues for a population growing yearly by immigration.

TOWN DEVELOPMENT

Urbanization not only spread westwards: it also changed in character. While the early cities were loosely structured, with little distinct separation of land uses (and social classes), great change accompanied the development of the industrial city; and as the cities grew, a pattern of land uses emerged. The central area became the location of religious, administrative, political functions, together with housing for the wealthy. The suburbs became less-desirable environments. (The word 'sub-urb' originally meant a settlement on the urban fringe: 'a place of inferior, debased, and especially licentious habits of life', according to the *Oxford English Dictionary* (Fishman 1987: 6).

As the economy grew, the inner parts of the cities became less-attractive places for living. Commercial activities increased in the center as central business districts began to develop. Workers were crowded into tenements in the surrounding zone, close to the factories (an efficient location since it minimized the walk to work). The factories themselves created noise and dirt, which was considerably increased with the coming of the railroads (which were not allowed to penetrate into the heart of the towns). As economic activity mushroomed, land values rose and land speculation increased, and those who could afford to escape to quieter surroundings did so. The opportunities for this were increased as 'transport' developed: horse-drawn buses in the 1830s, the first commuter train services in the 1840s, horse-drawn streetcars in the 1850s, a rapid growth of commuter trains in the 1860s, and, above all, the electric trolley routes of the late 1880s.

TRANSPORTATION FOR COMMUTERS

It was the electric trolley services which really let loose the forces of suburbanization. The growing middle class could now move well beyond the limits imposed by the earlier and poorer forms of transit, and they had the incomes to enable them to do so (leaving their former residences for multiple-occupation renting by working-class families). Suburbs grew at a remarkable rate along a narrow band parallel to the streetcar lines. The same happened with the commuter railroads which spread further out from the cities and served both middle- and upper-class housing markets. At the end of the century, the US 'had more miles of railway track than the rest of the world combined' (Jackson 1985: 91).

One interesting feature of the later suburbanization was the promotion of 'streetcar suburbs' by entrepreneurs who developed both transport and land. Subsidization of the streetcars to attract families to the suburbs enabled large profits to be made from the land. These land/transport developers often had little or no long-term interest in their transport operation, and services deteriorated after the housing development was completed. In some areas, the

services suffered from overextension and the collapse of subdivision plans; in others, short-term policies (and worse) caused services to flounder. Many continued only by municipal annexations and extensive subsidies. Later, of course, the private car released the constraints imposed by mass transit, thus still further worsening the difficulties of the latter.

The development of transport thus played a key role in urbanization (as it has continued to do); and each new technological advance made possible further extension of the suburbs, until in recent years the outward movement of a whole range of urban activities has transformed outer suburbs into totally new urban forms.

PUBLIC POLICIES AND SUBURBANIZATION

Housing

It is commonly held that suburbanization results from an innate desire on the part of Americans to own spacious single-family homes built at low density in pleasant, peaceful, green surroundings, separated from the bustle and problems of the city. Though there is no doubt on the compelling attractiveness of this idyllic image, there are other issues which are highly significant in the development of the suburbs. Foremost among these are public policies. Kenneth Jackson, in his classic book on suburbanization, *Crabgrass Frontier*, has gone so far as to state that 'suburbanization has been as much a governmental as a natural process' (1985: 11). One could go further and argue that the promotion of suburbanization has been among the most successful public policies ever pursued in the US. This, however, would be a distortion of the truth since suburbanization has typically been more a consequence than an objective of policy.

Thus, the most important contribution of public policy to the suburbanization process has been federal mortgage insurance. Before this was introduced in 1933, mortgages were commonly negotiated (for a third to a half of the value of the house) over a period of five to ten years, at the end of which the balance

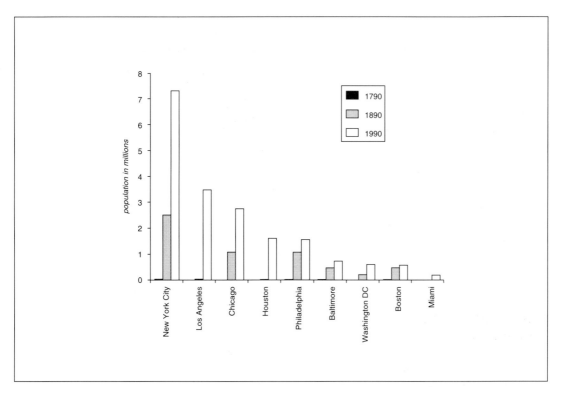

Figure 2.2 Population of Some Major Cities, 1790–1990
Source: Mills and Hamilton 1994: 74

was renegotiated at current rates (which, of course, were unpredictable and could be so high that the buyer might be unable to obtain a new mortgage and thus faced foreclosure). Buyers who could not afford the required deposit would take a second mortgage at a higher rate of interest. The system worked fairly well until the Great Depression, when its risky nature became all too apparent. Foreclosures became common, rising to a quarter of a million in 1932. In 1933, the rate was even higher – over a thousand a day (Jackson 1985: 193). The Depression seemed set to destroy completely the home-financing system and federal action became imperative.

The initial solution was the introduction of federal funds to home loan institutions with insurance against risk for depositors. The structure of home financing was changed: loans were made for a period of twenty years, with amortization over the life of the loan. This coped with part of the immediate problem, but it failed to bring about the hoped-for recovery of the housing market and extended provision was made in 1934 when the Federal Housing Administration (FHA) was established and mortgage insurance extended significantly. It worked; and it worked so well that it is still in operation today. Its impact on house production was substantial: housing starts rose from 93,000 in 1933 to 216,000 in 1935, and 619,000 in 1941 when the war sharply curtailed housebuilding. The new financing system considerably reduced the cost of house purchase and, in fact, it was often cheaper to buy than to rent. It should be noted, however, that though this system is seen as a major instrument of national housing policy, it was introduced for economic reasons: above all to reduce unemployment which reached 12 per cent in 1934 (and even higher in the construction industry).

Plate 5 A Distant Streetcar Suburb of Los Angeles, 1890
Courtesy Seaver Center for Western History Research, Natural History Museum of Los Angeles County

The changes brought about by the FHA were accompanied by – and accelerated – others, such as real estate finance and the organization and scale of the development process. Together, these changes provided a new base for large-scale housing development which, though held back by the war, burgeoned as soon as the war was over. The expansion of the federal role in the mortgage market was especially important. FHA policies were liberalized and, for example, allowed thirty-year mortgages with a mere 5 per cent deposit. Financing under the Veterans Administration (the GI Bill) was even more generous: it provided for mortgages without any deposit!

Most of these mortgages went to suburban houses, partly because it was in the suburbs that the majority of new houses were built, and partly because the federal agencies followed conventional business practices in relation to mortgages. These included favoring 'economically sound' locations over more-doubtful inner-city areas, owner-owned instead of rental dwellings, and racially homogeneous (i.e. white) districts. Attitudes such as these, which predominated in the private market, were shared by the public agencies: their interest was essentially in supporting the real estate and banking interests (from whose ranks their staff often came). All these factors favored the suburbs; and there were no official policies

Plate 6 Castello Plan, New Amsterdam, 1660
Courtesy I.N. Phelps Stokes Collection, Miriam and Ira D. Wallach Division of Arts, Prints and Photographs, The New York Public Library, Astor, Lennox and Tilden Foundations

which directed otherwise. Public action thus followed market forces.

Another benefit bestowed on home buyers by public policy came about almost by accident: favorable tax treatment. The home mortgage deduction is a significant and tangible assistance to home buyers, and it is buttressed by other advantages which are discussed in Chapter 13.

Though changes in the organization of housing production and its finance were extremely important in the suburbanization process, they were not sufficient in themselves to achieve the scale of suburbanization which took place. They provided the motive power; it was the highways programme which (literally) provided the track on which the suburbanizing engine could run.

Highways

The suburbs built before World War II were still part of the cities which they surrounded: they were only a short distance from the city center and they could take advantage of its facilities. They extended the city; and though they might keep it at bay (for example, by rejecting annexation), they did not threaten it. But a break had developed between the central city and the suburb and, to use Gelfand's analogy, this 'was one of the many fault lines' which would be apparent later (Gelfand 1975). The enormous suburban expansion of the post-war period was different: the suburbs grew in scale and character and, eventually, became a new type of urban settlement. This was more than peripheral to the city: it was increasingly independent of it and threatened its viability.

This change would not have been possible without the huge programs of highway building which have characterized the last half-century. Paradoxically, these began as an attempt to rescue the cities from decline: they were seen as providing easy access to the city. Unfortunately, it was not perceived that they might equally well provide easy exit. The initial programmes began with civic concern to adapt the cities to the changed transport era. If urban congestion was to be reduced, thus allowing cities to fulfil their traditional functions, radial freeways had to be constructed to a beltway around the urban area; these would distribute traffic to desired entrances to the central area which would be generously provided with the necessary car parks. In this way, so it was commonly thought, city streets would be freed of congestion and urban blight would be banished.

With hindsight, it is not easy to understand how such misconceived policies could ever have been adopted, let alone implemented on a scale which devastated so many acres of city land. Yet, for many years, there was widespread backing for the program: 'across America, superhighway proposals won broad popular support, for they seemed to promise greater mobility and a new boost for the central cities burdened by a horse-and-buggy street system' (Teaford 1990: 97). Local resources, state aid, and federal aid on a grand scale provided the funding for this and was later greatly increased by the Interstate Highway Act of 1956.

The interstate highway program (pressed with vigor by a well-organized lobby) was one of the greatest public works ever. It committed the federal government to an expenditure of $33,500 million for 41,000 miles of highway. It also established a Highway Trust Fund into which revenues from fuel taxes were automatically siphoned, thus providing a continual replenishment of the resources for new road building. (Diversion of some of these enormous funds into public transit was resisted until the Nixon era.)

The importance of this highway program is difficult to exaggerate: it confirmed the ascendancy of the car over all types of personal transport; it enabled the development of truck and (until its later replacement by cheap air travel) long-distance bus travel – and thereby accelerated the decline of rail; it provided unprecedented accessibility for employment, trade, leisure, and shopping. It was the ultimate in the extension of the possibilities of mobility for which Americans had long shown an addiction. Though it later suffered the excesses of its own achievements (which are discussed in Chapter 12) it transformed the urban scene. The suburbs were no longer peripheral: they had become the new center.

DECENTRALIZATION

In this remarkable transformation of urban America, population movement was both followed and led by the movement of employment and urban services. Industrial decentralization has a long history which pre-dates the rise of the truck – spurred first by rising central land prices, then by the greater locational freedom presented by the advent of both electrical power and the motor truck, and finally by the construction of the highway system. It was the truck and the highway system which had the most dramatic effect, but there had been considerable outward movement of industry very much earlier. Indeed, concern was being expressed about the effect of the 'decentralization of industry' on the cities as early as the first decade of the twentieth century – while others saw the resultant residential movement as a means by which the housing problem of the cities might be relieved (Scott 1969: 130).

Industrial decentralization (and warehousing) accelerated in the 1920s with the growth of both cheap electricity and transportation. Though most industrial traffic was by rail, truck traffic increased in importance as roads were built under the Federal Highway Act of 1916. (The number of trucks increased from 150,000 to 3.5 million between 1915 and 1930.) A major push for industrial suburbanization came with World War II and its requirements for huge plants that for reasons of safety (from the expected enemy attacks) and for logistics could not be located elsewhere. By the early 1960s, a half of industrial employment was in the suburbs; by the end of the 1970s, the proportion had increased to around two-thirds.

Retailing followed a similar pattern. Decentralization started with the shopping 'strips', aimed at the motorist-shopper. Suburban shopping centers catering for the new suburbanites came later. Sears and other major retailers built additional stores in suburban locations in the early 1920s, using the highly successful formula of generous parking space. The first modern-type shopping center was Country Club Plaza in Kansas City (built 1922–5). Its followers might well have been more numerous had it not been for the Depression (only eight had been built by 1946). The post-war boom, however, led to many more, and ever-bigger, shopping centers.

Eventually, the effect on many city stores was fatal. Perhaps the most symbolic was Hudsons of Detroit, the third-largest store in the United States (after Macy's in New York and Marshall Field in Chicago). As its customers moved out to the suburbs, Hudsons provided them with local stores until, by their success, the city store became redundant. It was no longer 'a simple fact that all roads in the Motor City led to Hudsons': its suburban branches succeeded all too well, and it closed in 1981 (Jackson 1985: 261).

Shops have long followed people, but a new feature of the huge enclosed malls which began to be developed in the late 1950s is that they became development catalysts themselves. A good example is Cherry Hill Mall in Delaware Township, New Jersey, developed by the Rouse Company in 1961. This provided a center for a center-less suburban spread and an identity for a diffuse area. It acted as such a catalyst to further building that the township changed its name to that of the shopping mall – a highly symbolic act. There are, however, many other malls which have become centers of activity in the suburbs, whether they followed the population or vice versa; it can sometimes be difficult to be sure which came first.

By the mid-1980s, there were 20,000 large shopping centers, accounting for almost two-thirds of the national retail trade (Jackson 1985: 259). Shopping centers of all sizes have become so much a part of American life and so tied in with suburbanization and personal car transport that they are now the norm: it is the thriving city center which is remarkable.

The suburbs have long ceased to be an appendage to the city. They have assumed a character of their own, as part of a regional mosaic of development which contains within its area most, if not all, of the functions formerly performed by cities (Palen 1995). So great has been the change that the term 'suburban' is no longer appropriate to many of these areas. They are not suburbs in the traditional sense of the term: they are a new type of decentralized city. Fishman (1987) has termed them 'technoburbs' but the term 'edge city', coined by journalist Joel Garreau (1991), has proved to more endearing.

The scale and speed of suburbanization is breathtaking. During the 1950s and 1960s, the suburban population increased from 35 to 84 million. Further increases in the following two decades resulted in the suburbs being home to virtually a half of the population in 1990.

BOX 2.1   EDGE CITIES

Edge Cities represent the third wave of our lives pushing into new frontiers in this half century. First, we moved our houses out past the traditional ideas of what constituted a city: this was the suburbanization of America, especially after World War II. Then we wearied of returning downtown for the necessities of life, so we moved our market places out to where we lived. This was the malling of America, especially in the 1960s and 1970s. Today, we have moved our means of creating wealth, the essence of urbanism – our jobs – out to where most of us have lived and shopped for two generations. That has led to the rise of Edge City.

Source: Garreau 1991: 4

Plate 7 Tysons Corner, Virginia, 1935
Courtesy of Fairfax County Public Library Photographic Archive

Had it not been for the Depression and World War II, the *speed* of suburbanization would have been slower (though it is hazardous to speculate whether its character would have been any different). The Depression caused a general and far-reaching slow-down in activity (including housebuilding and household formation) which meant that, when better conditions arrived, there was a pent-up demand to be met. This backlog was further increased by the war years which involved the diversion of enormous resources to war-related purposes. The end of the war released the pent-up demand (together with a baby boom) and led to a long period of growth and prosperity of which suburbanization was both the outcome and the hallmark.

Part of the speed of suburbanization was facilitated by changes in housebuilding techniques, of which the mass-production Levittown developments were pro-totypical. Levitt was the Henry Ford of the building industry: by means of prefabrication, pre-assembly, site planning, and scheduled delivery of 'components'

he transformed housebuilding into a manufacturing process and achieved an incredible rate of production. When it was in full operation this reached as many as 150 completed houses a week, at a rate of one every sixteen minutes. Levitt is the best known of the mass builders, but he was not unique. Many others experimented with various forms of inexpensive prefabrication. And, if the houses were basic, small, and cheap, they also met an urgent need and proved capable of adaptation to later changes in the incomes and aspirations of their owners – as Barbara Kelly's 1993 study has demonstrated.

CURRENT TRENDS

The people of the US are constantly on the move – to seek better homes, better jobs, better climate, or even just somewhere new. The amount of mobility is incredible. Over a sixth of the population moves every year. Many of these moves are for housing reasons:

Plate 8 Tysons Corner, Virginia, 1989
Courtesy of Fairfax County Public Library Photographic Archive

households move to another house to improve their housing conditions (or, less frequently, to obtain cheaper accommodation). This type of 'residential mobility' is conceptually different from movements caused by a change of job which necessitates a change of house, possibly in another county or state. In practice, the distinction is often not clear-cut. A change of job may follow a change of dwelling as well as vice versa; and sometimes a mixture of motives may be involved.

Statistically, most moves are short-distance (well over a half within the same county), but there are also huge long-distance movements, particularly over a period of time. The biggest of these has been the move to the West which, at various rates, has persisted since the frontier was breached. There were, however, other notable migrations. For example, the movement of blacks from the rural South to the urban North during and after World War II continued on a large scale a long-established pattern of migration – the theme song of the fast-paced jazz laments for the 'Double Diaspora' suffered by African Americans (Boorstin 1973: 299). About 200,000 blacks moved north between 1890 and 1910. The rate stepped up after the outbreak of World War I when immigration of unskilled laborers ceased, and Henry Ford and other northern manufacturers actively recruited southern blacks (even hiring special freight cars for their passage): half a million moved between 1914 and 1919.

Large movements also characterized the farming scene. The movement of the farming population in the 1920s was around 1.5 million. Since then there have been huge changes: between 1930 and 1970, the farm population fell from over 30 million to less than 10 million, and by 1990 it had fallen still further to 4.6 million – a mere 2 per cent of the population.

The movement of workers in World War II was immense: over a period of four years, some 20 million moved house in response to changes in the wartime economy, and another 12 million left home to join the armed forces (Brogan 1986: 584). The years following the end of World War II saw equally dramatic movements: from the farms; from older industrial areas to the new areas developing in response to government defense contracts (later to be dubbed the gunbelt); and to the rapidly growing suburbs.

These movements have added up to striking patterns of population redistribution. Demographers have identified three dominant patterns: the movement to the West which persists in its regional primacy; the redistribution from rural to urban places, and from non-metropolitan to metropolitan areas; and, within the metropolitan shifts, the movement 'up-the-size-hierarchy', with the largest metropolitan areas gaining the most (Frey 1989: 34).

There was a consistency in these patterns until the 1970s when changes occurred on such a scale as to give rise to the question whether there had been a 'turnaround' in migration trends. A major feature of this change was the growth of the non-metropolitan areas at a faster rate than that of the metropolitan areas. Moreover, some of the largest metropolitan areas, instead of continuing to exhibit the 'up-the-size-hierarchy' redistribution, actually experienced population loss. This 'counterurbanization' was a reversal of the secular trends of increased urban growth and western movement and, interestingly, it had its counterparts in many other countries. For a while, there was a spirited controversy on the reasons for this change and its likely continuation. Now, with the benefit of the passage of time, it is possible to evaluate the changes and to assess the likelihood of their continuation in the future. Figures from the 1990 census show that there was indeed a change in the 1980s, but it was not a simple return to the pattern of earlier years.

Three issues stand out. In the first place, the situation has become much more complicated (Frey 1994a; and Frey and Speare 1992). A major characteristic of the new situation is the speed and volatility of change. Employment in urban areas is being affected by unprecedented national and international forces. There is now a real sense in which one can speak of the global economy. The geographical impact has varied with the strength or diversity of economic structure and, as a result, differences in growth and decline have accelerated and widened. Unfavored areas included those with outdated manufacturing base economies

---

BOX 2.2   FEDERAL POLICY AND THE GUNBELT

Because of the size and singularity of the gunbelt, its rise ranks among the most powerful of changes in American settlement patterns in the post-war period, rivaling other momentous changes such as the continued movement from central city to suburb. A whole new set of industries, arrayed around aerospace production and including electronics, communication equipment, and computing, and populated by a set of insurgent firms, has led to an extraordinary shift in the nation's industrial center of gravity away from the heartland.

Labor pools have been built with ease around new emerging gunbelt cities – able to attract substantial new contingents of professional and technical labor. In other words, people follow jobs: new firms, industries, and military facilities fashioned, often very deliberately, the labor market institutions that would generate an ongoing supply of labor. The Pentagon facilitates this lopsided recruitment, out of the heartland and into the gunbelt, by paying for the relocations of scientific and technical personnel as a part of the 'cost of doing business'. Unintentionally, this mechanism has financed one of the greatest selective and for-profit population resettlements in the nation's history.

Source: Markusen *et al.* 1991

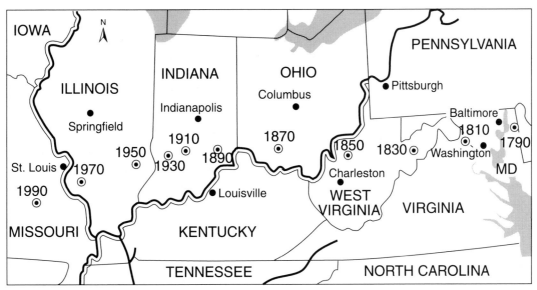

Figure 2.3 Center of US Population, 1790–1990
Source: US Department of Commerce

and those which suffered from cutbacks in mining or military expenditure. Favored areas were those which had growing financial, service, educational, health, and leisure centers. Some localities attracted very high rates of growth: for example, the resort and retirement areas such as Las Vegas and much of southern Florida.

Second, a dimension of increased importance is the growth and distribution of minority populations, especially blacks, Hispanics, and Asians. All have natural increases above the rate of whites, and immigration from Latin America and Asia has been significant. Of the 7.3 million immigrants in the decade 1981–90, 2.8 million came from Asia and 1.7 million from Mexico (US Bureau of Census 1994: Table 8). Though there has been greater 'dispersal' than in earlier decades, there is a marked concentration in certain areas. A striking example is the Los Angeles metropolitan area which had over a fifth of the total growth in minorities during the 1980s (numbering 2.8 million). As a result, the area housed 12 per cent of the entire country's minority population in 1990.

Third, the suburbs continue to be the favored location for new urban employment and residential growth. They have become growth areas in their own right, rather than simply recipients of people and jobs from the cities. Most suburbanites now work within the suburbs (though not necessarily the one in which they live), and the pattern of commuting has therefore changed and become more complex. It is not only car-based, but also car-dependent: public transport could not cope with the new patterns of commuting. Minority movement to the suburbs has grown significantly, particularly in the West which experienced the greatest increase in minority populations. The suburbs have become more differentiated by race and also by economic structure, and they 'represent the arena of future growth in most metropolitan areas' (Frey 1994a: 132). During the 1980s, suburbs grew on average at over twice the rate of the central cities.

These patterns are considerably complicated by regional differences which themselves are associated with the differential impacts of economic change. Regional restructuring plays itself out clearly in metropolitan areas. Thus while advanced service-based economies such as those of New York and Boston were able to build on these (offsetting some of the decline due to deindustrialization), areas heavily dependent on manufacturing, such as Detroit, Cleveland, and Pittsburgh, suffered further decline. (These cities are now exhibiting varying degrees of revitalization due in no small part to the efforts of city and state governments.)

In such ways has the pattern of urbanization become more complex. It now defies simple characterization, and it is both more volatile and unpredictable. This presents difficult problems of comprehension, let alone policy-making. When it is unclear at the local level what is happening and what forces are at work, it is not easy to forge relevant and workable policies. Yet there is nothing new in this, as history abundantly shows. The various phases of post-war urban policy have more frequently failed than succeeded, often because the underlying causes were not appreciated. But that there is scope for effective governmental action is apparent from the influence which this had had in shaping urbanization.

THE ROLE OF GOVERNMENT IN URBANIZATION

It is abundantly clear that the federal government has had a major influence on the scale and character of urbanization. Financial aid for housing, the building of a huge highway network, and the indirect funding of development in the gunbelt are three particularly important ways in which the federal government has at least facilitated, if not created, the modern suburbs. What is also clear is that this influence has been a consequence of a multiplicity of policies directed at other goals, and that there has been no policy related to urbanization. Though the federal government has been a major force in urbanization, it has not attempted to guide this – or even acknowledge it. As we shall see later, a few states and localities have attempted to influence the rate or nature of urbanization, though not with a great deal of success. The majority, however, have not even tried. Whether

they might have done, should have done, could have done, or might do in the future are questions to which we will return later.

FURTHER READING

There are many good general books on American history which deal well with urban issues. The three-volume *The Americans* by Boorstin, is very readable (and each volume has an extensive bibliography). The three volumes are: *The Colonial Experience* (1958); *The National Experience* (1965); and *The Democratic Experience* (1973). Equally enthralling reading is provided by Hugh Brogan's one-volume *The Pelican History of the United States* (1986).

The major histories of planning are Reps (1965) *The Making of Urban America: A History of City Planning in the United States*; and Scott's *American City Planning since 1890* (1969). Reps is concerned mainly with physical planning, and his volume contains a unique set of reproductions of town plans. Scott's book sets the development of planning within a wider social framework. Peter Hall's *Urban and Regional Planning* (1992) has a very useful chapter on planning in the US since 1945 as well as an analysis of the intellectual background in both the US and Europe (which is more thoroughly developed in his 1988 *Cities of Tomorrow: An Intellectual History of Urban Planning and Design in the Twentieth Century*).

Urban history is thoroughly dealt with by Glaab and Brown (1983) *A History of Urban America*. Sam Bass Warner's *The Urban Wilderness: A History of the American City* is a marvelous study of the historical roots of the major urban problems of today; first published in 1972, it was reprinted in 1995. See also his *Streetcar Suburbs* (2nd edn, 1978): this is a detailed study of the growth of Boston which gives a fascinating insight to the general process of suburbanization. A history covering the period from 1900 to 1990 is Teaford (1993) *The Twentieth-century American City* (this has an excellent bibliographic essay). Miller and Melvin (1987) *The Urbanization of Modern America: A Brief History* is an overview, with telling illustrations and guides to the relevant literature. Gelfand's *A Nation of Cities: The Federal Government and Urban America, 1933–1965* (1975) provides a particularly good account of the role of the federal government during the period covered. An excellent brief account is Gerckens (1988) 'Historical development of American city planning'.

The classic account of suburbanization is Jackson (1985) *Crabgrass Frontier: The Suburbanization of the United States*. Palen (1995) *The Suburbs* is a useful textbook which synthesizes material from a wide range of sources, with a mainly sociological orientation.

Fishman (1987) *Bourgeois Utopias: The Rise and Fall of Suburbia* is a study of the origins of suburbia in the United States and Britain and a discussion of 'the rise of the technoburb'. A journalistic discussion of the technoburb (with a more elegant term) is Garreau (1991) *Edge City*.

Studies of population change by Frey appear regularly; they provide detailed analysis of demographic change by region and urban area. Some are listed in the References.

Further reading on transport issues is given in Chapter 12.

QUESTIONS TO DISCUSS

1 **What are the main reasons for urbanization in the US? Have these changed over time?**

2 **To what extent can it be said that the suburbs have been created by federal policies?**

3 **Do you think it is appropriate to term the modern suburb 'a new urban form'?**

4 **Discuss the argument that in the post-war years 'cities had no realistic alternative but to embark on road building programs that were suicidal'.**

5 **Do you consider that current trends in decentralization are fundamentally different from those of earlier years?**

6 **Is it possible that edge cities will replace central cities as the 'heart' of American metropolitan areas? What would be the effects of such a shift?**

3

GOVERNING AND PLANNING URBAN AREAS

The 'city problem' in the United States was, as President A. Lawrence Lowell of Harvard University said, like a jelly fish. You could not pick up a part here and a part there and succeed. You had to lift it altogether.

Scott 1969: 110

BASIC NEEDS FOR GOVERNMENT

The original colonial cities were small both in terms of population and in area. They were intimate walking cities, usually less than a mile across. Their government was likewise modest, with the leading families taking control of matters in a natural way. Such services as were needed – for protection against hazards – were provided cooperatively. Life was based on family and community relationships, with little distinction between public and private enterprise. There was, however, a remarkable degree of regulation both by the English government (to keep the colonists under control) and by the new communities themselves (for their very existence). The former, of course, ultimately proved self-defeating when the onerous regulations (and 'taxation without representation') led to the Revolution and the birth of the United States. The latter, on the other hand, were self-imposed and had an acceptable rationale. Despite the abundance of land, the colonists quickly found that they had to plan and control the growing of certain crops. Without this, there was a danger that individual colonists, intent on maximizing their profits, would overproduce crops that were valuable for export and grow insufficient of the crops that were essential for local needs. Virginia restricted the growing of tobacco and

required 'each white adult male over 16 to grow two acres of corn, or suffer the penalty by forfeiting an entire tobacco crop. An Act of 1642 required the growing of at least one pound of flax and hemp, and an Act of 1656 required land owners to cultivate at least ten mulberry bushes per 100 acres in order to stimulate the production of silk' (Bosselman *et al.* 1973: 82).

In urban areas, the regulations were designed to promote health and safety. Following the great fire of Boston, laws were passed requiring the use of brick or stone in buildings. 'No dwelling house could be built otherwise, and the roof had to be of slate or tile upon penalty of a fine equal to double the value of the building.'

There are many other examples of the extensive amount of regulation in colonial times. Though the settlers had the freedom to use their land unfettered by the restrictive feudal-type controls which operated in the country which they had left, the sheer necessities of surviving in a strange, undeveloped country with its new hazards of climate, disease, and relationships with native Indians, forced a remarkable degree of self-regulation. There was also a religious strand which reinforced the dictates of the physical environment. Freedom to follow the religion of their choice was one of the reasons for the move to the new land, and each religious group had its own convictions and dogma. Settling in a new country meant building not only a new physical environment but also new communities in which religious beliefs could be practiced freely and in accordance with the observances which they demanded. The Puritan philosophy in particular stressed the divine joy of building a new society: the 'city upon the hill' to use the phrase immortalized by John Winthrop in the famous sermon delivered on board ship bound for the new world; but it was an orderly city which would serve God according to His rules.

As in any Utopia, the rules were not always followed, but neither were they ignored: they provided a framework for individual behavior and community action – and, unlike most Utopias, the city upon the hill *was* built. Similarly with William Penn's city of brotherly love: Philadelphia was a 'holy experiment' in the Quaker tradition which grew rapidly from the date of its foundation in 1683 as Friends emigrated from persecution in England (and later from Germany). The city was planned in detail (its legacy can still be clearly seen today), though Penn was unable to stem the pressures for land speculation which rapid growth engendered. Such speculation had no place in the earliest settlements: the problems of establishing and developing the community were primary. In addition to these matters of high principle, there were more practical considerations which the settlers had to consider. Since everything was to be provided from scratch, it was simple common sense to give some thought to the general layout of the settlement. At the least, it was necessary to decide upon the broad pattern of the streets and the location of public buildings. In fact, the early town charters listed requirements for such matters as the basic road layout, and the reservation of land for the church, the town house and the market place. Such charters continued to be drawn up in later years to guide the development of the western territories. Planning was therefore no stranger to early America, but, like other features of these times, the pace and character of growth brought about great change.

PRIVATISM

In particular, problems of survival gave way to problems of development, first of trading and later of manufacture. The towns became, above all, places for economic growth and money-making. This was an individualistic activity, in which government was seen as a facilitator of private enterprise rather than as a mechanism for order and control. The dominant philosophy became one of 'privatism': the free operation of individual initiatives in the search for private profit. Change took place as a result of private action and competition. Public controls were viewed as restraints on progress and, though they have developed significantly over time, privatism remains a powerful force in contemporary society. Indeed, much of the debate on current urban policy can be seen as a battle between philosophies of privatism and public planning. The philosophy of privatism has been well articulated in the writings of Sam Bass Warner (see Box 3.1). It is a distinctive, though not unique, feature of American urban development. The cities of the older world grew (as the very word 'city' indicates) as centers of civilization (Mumford 1961). They were places of religious, cultural, and political power where governments determined public policy. The American experience, born of its different history, was essentially entrepreneurial and disdainful of government.

THE GROWTH OF PUBLIC POWERS

In spite of this, issues arose which demanded the use of public powers. These were broadly of two sorts: economic and social. Economic development itself required governmental support. The building of roads, canals, and railroads involved large amounts of capital and were important in the competition between towns for key developments. The Erie Canal is the classic case of a huge public investment which reaped enormous benefits for New York. Even if private capital might have sufficed for many more modest ventures, there was a mutual interest of private and public bodies in the promotion of

---

BOX 3.1   PRIVATISM

Psychologically, privatism meant that the individual should seek happiness in personal independence and in the search for wealth: socially, privatism meant that the individual should see his first loyalty as his immediate family, and that a community should be a union of such money-making, accumulating families; politically, privatism meant that the community should keep the peace among individual money-makers and, if possible, help to create an open and thriving setting where each citizen would have some substantial opportunity to prosper.

Source: Sam Bass Warner 1968

---

economic development. Railroads were quick to see how important they were to the future success of individual cities, and they took good advantage of this in obtaining concessions and benefits. Entrepreneurship was a hallmark of city as well as private behavior.

The second type of problem was an outcome of the very success in the growth of the new towns. These gave rise to novel problems of public health and sanitation, overcrowding and congestion, public order, fire protection, education, and poverty. The problems were of an extreme nature (particularly as immigration expanded to incredible proportions) for which no ready solutions were apparent. One writer has suggested that these problems were on such an unprecedented scale as to create 'nearly irresolvable political and social tensions' (Judd 1988: 13) – an assertion which has a disturbingly contemporary relevance.

MACHINE POLITICS

The development of public policies has been an erratic one, beset with continual controversies about the relative roles of public and private action and of the organizational structures which are required. The early towns were 'omnibus' authorities, and new functions (if not undertaken by private or voluntary bodies) were simply added to the existing local government or, in many cases, given to *ad hoc* boards. (Philadelphia had thirty separate boards at one time.) The lack of a clear line between individual and public

enterprise led to rampant corruption which, in a curious way, oiled the city governmental machine. It did this by explicitly serving the political interests of the time through 'machine politics'. This grew as a response to the problems of the cities: the growth in population and economic activity required a huge provision of new streets, water and sewerage systems, transport and other public utilities; and when complete they needed many thousands of workers to run them. These were often under the control of local political machines which were as attuned to profit-making as any industrial entrepreneur: they needed a regular flow of income from graft to keep their machines running. The system worked and was accepted (if not acceptable); in the words of the infamous Boss Tweed of New York, 'This population is too hopelessly split up into races and factions to govern it under universal suffrage, except by the bribery of patronage or corruption.' The system also served the interests of business whose financial contributions bought influence in the awarding of contracts and franchises.

Machine politics is now looked back on as a corrupt form of government (and it led to a distrust of city government which has persisted), but it acted as a mediator between the multiplicity of conflicting interests in the cities. It provided direct support for ethnic groups and particularly for newcomers, who had to familiarize themselves with a culture which was foreign to them in so many ways. Today, with the large numbers of minorities and recent migrants in some cities, the question is being raised as to whether a contemporary version of machine politics would benefit these groups (Erie 1988).

THE REFORM MOVEMENT

The reform of municipal government and a later movement for municipal land use planning emanated from broader concerns for improving the character of urban life and the instruments of control. This was the 'Progressive Era' in US history, marked by a desire to bring about radical change both in the arrangements for planning and administering local affairs and in the worsening conditions under which so many of the urban population were now living. As with all such movements for reform, there were many strands. These included the revelations of successive inquiries and reports on urban living conditions (including the best-selling book by Jacob Riis, *How the Other Half Lives* (1890), and the massive six-volume Pittsburgh Survey (1907–8) – which was the most extensive of many such surveys). From the 1870s through to the outbreak of World War I, reports on social conditions appeared in profusion from religious organizations, welfare societies, researchers, settlement house workers, journalists, and many others. Even the federal government was moved to study some of the problems: the Commissioner of Labor produced reports in 1894–5 on slum housing (though funding restricted the investigation to only four cities: Baltimore, Chicago, New York, and Philadelphia), and on the experience in European cities of providing housing for 'the working people'. Better known was Upton Sinclair's *The Jungle* (1906) which depicted the horrors of Chicago's meat-packing industry; this was one piece of writing which led to clearly connected and tangible results – the establishment of the federal Food and Drug Administration in 1905. Sinclair's moving account of the wretched existence of Chicago's immigrants had no parallel effect, but it added to a growing concern about living conditions.

The overcrowded slums brought several responses, though public action was restricted by firmly held political beliefs. The direct provision of housing was seen as essentially a matter for market economics and, since incomes were low, housing standards were low. Though New York introduced its first tenement house law in 1867, following some disastrous fires,

building codes were not generally introduced until the last two decades of the century. Even then, they were primarily concerned with fire hazards, and not with other controls such as height (the absence of which later made possible the development of the office skyscraper). New York was again in the lead (in 1895) with legislation relating to slum clearance, but few other cities followed and little was achieved, though by the turn of the century it could be said that a national housing movement was under way (Lubove 1962).

PARKS

One of the powers granted by the 1895 New York legislation was for the provision of playgrounds and parks in crowded districts. This reflected another strand in the reform movement that was involved with landscape and parks. There were two overlapping concerns: one related essentially to natural and landscape beauty, the other to the desperate need of urban neighborhoods for some relief from their congested environment by way of the provision of recreational areas. The protagonists of both schools saw beautiful recreation areas as sources not only of rest from labor but also, more romantically, of moral rejuvenation.

New York's Central Park was (literally) the major landmark in the campaign for urban parks. The emphasis here was on natural beauty (considerably helped by human hands). The concern stemmed in part from the agrarian origins of the country and the high regard for the countryside and rural life, epitomized in Jeffersonianism, and later further idealized by Ralph Waldo Emerson, Henry Thoreau and others. This rural dream was by no means uniquely American but, as in Europe, it drew additional strength from the unlovely urbanization of the nineteenth century. Unlike European parks, however, the objective was not the landscape garden beloved by the upper classes of Europe but the preservation of wild land: Central Park dramatically demonstrated that this could be provided close to the developing urban areas. Similar parks, and even systems of parks, followed. The planning of these led to an elementary form of planning: judging how a park would relate to the development of the urban area and its transport provision. This was the route along which a number of landscape architects traveled to become planners.

Though aesthetic and romantic in origin, the parks movement was also concerned with meeting the needs for recreation and for relief from the over-crowding of city neighborhoods. The need for local parks assumed a high profile in the mid- and late-nineteenth century. Perhaps this was because the need was tangible, the provision relatively cheap, and the opposition weak. Certainly, the provision of neighborhood parks was less daunting than housing reform, where the hegemony of market principles made progress extremely slow. Be that as it may, the parks movement was a force of significance and, coupled with other concerns about the preservation of natural and landscape resources, it widened to embrace an attack on all forms of ugliness and the inefficiencies of contemporary cities.

THE CITY BEAUTIFUL

The widespread interest in parks constituted one of the elements which made up a concern for urban grace and beauty which became known as the City Beautiful movement. This reached its zenith with the World's Columbian Exposition: a 'plaster fantasy' which celebrated, in classical architectural terms, the four-hundredth anniversary of the landing of Columbus. It demonstrated the confidence of 'a nation grown rich by the development of its natural resources and its industries, a nation at last critical of its municipal institutions, and determined to remold them to serve broader public purposes'. Located on the shore of Lake Michigan, the 'white city' was the occasion of a national celebration (though, because of unavoidable delays, one year later than the anniversary). It was in some ways an irrelevance (an 'anachronistic symbol of accomplishment', in Mel Scott's phrase), but it represented in a tangible way the merging of a number of concerns and ideologies

which had developed over the last decades of the nineteenth century. It marked a desire to make American cities places of beauty, set in an artificial naturalistic landscape. Such dreams could not survive the realities of a growing industrial society: they were a reaction to it, not a solution for its problems. However, the Exposition has the important historical significance of being clearly placed in time and space as the marker of a coming together of numerous attempts to create a more humane and livable environment.

It was also responsible for a legacy of beautiful buildings which are to be seen throughout the country, in countless civic centers, boulevards, college campuses, railroad stations, banks, and other public buildings: all reflecting the Beaux Arts tradition which was embraced by the Exposition. (Their current fame often stems from their success in resisting re-development in the name of historic preservation – as with Penn Central Railroad's Grand Central terminal in New York which was the subject of a landmark zoning case, discussed in Chapter 9.)

The strands which joined together to create the City Beautiful Movement were many – some of which have already been touched upon. Following Wilson's analysis, seven of them can be singled out as being of particular importance. First, as the name suggests, there was a desire to make cities beautiful. The origins here were the landscape, park, and municipal art movements which, at least initially, were essentially aesthetic in conception, though they sometimes exhibited a degree of social awareness and even ideas of municipal efficiency. This was particularly apparent in a second strand: a perception of beauty which incorporated some concept of public or private profit. The idea that 'good design pays' is only a short step from the contention that 'good design is not more expensive than bad design' – a contention that is frequently heard in architectural circles. The argument went further, however, since it incorporated the idea that beautiful designs were more efficient. Black smoke pouring from a factory chimney was both ugly and inefficient (an idea which has achieved a new formulation in contemporary environmental policy). Similarly, a graceful design has palpable utilitarian features; an imposing boulevard has an effectiveness in accommodating traffic; an elegant road scheme is an efficient distributor of traffic. One enthusiast went so far with the conceptual marriage of beauty and utility as to coin the term 'beautility'. This mercifully failed to gain currency, but it was a neat epitome.

A third strand was the importance attached to expertise. Efficiency required experts, and there was a rapidly growing number of them at the end of the nineteenth century. It is not too much of an exaggeration to describe the time as 'the age of the expert'. The achievements of nineteenth-century capitalism had led to a belief in the great potential of business-like methods of production and control. This extended to the rapidly growing middle-class cadre of professionals: doctors, dentists, teachers, social workers, architects, and planners. The early beginnings of a technocratic society needed, and could afford, these new skills. Leading later to the conceptual transformation of the City Beautiful into the City Efficient, this belief in the expert, wedded to ideas of progress, had important implications for municipal government and planning.

There was a class element in this (which constitutes the fourth of our selected strands): the expanding middle class attracted a respect and achieved a position which gave them a power to influence the course of events that has been unsurpassed in later times. They were the high priests of the cult of expertism; they may not have had an answer to every technical problem, but they knew there was one to be found. They could also advise on what provisions should be made for the lower working classes: whether in the form of parks for recreational relief from the toil of everyday labor, or for beautiful landscapes to raise their spirits. The professional classes knew what was best, and they made some attempt to bridge the chasm between their ideals and the realities of the nineteenth-century city. One element in this paternalism was fear of open class conflict. Industrial strife was well known and there were fears that this might turn into something more sinister. But the prevailing philosophy was essentially confident and optimistic. This fifth strand in the current ideology prevailed over

fears of revolt, though not always easily. At least for New Yorkers, there was the vivid memory of the riots of 1863 'when the poor streamed out from their gloomy haunts to burn, murder and pillage' (Lubove 1962: 12).

More broadly, there were widespread concerns about the waywardness, the unruliness, the depravity of the working class. The belief in individual responsibility, the antagonism to socialistic ideas, the fears of immigrants were all too clear to see. These views played themselves out mainly in other arenas, but they impinged upon the City Beautiful movement by way of a belief in righteousness and reform. There was a fervor in this which might have belied deeper fears. Certainly, some of the language used was exaggerated, to say the least. Charles Mulford Robinson was perhaps the most florid in his *Modern City Art, or The City Made Beautiful*, where he foresaw:

> the adjustment of the city to its needs so fittingly that life will be made easier for a vast and growing proportion of mankind, and the bringing into it of that beauty which is the continual need and rightful heritage of men and which has been their persistent dream.
>
> (Robinson 1903: 375)

Every movement needs its poet. Robinson, however, was an effective poet, not only inspiring an awareness of the quality of what we would now call the environment, but also inspiring large numbers of people to do something about it. His 1907 book, *The Improvement of Towns and Cities*, was a best-seller, and stimulated the formation of large numbers of 'local improvement' societies.

Part of Robinson's beguiling effervescence stemmed from the sixth strand: the 'American rediscovery of Europe'. Though huge numbers of immigrants had forsaken the beauties of Europe for the more prosaic benefits of the New World, its architectural treasures were models to copy. So were some of its city governments: European cities were seen to work in a way which American cities did not. Frederick C. Howe's *European Cities at Work* (1911) extolled the superiorities of German expertise, though later others, more realistically, were critical of German enterprise-crushing bureaucracy.

Finally, there was the new acceptance of the

American City. With a heavy dose of wishful thinking, American cities were regarded as being capable of major improvement: all that was needed was the same dynamism in civic improvement that had proved so successful in the industrialization of the economy.

These and other influences that created the City Beautiful movement are of more than historical interest. The movement itself was only a name, not a concerted campaign; but its elements remain important not only for an understanding of the historical background to planning but also for an appreciation of the forces which still affect the conception and the operation of US planning.

MUNICIPAL REFORM

Some of the thinking which was associated with the City Beautiful movement also focused on city government and its inadequacies. This was a major plank in the wider scene of critical analysis which characterized the times. On this the 'muckrakers' had a field day. Lincoln Steffens' exposé of corruption, *The Shame of the Cities* (1904), was one of the most influential; but there were many others in the period from the 1880s to World War I who remorselessly attacked the blatant corruption of city governments. The issue figured continuously in the newspapers and magazines of these years. At the same time as the excesses of municipal corruption bred increasing discontent among businesses, the corruption of big business was itself a target of criticism. Both fed into the wider reform movement. Reform of city government thus attracted the support of those seeking to safeguard business profits as well as those who saw them as excessive, particularly in the context of widespread poverty. The forces of change were more complicated than this might suggest since, in supporting municipal reform, there was a common purpose among those concerned with beauty, with efficiency, with physical conditions, and with other concerns of what one historian (Brogan) has termed the 'Progressive Adventure'.

It was from this wide ferment of ideas that action emerged to rectify the inadequacies of municipal government. Here, visions of the City Beautiful merged with, and became dominated by, those of the City Efficient. A particularly important connecting thread was the promise offered by 'scientific management'. This had been perceived to be successful in fields as diverse as engineering and factory organization; and it was now seen as being equally relevant to the management and engineering of cities. The validity of 'the scientific method' became part of the religion of the age. The High Priest of this was Frederick Winslow Taylor, whose *Principles of Scientific Management* (1911) was, like all such texts, more widely quoted than actually read. So famous were his ideas that they gave rise to the eponymous creed of Taylorism. Taylor is remembered as the inventor of time-and-motion study, but the important aspect of his theories was the separation of planning from implementation: identifying the problems involved in a process, establishing a scientific way of resolving them, and then implementing the new system. Applied to cities, it spelled the separation of politics from administration, and the rule of the expert.

A streamlined commission type of municipal government seemed to be the answer to the 'weak mayor' form of government which was a notable feature of cities dominated by machine politics. The structural weakness lay in the division of responsibilities and financial power among many sectors of municipal government, and the lack of coordinated control. Accountability was confused, and the control of patronage was of greater importance than the good government of the municipality. Administrative fragmentation was both engendered by and supported machine politics. The system could not have provided a greater contrast to the businesslike management of industry (or at least the common image of this).

One popular remedy was the 'strong mayor' type of administration where, though elected, the mayor operated as a chief executive, coordinating all municipal government functions and maintaining an authority over departmental heads. An alternative was the commission form of government which was propelled into prominence by an accident of history: the devastating flood of 1900 which destroyed much of the town of Galveston, Texas, killing 6,000 of the town's 37,000 inhabitants. Previous efforts of the local business elite to reform the municipal government were now boosted, and local businessmen took over with a commission form of government which was approved by the state legislature. The commission consisted of five members who took responsibility for both policy and its administration. It was deemed remarkably successful in its restoration and improvement of the town and its services, and it was widely admired and copied. Nevertheless, it had its shortcomings, of which an important one was seen to be that it was not sufficiently businesslike: the commissioners were both policy-makers and administrators; and there was no guarantee that they would coordinate their respective fields of responsibility.

A solution to this perception of the problem was the transfer of all administrative functions and power to an expert city manager who would be able to coordinate the work of the separate departments. The city manager system also had the perceived advantage of further separating the government of municipalities from political 'interference'.

As experience was gained with these newer forms of municipal government, the commission system fell out of favor except in a few smaller municipalities (only 200 of the 6,700 cities now operate this system). Today, most municipalities work under the 'strong mayor' or the city manager system. City managers function most effectively in areas where the politics are not too divisive and where there is broad agreement on the significant policy issues. This is unusual (though not unknown) in the largest cities with more volatile politics, where the strong mayor system often works better. However, as with all generalizations about the US, such statements fail to cover all circumstances.

What is notable about much of the debate on the form of municipal government is that a preoccupation with controlling the excesses of corruption has marginalized the importance of the political process. It refuses, however, to be neutralized; and it often attains salience in debates about the planning issues which are the subject of this book. Structural changes do not avert the influence of politics: they simply alter the form in which this influence is allowed (or is not allowed) to flourish.

REFORM AND THE PLANNING FUNCTION

Urban planning emerged as a promising field of professional activity supported by public opinion and governmental capabilities in several countries in the early years of the twentieth century. The year 1909 stands out as a particular high point in this emergence: it was in this year that the first national conference on city planning was held. (It was also the year in which Wisconsin passed a state law providing powers for the creation of city planning commissions; Benjamin Marsh published what is arguably the first textbook on planning, *Introduction to City Planning*; Harvard introduced the first university course on city planning; Burnham completed his *Plan of Chicago* and a Chicago Plan Commission was established to implement it; and, illustrating the international character of the planning movement, Britain passed its first planning act.)

These were promising times, and in 1917 it seemed that the promise might be fulfilled: in that year the planning interests felt sufficiently bold to establish a new professional organization, the American City Planning Institute. The membership of this was, in all likelihood, the most diverse any professional body has ever witnessed: the fifty-two charter members included fourteen landscape architects, thirteen engineers, six attorneys, five architects, four realtors, two publishers, two 'housers', and an assorted group of writers, tax specialists, land economists, educators, and public officials (Scott 1969: 163). A common denominator was a recognition of the fact that cities had increasing problems with which existing institutional structures were unable to deal, or even comprehend. The spirit of reform was still in the air, but it wore a many-colored cloak.

It soon became apparent that more was needed than scientific management: the city planning commissions which were set up to give substance to the drive for efficiency were severely constrained in

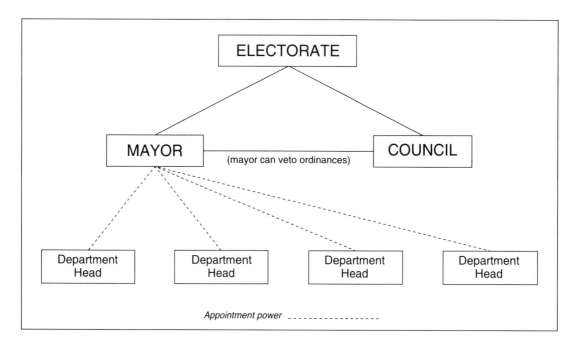

Figure 3.1 Council Structures
(a) Strong Mayor

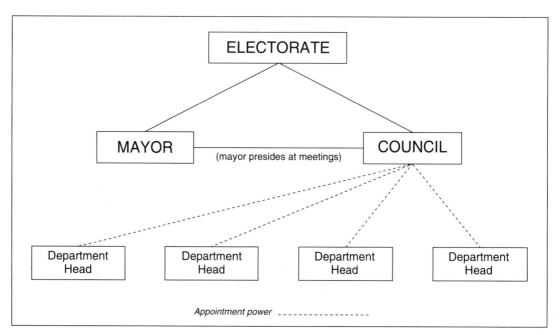

(b) Weak Mayor

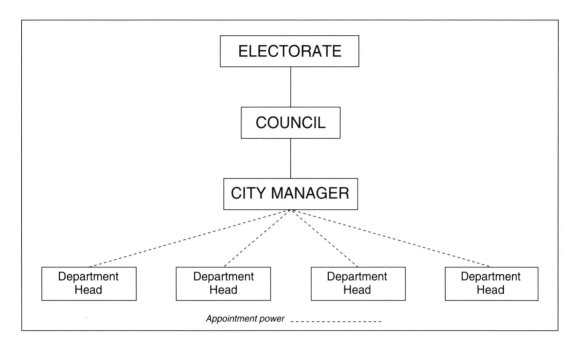

(c) Manager

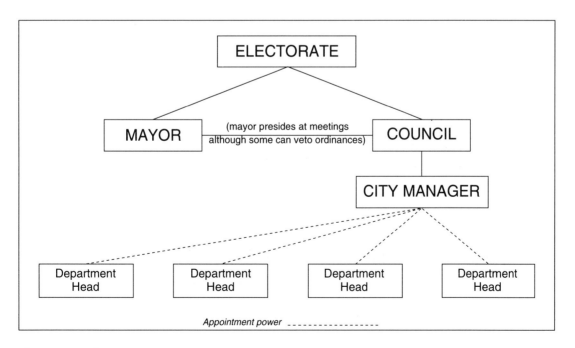

(d) Manager with Mayor

what they could actually do – despite the signs of purposive activity by such bodies as municipal information clearing houses and a burgeoning of planning courses, conferences, publications, and the like. The forces of privatism were too strong to be contained by public officials. Indeed, urban planners seldom did more than follow residential and commercial developers with transportation and sewerage systems. Despite a desire to imitate some of the trends which were emerging in Europe, the US planning movement was, in fact, unique. This stemmed from the uniqueness of the US itself: the dynamism of its urban system, the pace of its growth, the strength of its private enterprise, and the general reluctance to place fetters on the forces of development. In this, the gridiron plan played a significant role.

THE GRIDIRON PLAN

The foregoing discussion has outlined the character of the social framework within which a movement favorably disposed to planning was surfacing. Foremost, of course, was the need to facilitate urbanization: here the favored 'plan' had for long been established as the gridiron. The gridiron plan was particularly useful in land transactions and it was simple in the extreme. (Its disregard for topography sometimes made it literally 'extreme' – as was later so dramatically illustrated in San Francisco.) It provided for uniformity of lots, thus easing description for both legal deeds and for land sales (particularly when sight-unseen), and it enabled urban growth to proceed in an orderly fashion. New York, in 1811, explicitly accepted the 'decisive' advantages of the gridiron for the undeveloped part of the Manhattan Island. Commissioners appointed to propose a plan concluded:

> In considering the subject, they could not but bear in mind that a city is to be comprised principally of the habitations of men, and that straight sided and right angled houses are most cheap to build, and the most convenient to live in.
>
> (Glaab and Brown 1983: 252)

Even in towns where the gridiron was not wholly adopted (for example, in Buffalo, Indianapolis, and

Detroit), its influence was generally all too apparent and, even when natural areas were safeguarded as open spaces, market pressures often led municipalities to free them for development (Reps 1965). The gridiron also worked easily in the western urban promotions, since it was 'the natural tool of the land speculator' and fitted in neatly with the lines of the land ordinances (each township was divided into thirty-six square sections of a square mile: the checkerboard is still clearly visible to the air traveler).

However, the gridiron had its limitations (in addition to its blatant disregard for contours), especially when urban development spread over very large areas. Access became increasingly problematic, and land values in peripheral areas were affected. Difficulties were increased where one town ran into another. The problem clearly pointed to the need for some type of planning – but this raised the central problem with which planning always has to contend: how to balance the public interest against the rights of private property. The gridiron system did this in a way which was judged acceptable; an alternative was not to be readily found.

One approach which had only limited success was the establishment of boards of survey charged, as in Boston in 1891, with making 'plans showing the location of highways which present and future interests of the public require' (Scott 1969: 3). Sensible though this might seem, it adversely affected some property owners who were prepared (then as now) to seek a judicial remedy. In the Boston case, the Massachusetts Supreme Court invalidated the procedure on the ground that there had been a taking of property without due compensation. The attitude of the courts in most states was that property owners could not be required to conform to paper plans for new streets (as distinct from streets that were actually in existence). Though some states (such as Pennsylvania) took a more liberal line, it was some time before it was generally accepted that rights of way could be established by the use of the police power. In the meantime, the gridiron plan was almost universally accepted both in the peripheral extension of existing towns and in the establishment of new towns.

Plate 9 'A New and Accurate Plan of the City of New York,' 1797
Courtesy I.N. Phelps Stokes Collection, Miriam and Ira D. Wallach Division of Arts, Prints and Photographs, The New York Public Library, Astor, Lennox and Tilden Foundations

If such elementary planning as that concerned with the placing of new streets was problematic, it was hardly to be expected that more ambitious land use planning would be welcomed. At every turn, public action was hampered by the importance attached to property rights. Paradoxically, as more thought was given by planners to the subject of their emerging profession, it became increasingly apparent that planning was indeed a most difficult matter. Not daunted, however, an attempt was made in line with the spirit of the age to make it more scientific.

CITY PLANNING AS AN EXACT SCIENCE

City Beautiful plans were concerned above all with appearances. In this, they were precisely the opposite of the burgeoning industrial cities, where the over-riding object was production and profit. But to pit the beautiful against the ugly was not enough in this practical and increasingly 'scientific' age: what was needed was a plan which would increase the efficiency of the city. The new planners realized this and increasingly turned their attention to the physical workings of the city. Efforts to understand aspects of city life had become increasingly scientific. Studies such as those of the New York Council of Hygiene in the 1860s had carefully and methodically studied housing and sanitary conditions: they provided a model which other investigators copied and which set a standard to which planners aspired. But they faced difficulties which continue to beset those who attempt to make planning a scientific process: how to isolate and analyze the objects of study, how to coordinate the findings of separate studies, how to make planning 'comprehensive', and how to prevent political factors spoiling well-laid plans.

Few would be bold enough today to claim that the difficulties involved in such a task are surmountable: more likely it would be seen as a search for the Holy Grail. Some had an inkling of this even in the heady days before World War I. Speaking at the second national conference on city planning in 1910, Olmsted gave voice to a general apprehension about 'the complex unity, the appalling breadth and ramification, of real city planning'. The prospect of understanding, let alone controlling, the forces at work were daunting. Yet Olmsted was not overwhelmed by this prospect and, like many after him, considered that an attempt had to be made to understand 'the complex web of the city'.

Ironically, the search for a rational basis for planning led quickly away from the concerns of those who had shown how to study the social life of the city. Even a longstanding regard for housing waned as it became apparent that it was impossible to do anything significant within the existing social and political framework. What emerged was a preoccupation with physical controls by way of the separation of land uses: zoning became the focus of planning action. Plans were still commissioned, but they were typically superficial glossy productions which, while of some use for promotional purposes, were largely irrelevant to the problems which planners had initially glimpsed. Further, with its focus on legally enforceable uses of land, zoning lost the essential planning concern for future patterns of development. What was essentially a legal and administrative device for regulation (akin to a building or sanitary code) took the place of the vision which, even if remote from reality, inspired the plans of the City Beautiful era. In place of dreams of the future city came detailed regulation to prevent unwanted uses invading desirable residential uses. Thus the distinctive character of US land use planning was established.

REGIONAL PLANNING

Though zoning moved to center stage in the 1920s, wider concepts of planning were prominent in planning debates. Some of these were aimed at providing the machinery for dealing with the problems of servicing large areas undergoing subdivision or the protection of natural and recreational resources, and of coordinating the inherently limited capabilities of existing agencies. Such practical considerations led to the establishment of regional planning organizations such as the Los Angeles County Regional Planning Commission (the first of its kind) in 1922 and the Chicago Regional Planning Association in 1923. Such bodies had some success in road planning, particularly as growing car ownership dictated expanded road-building programs.

Alongside these efforts of practical persons were the visions of Henry Wright, Lewis Mumford, Clarence Stein, Catherine Bower, and others, who founded the Regional Planning Association of America in 1923. This espoused the cause of self-contained communities set in natural environments (what would today be called 'sustainable environments'). They achieved little success: their main practical experiment, the garden city of Radburn, New Jersey, fell victim to the Depression; and the governmental realization of their dream – the New Deal program of greenbelt towns – was axed by an antipathetic Congress in 1938 just as they were getting under way. Though Utopian, the ideas of these thinkers have persisted and form part of a tradition of planning thought which emerges from time to time as a vision of a comprehensive regional planning system.

Attempts to foster regional planning have a long history. Sadly, the story is not a thrilling one: with some notable exceptions (such as Portland, Oregon, and the Twin Cities of Minneapolis-St Paul) it is typically a succession of false starts and disappointed hopes. Nevertheless, some progress was made. The Tennessee Valley Authority was established as early as 1933, and a number of other economic planning agencies followed the Public Works and Economic Development Act of 1965 (such as the multi-state Appalachian Regional Commission). Regional physical planning bodies, such as the New York Adirondack Park Agency, New Jersey's Pinelands Commission, California's Coastal Commission, were set up by state governments. The 1954 Housing Act

Plate 10 Traffic Jam on Dearborn Street, Chicago, 1909
Courtesy Chicago Historical Society (ICHi-04191)

introduced federal financial assistance for metropolitan planning, and a large number of metropolitan
organizations were established. Additionally, special
acts were passed creating such agencies as the Twin
Cities Metropolitan Council (Minneapolis-St Paul).
These bodies differ enormously in function, power,
and performance, but all demonstrate a degree of
willingness to look at problems on a regional scale.

A thrust for creating a means of cooperation
between the constituent parts of metropolitan areas
came in President Kennedy's 1961 *Housing Message to
Congress* in which he argued that the old jurisdictional
boundaries were no longer adequate (see Box 3.2).
Without entering on a historical account, it is of

relevance to this chapter to give some indication of
the course of events.

Particular progress was made through the Bureau
of Roads, which required local governments to cooperate in a regional planning exercise as a condition
for highway construction grants. The system was
gradually extended and, in 1965, urban areas with a
population of more than 50,000 became ineligible
for federal grants for highway construction unless
they had a 'comprehensive transportation process for
the urban area as a whole, actively being carried on
through cooperative efforts between the states and
the local communities' (Advisory Commission on
Intergovernmental Relations 1964: 106).

BOX 3.2   PRESIDENT KENNEDY ON THE CITY AND ITS SUBURBS

The city and its suburbs are both interdependent parts of a single community bound together by the web of transportation and other public facilities and by common economic interests . . . This requires the establishment of an effective and comprehensive planning process in each metropolitan area embracing all activities, both public and private, which shape the community.

Source: President Kennedy's *Housing Message to Congress* 1961

The Urban Renewal Administration followed suit with its requirement that states and local governments produce comprehensive plans. Then, in 1968, the Bureau of the Budget issued Circular A-95 which sought to establish a 'network of state, regional, and metropolitan planning and development clearinghouses' to receive and disseminate information about proposed projects; to coordinate applicants for federal assistance; to act as a liaison between federal agencies contemplating federal development projects; and to perform the 'evaluation of the state, regional or metropolitan significance of federal or federally-assisted projects' (Mogulof 1971: 418; Elazar 1984: 186).

Between 1968 and 1970, the number of councils of government (COGs) increased from 100 to 220: almost all the 233 Standard Metropolitan Statistical Areas had regional councils of some type: COGs, economic development districts, regional planning commissions. These varied greatly in the extent to which they became involved in regional planning: many did as little as possible to meet the federal conditions. Others became actively committed, particularly after HUD introduced yet another

regional planning scheme: the comprehensive planning assistance program, popularly known (by reference to the relevant section in the Housing Act) as the 701 program (see Box 3.3).

Regional planning was not as effective as its protagonists had hoped. It was mainly a creature of federal initiatives; and frequently it did not receive more than nominal support from the member governments, who were apprehensive about the growth of an independent source of regional influence. Instead, they typically saw it as 'a service giver, a coordinator, a communications forum, and an insurance device for the continued flow of federal funds to local governments' (Mogulof 1971: 418). Moreover, though one of the major objectives was to ensure that individual federally funded projects were in harmony with metropolitan or regional plans, such plans often did not exist. Above all, there was no effective machinery for promoting and implementing them.

Weak though the COGs were, they constituted a point from which regional thinking could develop; hopefully, action would have followed. But in 1982 the Reagan administration rescinded Circular A-95

BOX 3.3   THE '701' PLANNING PROGRAM

HUD resources in the 701 planning program have become the institutional support for COGs, and HUD guidelines insist on the representation of a significant percentage of metropolitan area governments on COG policy boards. Additionally, it is HUD which has begun to prod 701 planning agencies with regard to 'citizen participation' in their policy structure. And it is HUD which has moved the COG into a new (and sometimes uncomfortable) concern with social problems by requiring that a housing element be a part of the 701 agency's regional planning.

Source: Mogulof 1971: 418

> **BOX 3.4   PORTLAND METRO: A DIRECTLY ELECTED REGIONAL GOVERNMENT**
>
> We, the people of the Portland area metropolitan services district, in order to establish an elected, visible and accountable regional government that is responsive to the citizens of the region and works cooperatively with our local governments; that undertakes, as its most important services, planning and policy making to preserve and enhance the quality of life and the environment for ourselves and future generations; and that provides regional services needed and desired by the citizens in an efficient and effective manner, do ordain this charter for the Portland areas metropolitan services district, to be known as 'Metro'.
>
> *Source*: Portland Metropolitan Services District Charter, 1992

and halted the system of federally funded regional clearing houses. In their place, states were encouraged to establish their own machinery. Some particular examples of this are illustrated at a number of points in this book. The indications are that, in response to increasing problems of environmental pollution and urban growth, and, above all in transportation, there is a reawakening of interest in forms of regional planning.

Recent studies such as those of David Rusk (1993) and Henry Cisneros (1995) have made persuasive statements of the need for regional planning. Even more eloquent is the establishment of a directly elected metropolitan government for Portland, Oregon (see Box 3.4). It is too early to pass judgment on these new endeavors, but the issues are discussed further in the chapter on growth management and in the final chapter.

FURTHER READING

Many of the titles recommended for the previous chapter are equally relevant here, particularly those concerned with nineteenth-century history.

The major historical study of the land use regulation issue is Bosselman *et al.* (1973) *The Taking Issue*.

On privatism, see Warner (1968) *The Private City: Philadelphia in Three Periods of its Growth*; and the more extensive discussion in Barnekov *et al.* (1989) *Privatism and Urban Policy in Britain and the United States*.

Good political science texts include Judd (1988) *The Politics of American Cities: Private Power and Public Policy*; Judd and Swanstrom (1994) *City Politics*; Ross *et al.* (1991) *Urban Politics: Power in Metropolitan America*; and Harrigan (1993) *Political Change in the Metropolis*.

For recent writings on metropolitan and regional planning, see Rusk (1993) *Cities Without Suburbs*, and Cisneros (1995) *Regionalism: The New Geography of Opportunity*. (Further references on state and regional growth management are given in Chapter 11.)

QUESTIONS TO DISCUSS

1 **What were the strands in the nineteenth-century reform movement?**

2 **Describe the City Beautiful Movement. How did it originate?**

3 **Discuss the benefits and problems of gridiron plans.**

4 **In what ways were the 'principles of scientific management' thought to be relevant to city planning?**

5 **Does the history of American city planning offer any lessons for the debate about the relative merits of comprehensive and incremental planning?**

PART II

LAND USE REGULATION

Planning arose from the need to protect property. Following the tradition of nuisance law, the favored technique since the 1920s has been zoning: the division of a local government area into districts which are subject to differing regulations regarding the use of land and the height and bulk of buildings that are permitted. A major reason for the introduction of zoning was the huge influx of immigrants into the cities (18 million in the thirty years from 1890 to 1920) and their impact on middle- and upper-class neighborhoods.

Chapter 4 summarizes the famous 1926 case of *Euclid*, in which the Supreme Court declared zoning to be constitutional, thus paving the way for its rapid spread across urban America.

The institutional and legal framework which developed is discussed in Chapter 5. Much if not most of the land use planning in the US is not planning but zoning and subdivision control. The former implies comprehensive policies for the use, development, and conservation of land. Zoning is the division of an area into districts with differing regulations; subdivision is the legal division of land for sale and development. In operating these controls, local governments can impose a range of conditions. Most of the discussion in this chapter relates to zoning, but distinctive aspects of subdivision control are dealt with separately at the end of the chapter.

Having described the way in which zoning has developed and the institutional and legal framework within which it operates, Chapter 6 examines in more detail the 'tools' of zoning which are available to land use planners. There are many of these, and the account is confined to the more important ones. Local governments sometimes impose conditions which are considered by developers to be onerous or unreasonable, and thus the courts are involved in settling disputes. Since most issues are dealt with at the state level, there can be a wide variation in judicial opinion on some issues. The case law is immense and sometimes incoherent. Only seldom does the Supreme Court establish a clear lead.

Among the conditions imposed on developers by municipalities are a range of charges and imposts. (Terminology varies, often confusingly: in Chapter 7 the inclusive term 'development charges' is used.) These charges have come about partly in response to increases in the costs of providing infrastructure; they emerged at the same time as municipal budgets began to come under severe pressure. Developers can shoulder these costs only in certain circumstances: when, for instance, market conditions allow them to pass the costs on to buyers, or they can accept a reduced rate of profit, or they can negotiate a lower price for land. Alternatively, a municipality may offer them an incentive or 'bonus' which improves their profitability. There is considerable scope in this area for ingenuity on the part of both municipalities and developers. This chapter describes some of these, and also raises the complex question of who, in the final analysis, bears the cost of charges.

4

THE EVOLUTION OF ZONING

Urban America was in something of a zoning crisis in the early 1920s. Like a patient who could endure his fever until he suddenly learned that there was a new remedy for it and who was then impatient to be cured, urban America was now sure that it would perish if it did not have zoning.

Scott 1969

THE NEED FOR PROPERTY PROTECTION

The history of land use controls is as old as history itself. In this chapter, after a short reference to colonial times, a rapid review is given of the foundations and early development of American controls from the nineteenth-century 'nuisance' cases up to the time of the classic *Euclid* case which laid upon zoning the imprimatur of the US Supreme Court. The full story of this case is a fascinating one, particularly with the virtually cliff-edge climax of the Supreme Court's deliberations. By way of introduction it is useful to list some of the more important factors which gave rise to zoning.

A major problem was public health, which grew rapidly as unbelievable numbers of immigrants crowded into cities totally unprepared to cater for their basic needs. Technological factors also played a major role: electricity increased the spread of the streetcar suburbs – the escape route of the middle class from the horrors of the insanitary and congested city. But they themselves contributed to this congestion. Even more did the two technological innovations of the steel frame and the elevator, which made towering skyscrapers both possible and practical (Goldberger 1981: 5). Central-city uses intensified as the middle

class sought semi-rural respite by new means of transport. Later the wizardry of Henry Ford escalated problems of traffic congestion to huge dimensions.

Other changes were in progress or in the wind: widespread regulation of election procedures, and the reform movement which was aimed at securing sound engineering-type solutions to problems of municipal administration. Even 'planning' was debated as a rational solution to the problems of the city. This started as a City Beautiful movement, but soon changed its character into a concern for the City Efficient. Neither got very far: they were, then as now, too long-term ('visionary' was the word) for practical men.

But one problem above all demanded attention: the safeguarding of the new suburbs from the blight which had stimulated their development. The solution was found in the extension of the law of nuisance to land uses, by way of zoning. Zoning provided long-term security against change: industry, garages, apartments, corner shops – indeed, anything which might threaten the sanctity of the single family dwelling suburb – could now be excluded. In Mel Scott's words: 'zoning was the heaven-sent nostrum for sick cities, the wonder drug of the planners, the balm sought by lending institutions and householders alike. City after city worked itself into a state of acute apprehension until it could adopt a zoning ordinance' (Scott 1969: 192). While few might understand what 'planning' involved, the protection provided by zoning was immediately apparent; and it spread at an incredible speed.

EARLY LAND USE CONTROLS

As already noted, land use controls have a long history in the US. Moreover, there were few problems with the taking of land for public purposes. Land was in abundance: so much so that questions of compensation hardly arose. Undeveloped land was perceived to be in such plentiful supply as to have no significant value. However, where developed, improved or enclosed land was physically acquired, compensation was normally payable. The power of eminent domain was accepted as an inherent power of government for which specific legislation was not required. The taking issue which became of such importance later received scant attention. Indeed, there is a paucity of evidence on the reasons why the taking clause became a part of the Constitution.

Matters changed dramatically with the adoption of the Constitution and the Bill of Rights, particularly when (under John Marshall) the Supreme Court claimed the singular power to determine the constitutionality of legislative acts. So far as the taking issue was concerned, it was accepted by both the federal and state courts that a regulatory action could not involve a taking. The term 'taking' was applied only to the physical acquisition of land by government – an approach encapsulated in the phrase: 'no taking without a touching'. Where the use of property was restricted by regulatory controls, no compensation was payable. This was so even if land owners were deprived of all use of their land, as is illustrated by an 1826 case involving a cemetery in New York City. Land which had originally been in the country well outside the urban area had been conveyed to the City for a church and cemetery. Over time, the City grew and surrounded the cemetery. A bylaw, passed by the City, which prohibited cemetery use was appealed by the cemetery. Since it

was generally believed at the time that burying the dead produced unhealthy vapors, the court held that it would be extremely unreasonable to endanger the public by the cemetery use, despite the terms of the lease. In such cases, since the physical property (as distinct from the property rights) had not been invaded, no compensation was appropriate.

In a much later case, that of *Mugler* v. *Kansas*, decided by the US Supreme Court in 1887, Mugler's brewery was made virtually worthless by a Kansas Act which prohibited the manufacture and sale of intoxicating liquor (Box 4.1). Mugler still retained his premises and could use them for any legal purpose – that is, excluding the formerly legal brewery use! (There must have been numerous Muglers in the US during the prohibition years.)

There were many such cases of the use of the police power. One further important example can be given here: the 1915 case of *Hadacheck* v. *Sebastian*. Hadacheck had owned and operated a brickworks in the open countryside since 1902; but in the following years residential development spread, and the area was annexed by the City of Los Angeles. The brickworks now became a nuisance to the local inhabitants, and the city passed an ordinance which effectively prohibited Hadacheck from continuing to operate his brickworks (which gave the land a value of $800,000), though he could use it for other purposes (value $60,000). The court held that 'vested interests' could not be asserted against the ordinance because of conditions which previously existed. 'To so hold would preclude development and fix a city forever in its primitive condition. There must be progress, and if in its march private interests are in the way they must yield to the good of the community.' In the court's view, the ordinance was a proper exercise of the police power.

Underlying these regulations was the English common-law concept of nuisance which held that no property should be used in such a manner as to injure that of another owner. These were largely 'negative' instruments, but gradually land use controls developed into more positive tools of planning. For instance, in 1867 San Francisco passed an ordinance which prohibited the building of slaughterhouses, hog

---

BOX 4.1   REGULATION IS NOT A TAKING

In *Mugler* v. *Kansas*, decided by the US Supreme Court in 1887, Mugler's brewery was made virtually worthless by a Kansas Act which prohibited the manufacture and sale of intoxicating liquor. Mugler still retained his premises and could use them for any legal purpose – that is excluding the formerly legal brewery use! In its judgment, the court argued:

> there is no justification for holding that the State, under the guise merely of police regulations, is here aiming to deprive the citizen of his constitutional rights; for we cannot shut out of view the fact, within the knowledge of all, that the public health, the public morals, and the public safety, may be endangered by the general use of intoxicating drinks . . . A prohibition simply upon the use of property for purposes that are declared, by valid legislation, to be injurious to the health, morals, or safety of the community, cannot, in any sense be deemed a taking or an appropriation of property for the public benefit.

Source: *Mugler* v. *Kansas* 1887

BOX 4.2   USE OF THE POLICE POWER

In the exercise of the police power, the uses in a municipality to which property may be put have been limited and also prohibited. Thus, the manufacture of bricks; the maintenance of a livery stable; a dairy; a public laundry; regulating billboards; a garage; the installation of sinks and water closets in tenement houses; the exclusion of certain business; a hay barn, wood yard or laundry; a stone crusher, machine shop or carpet beating establishment; the slaughter of animals; the disposition of garbage; registration of plumbers; prohibiting the erection of a billboard exceeding a certain height; regulating the height of buildings; compelling a street surface railroad corporation to change the location of its tracks; prohibiting the discharge of smoke; the storing of oil; and generally, any business, as well as the height and kind of building, may be regulated by a municipality under power conferred upon it by the legislature.

Source: *Lincoln Trust Co.* v. *Williams Building Corporation* 1920

---

storage facilities, and hide curing plants in certain districts of the city. Though clearly in the tradition of nuisance law, the ordinance was notable because it was 'preventive rather than after the fact and restricted land uses by physical areas of the city'; it thus 'set the stage for further evolution of land use zoning in the United States' (Gerckens 1988: 26). Such cases increased as the problem of urbanization escalated at a phenomenal rate.

IMMIGRATION AND URBANIZATION

In the decades from 1851–60 to 1871–80 migration into the US averaged 2.5 million. In the following decades it rose to 5.25 million (1881–90), to 3.66 million (1891–1900), and to nearly 9 million in 1901–10. In the single year 1907 it reached the staggering height of 1.25 million. Between 1890 and 1920, the population of the US rose by over 42 million. Urban areas grew at an incredible rate, quite overpowering the ability of city governments to provide basic public services. During the same three decades, the urban population of the US increased from 22 million to 54 million; the proportion of the population living in cities rose from 35 per cent to 51 per cent (Miller and Melvin 1987: 79); and the number of cities with a population of 50,000 or more rose from 50 to 144. The growth of individual cities was even more dramatic. Between 1880 and 1920, New York grew from 1,478,000 to 5,620,000;

Philadelphia from 847,000 to 1,823,000; Baltimore from 362,000 to 748,000; and Boston from 332,000 to 733,000. The difficulties created by these huge increases in population were exacerbated by the fact that the newcomers were different from previous immigrants:

> In the thirty years from 1890 to 1920, more than eighteen million immigrants poured into America's cities. These new immigrants were more 'foreign' than those who arrived before, coming mainly from Italy, Poland, Russia, Greece, and Eastern Europe. Overwhelmingly Catholic or Jewish, they came to cities that were already industrialized and class conscious. They made up the preponderance of the working force in the iron and steel, meatpacking, mining, and textile industries. They shared no collective memories of the frontier or the Civil War, much less of the American Revolution. Few spoke English, and many were illiterate even in their native language.
>
> (Judd 1988: 118)

They were therefore perceived as a threat to public health as well as to the sensibilities of middle- and upper-class residents of the outer city who had to pass the ghettoes on their way to work. Even more ominously, they threatened the entrenched urban political systems.

THE MOVEMENT FOR PLANNING

Planners in the first two decades of the twentieth century had few tools with which they could retune the urban system. Zoning, however, was one tool which offered great promise: it had a particular appeal which extended beyond those whose essential concern was with planning (and who saw zoning merely as an instrument of planning). The crucial feature of zoning, then as now, is its utility in excluding unwanted neighbors.

There was one major difficulty: it was unclear whether zoning would be accepted by the courts as constitutional. That the fears were justified was clearly illustrated several years later, after the passing of the New York Ordinance, by the rejection of the *Euclid* ordinance by the lower court. Considerable effort and skill was employed by planners and lawyers

in drafting ordinances which would stand judicial scrutiny. The battle – the word is appropriate – was between those who saw zoning as 'a protection of the suburban American home against the encroachment of urban blight and danger', and those who saw it as 'the un-restrained caprice of village councils claiming unlimited control over private property in derogation of the Constitution' (Brooks 1989: 7). However, the first major zoning ordinance emerged in New York where the forces in favor of zoning were exceptionally strong.

THE NEW YORK CITY ZONING ORDINANCE OF 1916

The 1916 New York City zoning ordinance is usually regarded as the first comprehensive zoning ordinance in the US. It was the successful outcome of an open campaign to stop changes that were taking place on Fifth Avenue. It was a war on two fronts, one between carriage-trade merchants and the invading garment industry, the other between wealthy residents and the invading retail trade. In Toll's words: 'If this was war of sorts, it was in truth a double war: garment manufacturers fighting retail merchants fighting wealthy residents. The entire conflict was much closer in spirit to social Darwinism than to the Geneva Convention. There were no rules and only one objective, survival by any means' (Toll 1969: 110.) But it was the encroachment of the Jewish garment-makers and their immigrant workers which formed the central issue. Property values fell by a half in the five years up to 1916 (Feagin 1989: 81).

A Commission on Heights of Buildings report in 1913 recommended that height, area, and use should be regulated in the interests of public health and safety, and that the regulations should be adapted to the varying needs of the different districts – a radical innovation. The city and the state legislature accepted these proposals, and the city charter was amended to include 'districting' provisions. In 1916, a comprehensive zoning code was adopted for the whole city.

Plate 11 Suburban Housing, Tampa, Florida
Courtesy Alex MacLean/Landslides

There was another difference between the new zoning controls and the well-established police power regulation of buildings and factories. Whereas the latter was intended to solve existing problems and to promote health and safety, zoning applied only to new development. So far as Fifth Avenue was concerned, further incursion by the garment industry was preventable, but nothing could be done about the changes that had already happened, at least not through zoning itself: political action was another matter. Existing uses were hallowed as 'nonconforming': 'In any building or premises any lawful use existing therein at the time of the passage of this resolution may be continued therein, although not conforming to the regulations of the use district in which it is maintained.' Thus, ironically, the very problem which gave rise to the zoning ordinance remained untouched. The reason is not far to seek:

there was so much concern that the newfangled zoning system would reduce property values that the Commission was most anxious to allay the fears. Indeed, huge areas were zoned for business and industrial use. Protection of 'investments' was, and remains, a major objective of zoning.

Successful though the campaign for the New York ordinance appeared, it was, in one crucial respect, a dismal failure: in contrast to the hopes of the proponents of the planning movement, it lacked any 'planning' component. It was a substitute for a plan: it was concerned with protecting existing property interests rather than with providing for future needs. Nevertheless, it was rapidly copied by numerous cities throughout the United States. But the most significant indication of progress was the appointment, by Secretary of State Herbert Hoover, of the Advisory Committee on Building Codes and Zoning. This

---

BOX 4.3   THE ATTRACTION OF ZONING

Nothing appeared so destructive of urban order as garages and machine shops in residential areas, or loft buildings in exclusive shopping districts, or breweries amid small stores and light manufacturing establishments. Nothing caused an investor so much anguish as the sight of a grocery store being erected next door to a single family residence on which he had lent money. Nothing made whole neighborhoods feel so outraged and helpless as the construction of apartment houses when the private deed restrictions expired and there was no zoning to prevent vacant lots from being used for multifamily structures. Zoning was the heaven-sent nostrum for sick cities, the wonder drug of the planners, the balm sought by lending institutions and householders alike.

Source: Scott 1969: 192

BOX 4.4   THE NOVELTY OF ZONING (1916)

The novel feature of zoning as distinguished from building code regulations, tenement house laws, and factory laws was that suitable regulations for different districts were established. We have become so accustomed to zoning regulations that it is difficult to understand how fixed the popular notion was that all land should be regulated in the same way throughout a municipality. On this account imposing different regulations on different areas appeared to many to be a discriminatory, arbitrary, and therefore an unlawful invasion of private rights. To counteract this impression it was considered important that the regulations within each district should be uniform for the same kind or class of buildings. A provision to this effect was placed in the original zoning clauses of the charter of New York city, and there can be no doubt that courts which early passed upon these regulations were to a considerable extent persuaded to favor them on account of this requirement of uniformity. If it had been possible to make different regulations for the same sort of buildings in different parts of the same district, it is unlikely that zoning would have received the court approval that it now has.

Source: Bassett 1940: 26

---

committee drafted a Standard State Zoning Enabling Act (SSZEA) which rapidly became the model for a large number of zoning ordinances.

THE STANDARD STATE ZONING ENABLING ACT

Zoning was a part of the scientific management movement which was sweeping America in the first quarter of the twentieth century. Herbert Hoover was an important figure on this stage. He was instrumental in creating a new area of federal responsibility: one which Christine Boyer (1983) has termed the 'cooperative state'. Central to this was scientific study of the facts (and the collection of scientific data), and the establishment of 'a central clearinghouse for social and economic reforms'. Boyer documents some of the areas in which Hoover applied this philosophy; these included the standardization of industrial parts and of the plans, designs, materials, and structural elements of houses; and the coordination of information concerning the housing market. Zoning clearly fell into this kind of thinking (see Box 4.5).

Hoover's philosophy was that the role of the state was not to interfere with market forces, but to make them more efficient by, for example, facilitating the production of better market information, advancing

---

BOX 4.5   ZONING: A NEW SYSTEM OF ORDER

In the search for a new order to the American city, the division of land uses and regulations restricting building heights and bulk became tactical rearrangements . . . Zoning, it was claimed, embodied and exemplified the idea of orderliness in city development; it encouraged the erection of the right building, in the right form, in the right place. 'What would we think of a housewife who insisted on keeping her gas range in the parlor and her piano in the kitchen?' Yet these were commonplace anomalies in the American city of the 1920s: gas tanks next to parks, garages next to schools, boiler shops next to hospitals, stables next to churches, and funeral parlors next to dwelling houses.

Source: Boyer 1983: 155

---

the acceptance of standardization, and (in the area of housing and urban development) assisting with the introduction of a system for orderly development which would be safe as an investment for both lenders and borrowers. Zoning was seen as the instrument for providing the necessary security against both unwanted development and legal challenge. In particular, it provided protection to home owners from uncongenial neighboring uses which would affect both amenity and market value.

The SSZEA gave state legislatures 'a procedure, based upon an accepted concept of property rights and careful legal precedent, for each community to follow' (Boyer 1983: 164). A crucial element in the rationale here was the belief that a single legal code would pass legal muster in a way which a multiplicity of individual local ordinances would not. A carefully crafted ordinance, based on this universal model and embodying the fruits of planning expertise, and supported by local citizens, would provide a defensible framework for an extension of the hard-to-define limits of the police power.

The model act was hugely popular. The first edition, published in 1924, became a best-seller with sales of more than 55,000 copies. Within a year, nearly a quarter of the states had passed enabling acts which were modeled substantially on the Standard Act.

And so, though planning languished, zoning boomed: by 1926, forty-three of the (then forty-eight) states had adopted zoning enabling legislation; some 420 local governments containing nearly a quarter of the population had adopted zoning

ordinances, and hundreds more were in process of preparing them (Mandelker and Cunningham 1990: 166). By 1929, 754 local governments had adopted zoning ordinances: these contained about three-fifths of the urban population of the country (Hubbard and Hubbard 1929: 166).

The list of purposes of zoning set out in the Standard Act (see Box 4.6), which is constantly paraded before the courts, omits the one which is by far the most important: the exclusion of unwanted people or uses, and thus the preservation of the status quo. These exclusionary objectives are seldom much below the surface, even when they are not explicit. Of course, all zoning is exclusionary; by definition zoning excludes *some* uses. The one exception (now rarely used) is where a zone is 'unrestricted'. Thus, in the 1916 New York ordinance, areas not zoned residential or business were unrestricted. The village of Euclid also had an unrestricted zone, as does any similar 'cumulative' zoning system. *Euclid* provided, first, for an exclusively single family house zone. The second use zone provided additionally for two-family houses; the third further included apartment houses; the fourth offices and shops; and so on until the final zone could accommodate all uses (and was therefore, in effect, an unrestricted zone). In each zone, development could take place that accorded not only with its specific categorization but also with all 'higher' uses. Under this system, a single family house could be built in any zone, but elsewhere only the uses specified for the particular zone *and* all higher zones were permitted.

BOX 4.6   THE PURPOSES OF ZONING

The Standard State Zoning Enabling Act listed the following purposes of zoning:

- to lessen congestion in the streets;
- to secure safety from fire, panic, and other dangers;
- to promote health and the general welfare;
- to provide adequate light and air;
- to prevent the overcrowding of land;
- to avoid undue concentration of population;
- to facilitate the adequate provision of transportation, water, sewerage, schools, parks, and other public requirements.

THE *EUCLID* CASE

Despite the growing popularity of comprehensive zoning, it was not until 1926 that the Supreme Court dealt with its constitutionality. In that year, in the *Euclid* case, the Supreme Court upheld the constitutionality of the Euclid zoning ordinance, and thus put its seal of approval on comprehensive zoning. This represented a significant extension of the police power in that it enabled a municipality to prohibit uses which were not 'nuisances' in the strict sense of the term. In particular, shops, industry, and apartments were excluded from single family zones. Apartments in particular were greatly feared by home owners:

> Once a block of homes is invaded by flats and apartments, few new single family dwellings ever go up afterwards. It is marked for change, and the land adjoining is forever after held on a speculative basis in the hope that it may all become commercially remunerative, generally without thought for the great majority of adjoining owners who have invested for a home and a home neighborhood only.
>
> (Cheney 1920)

The *Euclid* decision had the important social implication that apartment living could be a 'use' category for the purposes of land use planning. Thus another dimension was added to the exclusionary nature of land use regulation.

THE NARROWNESS OF ZONING

Initially, zoning was concerned essentially with 'districts'. Though lip service was paid to comprehensive plans (which were supposed to form the rational basis for the operation of zoning) this did not amount to much in practice. The term 'comprehensive' came to mean little more than a zoning provision which covered all or most of the districts in a local government area. However, as we shall see later, the overriding concern with exclusion led to zoning policies which were concerned with the whole of an area. In this way, the term 'comprehensive' took on a new meaning: safeguarding the status quo of a neighborhood was simply writ large.

It is important to stress that zoning is an inherently rigid instrument. This remains so in spite of the extraordinary ingenuity which has been displayed in adapting it to the real moving world; and in this rigidity lies its enormous popular appeal. The planning ideal of flexibility is anathema to protectionist home owners. Rigidity provides a degree of certainty and security. But zoning is not planning: it is a restricted instrument for districting.

FURTHER READING

There is some fascinating reading on the *Euclid* case which involved far more than appears at first sight: a major battle on the desirability and legality of zoning raged behind the scenes. The fullest and most accessible account is given in Haar and Kayden (1989a) *Zoning and the American Dream*, but see also Toll (1969) *Zoned American*, and McCormack (1946) 'A law clerk's recollections'; and Flack (1986) '*Euclid* v. *Ambler*: a retrospective'. A brief overview is given in Scott (1969) *American City Planning since 1890*.

More generally, in addition to Haar and Kayden, a succinct account of the 'Historical development of American planning' is Gerckens (1988). Also recommended is Boyer (1983) *Dreaming the Rational City: The Myth of American City Planning*.

QUESTIONS TO DISCUSS

1   Describe the police power; what does it have to do with zoning?

2   Discuss the constitutionality and fairness of the Mugler and Hadacheck cases.

3   In what ways was zoning unique?

4   What are 'standard state enabling acts'? Why were they important?

5   Why did New York pass a zoning ordinance in 1916? Why did it have implications nation-wide?

6   Consider the fairness of the *Euclid* decision.

7   'Zoning is not planning.' Discuss.

5

THE INSTITUTIONAL AND LEGAL FRAMEWORK

American cities seldom make and never carry out comprehensive plans. Plan making is with us an idle exercise, for we neither agree upon the content of a 'public interest' that ought to override private ones nor permit the centralization of authority needed to carry a plan into effect if one were made.

Banfield 1961

PLANNING AND ZONING

Zoning is an exercise of the police power: the inherent power of a sovereign government to legislate for the health, welfare, and safety of the community. The Constitution confers the police power upon the states which in turn delegate it to the local governments. All fifty states have passed legislation enabling municipalities (and often counties) to operate zoning controls. Most are based on the Standard State Zoning Enabling Act (SSZEA) issued by the Department of Commerce in the mid-1920s.

Paradoxically (in view of what was said above about the distinctions between planning and zoning) another section provides that zoning regulations 'shall be made in accordance with a comprehensive plan'. In fact, zoning was conceived (at least by planners, if not by lawyers) as a tool of planning. But generally the part became the whole, and (with notable exceptions considered later) practice does not follow the text of the Act.

Some states require zoning to be consistent with a comprehensive plan, though consistency is rarely defined; even when it is, the machinery for enforcing it is generally weak. There are some exceptions, particularly in states which have developed a strong growth management policy. These are discussed at length in Chapter 11. For the most part, however, 'in accordance with a comprehensive plan' does not mean what the words suggest; instead, it means that zoning should be carried out comprehensively rather than in a piecemeal manner. In some states, the requirement has come to mean little more than that the zoning laws shall be reasonable. Moreover, where there is a separate comprehensive plan, it is the zoning ordinance which usually carries the force of law, not the plan. The judgment in a 1987 Maryland case captures the essence of the matter (see Box 5.2) Nevertheless, increasing use is being made of comprehensive plans (or master plans, or general plans: the terms are used interchangeably) by both local governments and the courts. Zoning decisions are much easier to defend before the courts if a strong planning framework can be demonstrated.

ZONING AS A LOCAL MATTER

It is important to appreciate that US zoning is essentially a *local* matter. Even the decision on whether to operate a zoning system is usually a local one. Some localities have highly sophisticated zoning systems; some have none at all. But however complex a zoning system may be, it typically remains what it always

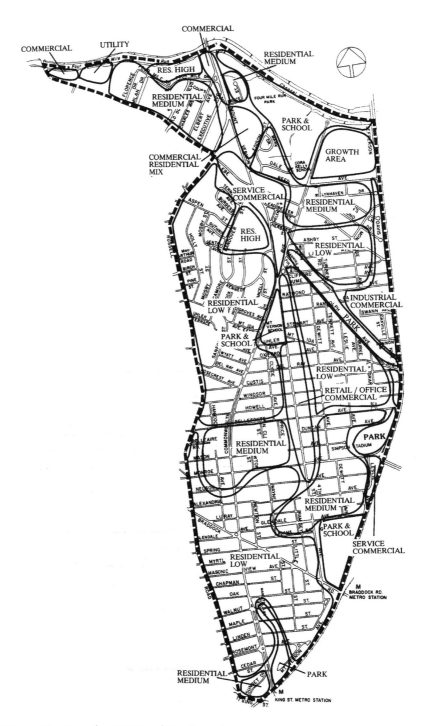

Figure 5.1 Potomac West Area Plan 1992: Land Use Concept

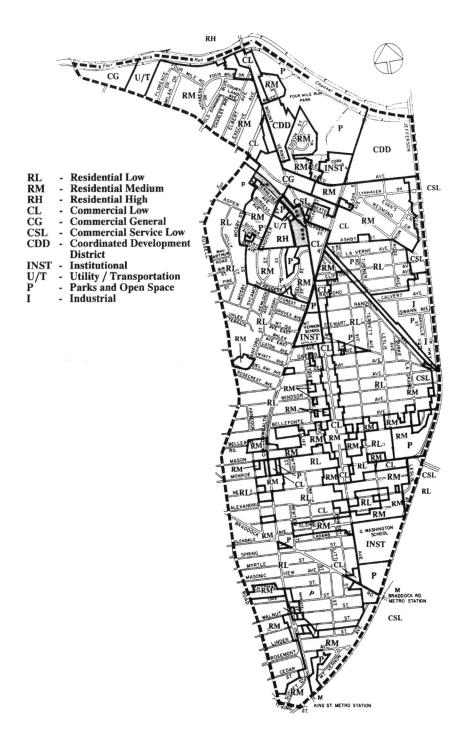

RL - Residential Low
RM - Residential Medium
RH - Residential High
CL - Commercial Low
CG - Commercial General
CSL - Commercial Service Low
CDD - Coordinated Development
       District
INST - Institutional
U/T - Utility / Transportation
P   - Parks and Open Space
I   - Industrial

Figure 5.2 Potomac West Area Plan 1992: Proposed Land Use

BOX 5.1   STANDARD STATE ZONING ENABLING ACT

1 For the purpose of promoting health, safety, morals, or the general welfare of the community, the legislative body of cities and incorporated villages is hereby empowered to regulate and restrict the height, number of stories, and size of buildings and other structures, the percentage of lot that may be occupied, the size of yards, courts, and other open spaces, the density of population, and the location and use of buildings, structures, and land for trade, industry, residence or other purposes.
2 For any or all of said purposes the local legislative body may divide the municipality into districts of such number, shape and area as may be deemed best suited to carry out the purposes of this act; and within such districts it may regulate and restrict the erection, construction, reconstruction, alteration, repair, or use of buildings, structures, or land. All such regulations shall be uniform for each class or kind of building throughout each district, but the regulations in one district may differ from those in other districts.

BOX 5.2   COMPREHENSIVE PLANS AND ZONING

Comprehensive plans . . . represent only a basic scheme generally outlining planning and zoning objectives in an extensive area, and are in no sense a final plan; they are continually subject to modification in the light of actual land use development and serve as a guide rather than a strait jacket . . . The zoning as recommended or proposed in the master plan may well become incorporated in a comprehensive zoning map . . . but this will not be so until it is officially adopted and designated as such by the District Council.

Source: West Montgomery County Citizens Association v. *Maryland National Capitol Park and Planning Commission* 1987

---

has been: 'a process by which the residents of a *local* community examine what people propose to do with their land, and decide whether or not they will permit it' (Garner and Callies 1972: 305).

The distinction between the ideal of planning and the reality of zoning is an important one. Planning is concerned with the long-term development (or preservation) of an area and the relationship between local objectives and overall community and regional goals. Zoning is a major instrument of this; but it is more. Indeed, it has taken the place of the function to which it is supposedly subservient. One of the reasons for this is that responsibility for land use controls has been delegated to the lowest level of local government. These local authorities have traditionally been concerned with attracting development to their areas, but since the 1970s there has been increasing pressure from electors for their communities to be preserved as they are (or at least safeguarded from unwelcome uses such as industry, apartments, and low-income

housing). The powers of zoning provide a very effective tool for this – a tool which can be wielded with a skill that thwarts judicial action. The contrast with planning is a sharp one: a comprehensive plan would deal not only with the needs of the existing inhabitants of an area, but also with its role in meeting the needs for housing newcomers, whatever their income or color. Additionally, it would make provision for such undesirable land uses as power stations, landfills, and a host of other uses which have given rise to the acronym NIMBY: 'not in my back yard', and its more recent progeny NIMTOO: 'not in my term of office'. This can be, and is, done by some local governments; but there are many more who use zoning as a means of precluding comprehensive planning. Some flavor of the action at the local level is given in the 1991 report of the Advisory Commission on Regulatory Barriers to Affordable Housing, more popularly known as the NIMBY report (see Box 5.3).

BOX 5.3   NIMBY

In addition to lobbying elected officials, NIMBY groups regularly participate in the regulatory process through vocal input at public forums and hearings dealing with land use and development issues. Unlike the strict rules governing judicial proceedings, many localities have no specific rules regarding who can testify at public hearings or what rules of evidence apply. Participants often represent *ad hoc* groups that coalesce around a particular development issue. They can be very effective at packing hearing rooms and leaving the impression that public opinion is strongly against whatever project they oppose.

THE LOCAL MANAGERS OF ZONING

Local governments carry out their zoning and planning powers within the framework of powers conferred on them by the individual states, either by constitutional home rule authority or by a specific enabling Act. There are thus fifty different systems of local government – which fortunately it is not necessary to analyze here. What has to be said, however, is that though some states exercise varying degrees of control over local governments, most do not.

The fifty states contain 80,000 local governmental units. Over half of these are school districts and other 'special districts' for particular functions such as natural resources, fire protection, and housing and community development. In 1987, there were 3,042 counties, 19,200 municipalities, and 16,691 townships (see Box 5.4).

The variation among states is exemplified by a few statistics from the *Census of Governments*. Of the 19,200 municipal governments, 183 have populations of 100,000 or more population, while nearly half (9,369) have less than 1,000. (In terms of inhabitants, 62 million live in the former, while only 40,000 live in the latter.) Illinois, with 1,279 such governments, has more municipalities than any other state; Texas has 1,156, and Pennsylvania 1,022. At the other extreme are states such as Connecticut and Massachusetts which have fewer than fifty municipalities. Organized county governments are common, though their power and functions vary. Twenty states have 'townships' which have powers similar to those of municipalities, except that their boundaries are defined without regard to the concentration or distribution of population. Counties play a role in zoning and planning in parts of the country, though the nature of this differs widely.

The doctrine of the separation of powers is an important feature of the US system of government. In brief (and therefore ignoring deviations from the normal rule), a zoning ordinance is passed by the legislative body (e.g. a municipality); applications for rezoning or variances are reviewed by an independent commission (the planning or zoning commission/ board); and appeals are to a board of adjustment (or appeals), and sometimes to the legislative body, and finally – on legal or constitutional grounds – to the normal courts. Furthermore, the role for discretion is severely limited (in theory at least). Indeed, zoning was originally conceived as being virtually 'self-executing': the zoning ordinance and the zoning map would spell out the permitted land uses in such clarity and detail that there would be little room for doubt or discretion. Thus 'policy' is seen firmly as the responsibility of the legislative body, while the commission deals with its execution through the issuance of permits and the occasional variance or exception.

As previously indicated, 'policy' is usually a matter only for the local government: no higher level of government is generally involved. The courts hear appeals against local decisions and therefore, in one sense, act as a type of policy-imposing body. However, policy enters into the courts' deliberations only to the extent that they do or do not defer to the legislative judgment of municipalities: what is termed the 'presumption of validity' (or 'judicial deference'). The courts are concerned, not with planning policy issues, but with legal and constitutional matters. As will become apparent, this neat division between policy and law does not work in practice.

> NOW TOTALLY DESTROYED. LAWYERS (OFFICERS OF THE COURTS) OCCUPY ALL THREE BRANCHES OF GOVT.

BOX 5.4   UNITS OF GOVERNMENT

Type of government	Number of units	
Total	83,237	
US Government		1
State governments		50
Local governments	83,186	
County		3,042
Municipal		19,200
Township		16,691
School district		14,721
Special district		29,532

Source: US Bureau of the Census, 1987 Census of Governments

THE CONSTITUTIONAL FRAMEWORK

Both the federal and state constitutions include provisions which are binding on municipalities. One of the most important of these is the protection of property rights. The Fifth Amendment to the Constitution provides: 'nor shall private property be taken for public use without just compensation' (see Box 5.5). The 'taking issue' (alternatively known as the 'just compensation issue') is at the heart of the major problem facing zoning: when does the exercise of the police power over land use constitute such an infringement of the property right as to become a 'taking'? The crucial matter, of course, is the definition of a 'taking'.

An enormous amount of thought, effort and scholarship has been applied to this question, yet the position is not clear. Postponing fuller discussion until later, here it suffices to note that (to quote the famous words of Justice Holmes in the 1922 case of *Pennsylvania Coal Company* v. *Mahon*) 'the general rule . . . is, that while property may be regulated to a certain extent, if regulation goes too far it will be recognized as a taking'. However, this is not very helpful since we are still left with the puzzle as to where the dividing line is between zoning decisions which are acceptable and those which go 'too far'. In truth, there is none: the Supreme Court has taken the view that (as with the question of obscenity) no generally applicable definition is possible, and each case must be decided upon its merits. The classic statement on this was made in the *Penn Central* case (see Box 5.6).

The Fifth Amendment also includes what is termed 'the public use doctrine': that property can be 'taken' only for a public use. The interpretation of this doctrine has changed significantly in recent decades (illustrating the changes that can take place in the constitutional framework). Until the early 1950s, it was conservatively interpreted as meaning that property which was taken had to be literally used by a public body. It could not be taken for a joint public–private venture, and still less for a private use. Short shrift was made of this in the 1954 *Berman* v. *Parker* case in which it was held that the public purchase of a slum area and its leasing for redevelopment by private enterprise constituted a public use. The court went further: in magisterial terms it declared that the public use requirement of the Constitution was 'coterminous with the scope of a sovereign's police powers'. It also declared that the concept of public welfare was so broad that it could encompass aesthetic matters: 'it is within the power of the legislature to determine that the community should be beautiful as well as healthy, spacious as well as clean, well-balanced as well as carefully patrolled'.

BOX 5.5   CONSTITUTIONAL PROTECTIONS

Amendment V

No person shall be held to answer for a capital, or otherwise infamous crime, unless on a presentment or indictment of a Grand Jury, except in cases arising in the land or naval forces, or in the militia, when in actual service in time of war or public danger; nor shall any person be subject for the same offense to be twice put in jeopardy of life or limb; nor shall be compelled in any criminal case to be a witness against himself, nor be deprived of life, liberty, or property, without due process of law; nor shall private property be taken for public use without just compensation.

Amendment XIV

All persons born or naturalized in the United States, and subject to the jurisdiction thereof, are citizens of the United States and of the State wherein they reside. No State shall make or enforce any law which shall abridge the privileges or immunities of the United States; nor shall any State deprive any person of life, liberty, or property, without due process of law; nor deny to any person within its jurisdiction the equal protection of the laws.

BOX 5.6   THE TAKING ISSUE – THE *PENN CENTRAL* CASE

The question of what constitutes a 'taking' for the purposes of the Fifth Amendment has proved to be a problem of considerable difficulty. While this Court has recognized that the 'Fifth Amendment's guarantee . . . [is] designed to bar Government from forcing some people alone to bear public burdens which, in all fairness and justice, should be borne by the public as a whole', this Court, quite simply, has been unable to develop any 'set formula' for determining when 'justice and fairness' require that economic injuries caused by public action be compensated by the Government, rather than remain disproportionately concentrated on a few persons. Indeed, we have frequently observed that whether a particular restriction will be rendered invalid by the Government's failure to pay for any losses proximately caused by it depends largely 'upon the particular circumstances [in that] case'.

Later cases have further extended 'the public purpose'. One particularly well-known and controversial case was the clearance of the Poletown neighborhood of Detroit for the purpose of accommodating a new General Motors plant. The city, faced with the prospect of General Motors moving out of the area, condemned some 465 acres of land and conveyed it on favorable terms to GM. The Supreme Court of Michigan, in *Poletown Neighborhood Council* v. *Detroit*, ruled that

the power of eminent domain is to be used in this instance primarily to accomplish the essential public

purpose of alleviating unemployment and revitalizing the economic base of the community. The benefit to a private interest is merely incidental. The new factory led to the destruction of 1,021 homes and apartment buildings, 155 businesses, churches and a hospital, displaced 3,500 people, and all but obliterated a more or less stably integrated community embodying a century of Polish cultural life.

(Hill 1986: 111)

Other relevant constitutional provisions require that land use controls be operated by 'due process': the Fourteenth Amendment states that 'no person . . . shall be deprived of life, liberty, or property without due process of law'. The due process clause applies both substantively (is the action legitimate?) and procedurally (is it administered fairly?). Substantive due process requires that controls serve a legitimate governmental interest such as the public health, safety, and general welfare. (A zoning ordinance which excluded low-income families could be challenged on substantive due process grounds.) Procedural due process requires that fair and proper procedures are followed in relation, for example, to public notice of and hearings on zoning ordinances. It further requires that an ordinance be clear and specific: a property owner must be able to ascertain what he may or may not do with his property. If the ordinance is not clear, it can be challenged as being 'void for vagueness'. For example, a provision that authorized a planning commission to permit development on criteria which 'include but are not limited to' those set out in the provision would be void since it would allow the commission to consider unspecified, alternative criteria.

The Fourteenth Amendment also provides that no state 'shall deny to any person within its jurisdiction the equal protection of the laws'. An ordinance which involved racial considerations clearly denies equal protection. However, as so often with zoning matters, cases are often not at all clear. Unequal results can be obtained by devious mechanisms such as the prohibition of multi-family dwellings and the imposition of large minimum lot sizes or large minimum dwelling sizes, or even the regulation of laundries.

THE ROLE OF THE COURTS

Recourse to the courts is a marked feature of the American system of government. As Tocqueville noted 150 years ago, 'there is hardly a political question in the United States which does not sooner or later turn into a judicial one'. Constitutional safeguards can transform a small administrative matter into a major judicial issue. It is therefore not surprising that the courts play a major role in the land use planning process. The role is, moreover, an 'active' one: decisions change over time in the light of

changing economic and social conditions, and also the political complexion of the court. The Supreme Court of the United States, as its name suggests, is the final arbiter; but it does not stand alone. There are over a hundred federal courts, and each of the fifty states has its own system of courts, including a supreme court. Decisions at the state level stand unless overturned by the US Supreme Court. There are important implications of this. First, until a matter is settled by the US Supreme Court (and few cases reach this level), the law can differ among the states. At the extreme, it is theoretically possible for there to be fifty different interpretations of a legal issue. This is particularly important in land use planning since the majority of zoning cases are dealt with at the state level. As a result, judgments vary considerably.

Another issue on the role of the courts needs to be made here. Their function is to ensure that municipalities are acting in a constitutional manner. It is not their role to act as a 'super board of adjustment' or 'planning commission of last resort'. There is a 'presumption of validity' in the actions of a municipality to which the courts give 'judicial deference'. This is nicely illustrated by a Missouri case (*City of Ladue* v. *Horn* 1986). In the zoning ordinance of this city, a family is defined as 'one or more persons related by blood, marriage or adoption, occupying a dwelling unit as an individual housekeeping organization'. The case concerned two unmarried adults who were living, along with their teenage children, in a single family zone. Clearly they offended the zoning ordinance, but was the ordinance constitutional? Certainly, concluded the court: the city had seriously considered the matter and had come to a decision which was within their competence (see Box 5.7).

Closely related is the 'fairly debatable' concept. This holds that if a decision is a matter of opinion, i.e. open to fair or reasonable debate, the court will not substitute its judgment for that of the responsible legislative body. The role of the courts is not to sit in judgment on the wisdom of a local government's legislative actions: that is the function of the political process. The judicial role is circumscribed. Typically, it can overrule a legislative body only if its actions are shown to be clearly arbitrary, capricious, and

---

BOX 5.7   PRESUMPTION OF VALIDITY

The stated purpose of Ladue's zoning ordinance is the promotion of the health, safety, morals and general welfare in the city. Whether Ladue could have adopted less restrictive means to achieve these goals is not a controlling factor in considering the constitutionality of the zoning ordinance. Rather, our focus is on whether there exists some reasonable basis for the means actually employed. In making such a determination, if any state of facts either known or which could reasonably be assumed is presented in support of the ordinance, we must defer to the legislative judgment. We find that Ladue has not acted arbitrarily in enacting its zoning ordinance which defines family as those related by blood, marriage or adoption. Given the fact that Ladue has so defined family, we defer to its legislative judgment.

Source: *City of Ladue* v. *Horn*, 1986

---

unreasonable. In short, 'a court does not sit as a super zoning board with power to act *de novo*, but rather has, in the absence of alleged racial or economic discrimination, a limited power of review' (Wright and Gitelman 1982: 527).

While this is the traditional view, there is no doubt that the local zoning process is frequently subject to irresistible pressure, and decisions are often taken which serve narrow interests. Rezonings, for example, are often made in defiance of the policy enshrined in the zoning ordinance. Some state courts have held that zoning decisions can be administrative rather than legislative in character, i.e. they constitute the *application* of policy as distinct from the *making* of policy. This is particularly so where *ad hoc* decisions are taken on rezoning. Where this is held, there is no presumption of validity (which applies only to legislative acts), and the court requires to be satisfied that the rezoning is needed in the public interest. The state which has been particularly aggressive on this matter is Oregon, and a few states have followed its lead – but most have not. If this seems confusing, that is because it is. There is no generally accepted way of distinguishing between legislative and administrative decisions. Without in any way denying the importance of constitutional issues in land use regulation, it is important to note that they normally operate as a backcloth to local decision-making, rather than being on the front line.

Constitutional issues may arise at any time (often unexpectedly), but it is easy to be misled about their primacy. The voluminous legal text books on land

use planning contribute to this incorrect impression. The very size of these texts is due at least in part to the fact that the courts frequently differ among themselves or refuse to clarify principles which planning authorities can follow. The reader has to digest the cases (which, thanks to the writers of the text books, are reduced to manageable length) and try to establish how the particular issues in which he or she is interested might be treated. Legal texts are misleading also in that they give the impression that constitutional and legal matters are all-important in land use planning. In fact, legal issues are normally in the background, and their influence on local governments is typically limited. The proportion of cases that reach the courts is very small – though there is always a danger of this happening where an aggrieved person has the time and money to embark on a legal challenge. But this is (statistically) unusual; and the general experience of those who come into contact with land use planning is of a bureaucratic rather than a constitutional nature.

SUBDIVISION CONTROLS

While zoning is concerned with the use of land, subdivision regulations relate to the division of land for sale. Originally designed to keep track of the legal ownership of land and to facilitate the establishment of clear titles (thus simplifying transactions), it has grown into a formidable tool of land use planning. The Standard City Planning Enabling Act defines

Plate 12 Two Housing Developments, Laguna Beach, California
Courtesy Alex MacLean/Landslides

subdivision as 'the division of a lot, tract, or parcel of land into two or more lots, plats, sites, or other divisions of land for the purpose, whether immediate or future, of sale or of building development'. (This Act is not to be confused with the Standard State Zoning Enabling Act which, as its title indicates, is concerned with zoning.)

Though subdivision and zoning are quite distinct in origin, they have come to share some important control features. With zoning, these are built into the zoning ordinance or imposed in the administration of the ordinance. Subdivision has acquired similar features (though it is usually applied only to residential development). The first controls were restricted to matters relating to roads. These ensured, for instance, that any streets built in a subdivision would be aligned with existing streets. These controls were extended to deal with the width of streets and side-walks, setbacks, and such like. This enabled local governments to prevent the creation of lots that were unacceptably small or badly configured. But it also gave them the scope to impose conditions relating to 'improvements'. It was not a big step, politically, to move from requiring that roads be a certain width to making the actual provision of the roads a condition of subdivision. Many subdivision enabling acts provided for the dedication of these roads – and also sewers, water mains, and other public facilities. As a result of these extensions to subdivision control, 'the subdivision ordinance was well on its way to becoming a development code by the 1950s' (Callies and Grant 1991).

Extensions of control continued in later years: to the provision and dedication of schools, police and fire stations, parks, and similar on-site facilities. Later, on the logical argument that new development

---

BOX 5.8   SUBDIVISION CONTROL

Subdivision, as the term suggests, is the division of raw land into smaller parcels for the purpose of sale or/and development. The granting of subdivision approval is subject to regulations which establish requirements for streets, lot lines, etc. In short, subdivision is a land use control very similar to zoning: the difference is that while both deal with the physical development of a lot, zoning deals also with the *use* of the land.

A traditional definition of subdivision is 'the division of land, lot, tract, or parcel into two or more lots, parcels, plats, sites, or other divisions of land for the purpose of sale, lease, offer, of development, whether immediate or future'.

Controls over subdivisions preceded comprehensive zoning, but it was not until the years after World War II that subdivision ordinances grew into development codes. These have imposed increasing conditions such as exactions, dedications, and impact fees.

---

had 'impacts' beyond the site being developed, conditions were imposed relating to 'off-site' facilities or payments in lieu. Thus controls originally designed to secure orderly development have been transformed into a complex system of dedication, exactions, and impact fees (see Chapter 7). There is no significant difference in principle between such impositions and those levied under the umbrella of zoning.

The relationship between subdivision and zoning is thus somewhat blurred; but there is an important distinction. Subdivision controls must comply with the zoning ordinance. They cannot be used to amend the zoning ordinance.

FURTHER READING

There are several good texts on American planning law, including Mandelker (1993) *Land Use Law* which contains much more discussion than is general, and omits the extracts from cases which characterizes many law books; the second edition (1994) of *Cases and Materials on Land Use* by Callies *et al.* is explicitly more manageable and user-friendly than its predecessor. A volume in the West *Nutshell* series is succinct and makes the subject seem surprisingly comprehensible: Wright and Wright (1985) *Land Use in a Nutshell*.

For a discussion of 'public purpose', see Merrill (1986) 'The economics of public use'. A blistering attack on the

judicial history of 'public use' is to be found in Paul (1987) *Property Rights and Eminent Domain*.

The tragic story of the destruction of Poletown is set out in Wylie (1989) *Poletown: Community Betrayed*; see also Hill (1986) 'Crisis in the Motor City: the politics of economic development in Detroit'.

The role of the courts is discussed at length in Waltman and Holland (1988) *The Political Role of Law Courts in Modern Democracies*. See also Goldman and Jahnige (1985) *The Federal Courts as a Political System*. An interesting insight into the operation of the Supreme Court is given in Tribe (1985) *God Save This Honorable Court: How the Choice of Supreme Court Justices Shapes Our History*. See also Haar and Kayden (1989b) *Landmark Justice: The Influence of William J. Brennan on America's Communities*.

Subdivisions are discussed within a legal framework in Mandelker (1993) *Land Use Law* (ch. 9). Callies and Grant (1991) provide a comprehensive picture in 'Paying for growth and planning gain'. See also (for California), Fulton (1991) *Guide to California Planning*.

QUESTIONS TO DISCUSS

1   How does a local government obtain its powers of zoning?

2 In what way do the courts 'presume' that a municipality's judgment is valid? Why do they do this?

3 What restrictions are there on the zoning power?

4 Describe the differences between zoning and subdivision.

6

THE TECHNIQUES OF ZONING

A quiet place where roads are wide, people few, and motor vehicles restricted are legitimate guidelines in a land use project addressed to family needs . . . The police power is not confined to the elimination of filth, stench, and unhealthy places. It is ample to lay out zones where family values, youth values, and the blessings of quiet seclusion, and clean air make the area a sanctuary for people.

Belle Terre v. *Boraas*, 1974

THE TRADITIONAL TECHNIQUES OF ZONING

Zoning is the division of an area into zones within which uses are permitted as set out in the zoning ordinance. The ordinance also details the restrictions and conditions which apply in each zone. Thus, the ordinance for the city of Newark, Delaware, has seventeen classes of district including residential, business, and industrial. There are seven classes of residential districts which are distinguished by house type and density. For example, one classification provides for districts with single family, detached houses having a minimum lot area of a half acre, a minimum lot width of 100 feet, a building setback of 40 feet, a rear yard of 50 feet, and two side yards with an individual width of at least 15 feet (and a combined width of 35 feet). Two other one-family detached residential districts have somewhat lower standards; similarly with one-family semi-detached residential districts. In the three detached districts, the taking of boarders is restricted to not more than three in any one-family dwelling. For a one-family dwelling in which the owner is non-resident the limit is reduced from three to two.

Other residential districts are garden apartments up to three stories in height, high-rise apartments of more than three stories with an elevator, and row or town houses. Certain uses are permissible by 'special use permit'. These include police and fire stations, golf courses, professional offices in a residential dwelling, 'customary home occupations', day-care centers, and private non-profit swimming clubs. The zoning ordinance also provides for a Board of Adjustment to which appeals can be made against the decision of the building inspector in enforcing the ordinance, or for a variance from the provisions of the ordinance where a literal enforcement would result in unnecessary hardship.

THE SINGLE FAMILY ZONE: WHAT IS A FAMILY?

Since the protection of the single family home is a major reason for (and a major objective of) zoning, it is clearly necessary to define 'family'. Without a definition it would be possible for a group of un-related students to live 'as a family' and introduce discordant elements into a single family zone. But definitions can raise as many problems as they solve; and so it is in this case. The first difficulty is whether

> ## BOX 6.1   STANDARD STATE ZONING ENABLING ACT
>
> The Standard State Zoning Enabling Act includes the following provisions:
>
> (i) *Grant of power* For the purpose of promoting health, safety, morals, or the general welfare of the community, the legislative body of cities and incorporated villages is hereby empowered to regulate and restrict the height, number of stories, and size of buildings and other structures, the percentage of lot that may be occupied, the size of yards, courts, and other open spaces, the density of population, and the location and use of buildings, structures, and land for trade, industry, residence, or other purposes.
>
> (ii) *Districts* For any or all of said purposes the local legislative body may divide the municipality into districts of such number, shape, and area as may be deemed best suited to carry out the purposes of this act; and within such districts it may regulate and restrict the erection, construction, reconstruction, alteration, repair, or use of buildings, structures or land. All such regulations shall be uniform for each class or kind of building throughout each district, but the regulations in one district may differ from those in other districts.

it is constitutional to 'penetrate so deeply . . . into the internal composition of a single housekeeping unit'. The answer seems to be in the negative except in a few states. In the notorious 1974 *Belle Terre* case, the US Supreme Court upheld a definition that required a family to consist of persons related by blood, adoption or marriage, or a maximum of two unrelated people. The court held that 'the regimes of boarding houses, fraternity houses, and the like present urban problems. More people occupy a given space; more cars rather continuously pass by; more cars are parked; noise travels with crowds.' It could have been objected that the same would result from a family with four teenage children, but the court was carried away by its respect for judicial deference, its overwhelming concern for the archetypical sub-urban family – and the poetry of its own words, a further oft-quoted sample of which is given at the head of this chapter.

The matter did not end there, however, since a later case (*Moore* v. *City of East Cleveland*) concerned an embarrassingly nonsensical outcome. The city of East Cleveland, Ohio, had a complex definition of a family which had the result of making one owner's occupancy of her house illegal. The oddity was that all the occupants were related by blood, but the degree of relationship was insufficient to satisfy the ordinance: the family consisted of Mrs Moore, her son, and two grandsons who were first cousins rather than brothers. Mrs Moore received a notice of violation from the city stating that one of the grandsons was an 'illegal occupant'. Mrs Moore refused to remove him, and the city filed a criminal charge. Upon conviction she was sentenced to five days in jail and a $25 fine. The city argued before the Supreme Court that its decision in *Belle Terre* required it to sustain the ordinance. The usual case was made about the need to prevent overcrowding, to minimize traffic and parking congestion, and to avoid an undue financial burden on East Cleveland's school system.

Surprisingly, at least to those who are unfamiliar with the element of unpredictability which is to be found in the workings of the Supreme Court, the justices had great difficulty with this case. However, the majority concluded that the ordinance was an 'intrusive regulation of the family', and that it was distinguishable from *Belle Terre* in that the latter case dealt with *unrelated* persons.

The issue is not, however, settled; and perhaps, like many other zoning matters, it may never be. In fact, generally there seems to be a trend to liberalize the meaning of the term 'family' to take into account the freer modes of conjugality that are now more common. The test appears to be whether there is 'a legitimate aim of maintaining a family style of living' (Wright and Wright 1985: 179).

GROUP HOMES

A similar test has been applied to group homes for foster children, the mentally retarded, and other groups to which neighbors may object. The rationale here is that the essential purpose of a group home is to provide a family-like environment (in contrast to the custodial character of an institution). The situation is clear where a foster home consists of a married couple and their children, plus foster children. The issue is more difficult when a home is staffed by professionals, and court decisions in such cases are conflicting.

In recent years, many states have passed legislation to prevent the exclusionary zoning of group homes. The nature of the legislation varies: some measures are restricted to certain types of home while others are much broader. Some designate group homes as a 'special exception' under the zoning ordinance; others classify group homes as a separate use to which special standards apply.

THE SINGLE FAMILY HOUSE: SHOULD THERE BE A MINIMUM SIZE?

Photographs of insanitary, tiny, crowded tenements leave one in no doubt that there are standards below which society will not, in all conscience, wish families to live. These standards vary over time and space. What is considered intolerable at the end of the 1990s is very different from what was so considered in the 1790s. Similarly, contemporary standards in the US are very different from those in Bangladesh. Every society has to define for itself the standards at which it expects (and will assist) its people to live. There is nothing scientific about this: it is a matter for judgment and political decision. Yet the zoning system frequently brings these matters before the courts for adjudication; for example, is the minimum lot size or the minimum floor area prescribed by a zoning ordinance acceptable? Unfortunately, the question is more narrowly conceived than this since the courts commonly operate on 'the presumption of validity': that an act of a legislative body cannot be challenged unless it is blatantly unfair. (This was discussed in the previous chapter.) This makes it difficult to challenge minimum area requirements because the onus of proof is on the plaintiff to show that the provision could not have had a valid purpose. As a result, the argument before (and of) the court is usually couched in terms of a dispute between an individual developer and the local inhabitants: wider issues of exclusion and of regional housing needs tend to be pushed into the background, even if they surface at all.

Two cases illustrate the issue. In a 1953 New Jersey case (*Lionshead Lake*), an ordinance provided that residential areas should have a minimum square footage of 768 for a one-story dwelling, 1,000 for a two-story dwelling having an attached garage, and 1,200 for a two story dwelling not having an attached garage. Despite so-called 'expert' testimony which maintained that there was scientific evidence on the effect of living space on mental and emotional health, the trial court concluded that the requirements 'were not reasonably related to the public health, were arbitrary and unreasonable, and not within the police powers' of the township. The New Jersey Supreme Court disagreed, and held that

> it is the prevailing view in municipalities throughout the state that such minimum floor area standards are necessary to protect the character of the community . . . In the light of the constitution and of the enabling statutes, the right of a municipality to impose minimum floor area requirements is beyond controversy.

The court soon had cause to regret these words: its decision gave rise to extensive academic discussion which caused it to rethink its position in later cases. In a 1979 case, it gave prominence to the issue of 'economic segregation'. It noted that, in the quarter-century since *Lionshead Lake*, changes had taken place which were reflected in legislative and judicial attitudes: 'once it is demonstrated that the ordinance excludes people on an economic basis without on its face relating the minimum floor area to one or more appropriate variables, the burden of proof shifts to the municipality to show a proper purpose is being served'.

LARGE LOT ZONING: MAINTAINING COMMUNITY CHARACTER

Large lot zoning has the ostensible purpose of safe-guarding the public welfare, for example by ensuring that there is good access for fire engines, that roads do not become unbearably congested, or that there is adequate open space. These and similar worthy objectives appear frequently in zoning cases, as does an alternative formulation: to keep out undesirable (that is different) people, and to maintain the social and economic exclusiveness of an area. A leading case arose in the Boston suburb of Needham. To control the amount of development in the area, the town passed an ordinance which provided for a minimum lot size of one acre over much of the area. Though declaring that insular interests must give way to the wider good, the court held that the zoning was valid and reasonable. It was swayed by the fact that 'many other communities when faced with an apparently similar problem have determined that the public interest was best served by the adoption of a restriction in some instances identical and in others nearly identical with that imposed' by Needham.

Other apparently similar cases have been decided differently, but no selection of decisions is necessarily representative. On the contrary, as a leading legal digest expresses the matter: 'the validity of large lot zoning is likely to vary depending on the size of the lot, the circumstances of the community or area involved, and the hostility or lack of it to large lot zoning in a particular jurisdiction' (Wright and Wright 1985: 183).

APARTMENTS AND MOBILE HOMES

The reader who has come this far will not be surprised to find that apartments and mobile homes (often far from 'mobile') are the targets of particularly explicit exclusionary practices. However, courts differ in their attitudes to these. Some have gone so far as to approve the restriction throughout an entire jurisdiction of all uses except single family dwellings. By contrast, other cases have ruled that municipalities must allow all types of dwellings in their area. A leading case is *Girsh*; this invalidated a zoning ordinance that totally excluded apartments from the Philadelphia suburb of Nether Providence, Delaware County. The Supreme Court of Pennsylvania ruled that: 'Nether Providence Township may not permissibly choose to only take as many people as can live in single family housing, in effect freezing the population at near present levels.' Unfortunately, the court did not indicate what rights the owners of the Girsh property had as a result of its decision. Nether Providence subsequently zoned several pieces of land – but not the Girsh property – for apartments, claiming that it had thereby complied with the court's decision. The Girsh property owners disagreed, and after two years won a clarifying order from the court which directed the township to grant the permits required for the development of their site. In the meantime, the township had begun procedures to condemn the Girsh property for a public park. This is a typical example of the way in which the drama of land use disputes is played out.

One obvious question which arises with mobile homes is a definitional one: is not a mobile home a single family dwelling? Certainly, modern well-equipped mobile homes in an attractive park may be difficult to distinguish from the stereotypical single family home which, in fact, nowadays can be largely factory produced. The point becomes one of particular significance with 'manufactured housing' intended for permanent siting. This type of housing has been built since 1976 under a national code of health and safety requirements. An observer might have thought that locating an immobile manufactured house on a permanent site would have translated a 'mobile home' into a 'single family dwelling'. Not so, for example, in the village of Cahokia, Illinois, where the zoning ordinance not only restricted manufactured housing to mobile housing parks, but also prohibited such housing from being permanently fixed in such a way as would prevent its removal. The Illinois Supreme Court upheld the ordinance on the grounds that a mobile home might be detrimental to the value of adjacent conventional single family homes, stifle development in the area, or create potential hazards to public health.

There are innumerable such cases. Some reveal remarkable ingenuity on the part of municipal governments in devising methods for excluding mobile homes: a minimum width for all dwellings; a three-acre minimum lot size; a minimum of 'core living space' for all dwellings of 20 × 20 feet.

Prior to the Fair Housing Amendment Act of 1988, it was common for local governments to restrict mobile homes to adults and seniors only. This Act makes it unlawful to discriminate against families in the sale, rental or financing of housing (with some exceptions in the case of housing communities for senior citizens). With changes in design and layout, mobile housing (now more commonly termed manufactured housing) has become more acceptable in recent years. Indeed, it is often difficult to identify what is and what is not 'manufactured'.

CONDITIONAL USES

There are some uses which, though permissible (and necessary), require review to ensure that they do not have an undesirable impact on an area. Hospitals, schools, day-care centers, and clubs, for example, are needed in a community, but their specific location may give rise to traffic congestion and dangers, or to severe parking difficulties. Similarly with gas stations in commercial districts, and multi-family dwellings in a single family district. Zoning ordinances typically make specific provision for such developments which require special restrictions. Though terminology varies among municipalities, these are appropriately termed 'conditional uses'.

VARIANCES

While a conditional use is one which is permissible under the conditions of the zoning ordinance, a variance involves a relaxation of the provisions of the ordinance. The Standard State Zoning Enabling Act confers on the board of adjustment the power 'to authorize upon appeal in specific cases such variance from the terms of the ordinance as will not be contrary to the public interest, where, owing to special conditions, a literal enforcement of the provisions of the ordinance will result in unnecessary hardship, and so that the spirit of the ordinance shall be observed and substantial justice done'.

Variances are of two types: 'area' (or 'bulk') and 'use'. The former involves a departure from the requirements of the ordinance in relation to such matters as lot width, lot area, setback and the like. By contrast, a use variance allows the establishment (or continuation) of a use which is prohibited by the ordinance. Allowing a house to be built closer to the lot line laid down in the variance would be an area variance; allowing a multi-family house in a single family district would be a use variance. In many states, the distinction is of no consequence since the same conditions have to be met (as is the case with the SSZEA provisions). In others, the distinction is crucial since use variances are totally prohibited.

The hardship theoretically has to be one which applies to a particular property, not to the personal circumstances of the owner. The rationale for this is that the matter for consideration is the relationship between the particular plot and the wider area. Any effect which a variance has on this wider area will persist after a change of ownership, or even if the hardship ceases. In fact, many variances are given precisely because of personal hardship. One board had an explicit policy of allowing any use variance requested by a disabled veteran – including automotive repair and body work at homes in a residential area, and the sale of groceries in the front room of a residence. This may be unusual, but there is plenty of evidence to show that boards frequently do consider personal circumstances. One board permitted home occupations in cases of personal hardship on the ground that the harm to the particular neighborhood was far outweighed by the economic hardship to the applicant.

The tests set out in the Otto case (Box 6.2) have been widely, though certainly not universally, adopted. In particular, the requirement that there be an inability to make a reasonable return has become a standard requirement for variances – though 'reasonable' should not be interpreted to mean 'maximum'.

Plate 13 Mixed Use: Church and Gas Station
Courtesy of Jay Hicks

Plate 14 Variance Notice
Courtesy of Jay Hicks

BOX 6.2   VARIANCES – THE HARDSHIP TEST

The classic statement of the hardship test appears in the 1939 New York case of *Otto* v. *Steinhilber*:

> Before the board may . . . grant a variance upon the ground of unnecessary hardship, the record must show that (1) the land in question cannot yield a reasonable return if used only for a purpose allowed in that zone; (2) that the plight of the owner is due to unique circumstances and not to the general conditions of the neighborhood which may reflect the unreasonableness of the zoning ordinance itself; and (3) that the use to be authorized by the variance will not alter the essential nature of the locality.

Many cases could be quoted, but even a long list would be misleading since the differences on the issue among (and even within) the states are great.

It was the original intention that variances would be exceptional. It has not worked out that way. The variance is a popular tool of the boards of appeal who see themselves as brokers for the hard-pressed citizen against the harshness of the law. One writer has suggested that the board of appeal operates as a kind of jury, dispensing rough justice in its hearings of variance applications, resulting in decisions which 'are very apt to reflect the conscience of the community – a close approximation of what most people in the community would think the proper course of action'. Various studies have convincingly shown that boards of adjustment commonly operate according to their own sense of what is right, with little regard for the law or even their local planning department. Most applications are in fact approved.

Since the evidence suggests that illegal use of variances is widespread, it has been proposed that variances should be abolished. It has been argued that this would lead to better and more carefully drafted zoning ordinances. It has also been suggested that variances should be subject to review by a higher authority such as a state review board or specialized courts having metropolitan jurisdiction, or that the power to grant variances be taken away from boards of appeal. But however much lawyers attack the legal deficiencies of the variance, its popularity at the local level assures its continuance as a major feature of the zoning system. There is more to planning than law.

SPOT ZONING

'Spot zoning' is the unjustifiable singling out of a piece of property for preferential treatment. It is not a statutory term: it is a judicial epithet signifying legal invalidity. In a Connecticut case, the court warned that an amendment to the zoning map 'which gives to a single lot or a small area privileges which are not extended to other land in the vicinity is in general against sound public policy and obnoxious to the law'. Such spot zoning is frowned upon by the courts, but if a planning commission decides 'on facts affording a sufficient basis and in the exercise of a proper discretion, that it would serve the best interests of the community as a whole to permit a use of a single lot or small area in a different way than was allowed in surrounding territory, it would not be guilty of spot zoning in any sense obnoxious to the law'.

In this particular case, a land owner requested a rezoning of a small piece of land near (but not adjacent to) some new development. The existing zoning was for residential use, but the owner saw a need for some shops, and he proposed to erect a drug store, and hardware and a grocery store, a bakeshop, and a beauty parlor. He requested an appropriate change of zoning which the planning commission granted. On appeal, the trial court concluded that the requested change amounted to spot zoning, but the appeal court reversed. Its argument was that, on the facts of the case, there was by no means unanimous opposition from surrounding owners to the proposed development and that, even had there been, 'it was the duty of the commission to look beyond the effect of the change upon them to the general welfare of the community'. This the planning commission had done in deciding to support the rezoning: there was a need for additional stores in the area; and it was the policy of the commission 'to encourage decentralization of business in order to relieve traffic congestion and that, as part of that policy, it was considered desirable to permit neighborhood stores in outlying districts'.

DOWNZONING

While an upzoning may well raise the wrath of the neighborhood, an amendment to rezone to a use of lower intensity – a 'downzoning' – is often the result of neighborhood pressure. Since a downzoning is likely to reduce the value of undeveloped land, an objection is likely on the part of the land owner.

A good illustration is a 1983 Iowa case, where a city downzoned some six acres of land on which the owner was intending to build a federally subsidized housing project. The downzoning took place after a public outcry, though ostensibly on the ground that the city's electrical, water, sewerage and road systems were inadequate for a concentration of multi-family dwellings in the area. Not surprisingly, the owner claimed that the reasons given were mere pretext, and that the downzoning was racially motivated. The court, however, held that the city's decision had been taken for valid reasons, i.e. the inadequacy of the utility systems. Furthermore, there was no evidence that the city had a discriminatory purpose.

Thus, applying the 'fairly debatable' rule, the down-zoning was upheld. The court added:

> zoning is not static. A city's comprehensive plan is always subject to reasonable revisions designed to meet the ever-changing needs and conditions of a community. We conclude that the council rationally decided to rezone this section of the city to further the public welfare in accordance with a comprehensive plan.

By contrast, a Connecticut case was decided the opposite way. The court rejected a downzoning which affected the whole of one of the two districts into which the town was divided. It noted that the downzoning was 'made in demand of the people to keep Warren a rural community with open spaces and keep undesirable businesses out'.

Cases on downzoning abound but, because of their great variety and lack of consistency, it is difficult to make any general sense of them. However, one point can be made with a moderate degree of certainty: piecemeal downzonings are likely to be examined much more carefully by the courts, without the usual assumption of validity.

CONTRACT ZONING AND SITE PLAN REVIEW

Zoning theoretically requires uniform conditions within districts (see the last sentence in Box 6.1). Uniformity, however, can lead to undesirable rigidity, and it may be to the benefit of both the owner and the community to depart from a uniform regulation. It is here that contract zoning can be useful.

Essentially, contract zoning is, as the term suggests, the rezoning of a property subject to the terms of a contract. Typically, the terms are negotiated between the owner and the local government following a specific proposal by the owner. There is much learned discourse on the validity and the desirability of contract zoning. The argument in favor holds that conditions can render acceptable a use which otherwise would be unacceptable. The contrary argument is that the police power cannot be subject to bargaining, that conditional rezoning is illegal spot zoning, and that local governments have no power to enact contract zoning amendments. Courts differ widely in their attitudes, and overall the position is confused, to say the least.

Many of the conditions that have been imposed are now normally included in 'site plan review'. This is the preparation of a site plan for approval by the planning board. Such a review can be a normal zoning requirement, or a special requirement for particular types of development such as cluster zones and planned unit developments.

CLUSTER ZONING AND PLANNED UNIT DEVELOPMENT

Traditional zoning is based on the assumption that residential development will take the form of single family houses on individual lots. New patterns of development emerged in the post-war years which require much more flexibility. This is provided by cluster zoning which involves the clustering of development on one part of a site, leaving the remainder for open space, recreation, amenity, or preservation. The overall density of the site is unchanged but, of course, the density of the developed part is increased. This has a number of advantages: the cost of paving and of supplying utilities is reduced; attractive landscape features (or wetlands) can be protected; open space can be provided for recreation; and housing can be provided of a type suitable for 'non-traditional' households who do not want the bother of maintaining a large lot.

The basic idea, of course, is not a new one. It goes back to Clarence Stein, Ebenezer Howard, Clarence Perry, and Frederick Law Olmsted. Perhaps its most famous prototype is Radburn, New Jersey: a 149-acre development with a strict separation of road systems, traffic-free residential culs-de-sac, and a continuous inner park.

A refinement of the cluster concept is the 'planned unit development', affectionately known as a PUD. This differs from cluster zoning in that it is more than a design and planning concept: it is also provides a legal framework for the review and approval of development. It also can incorporate (or even be

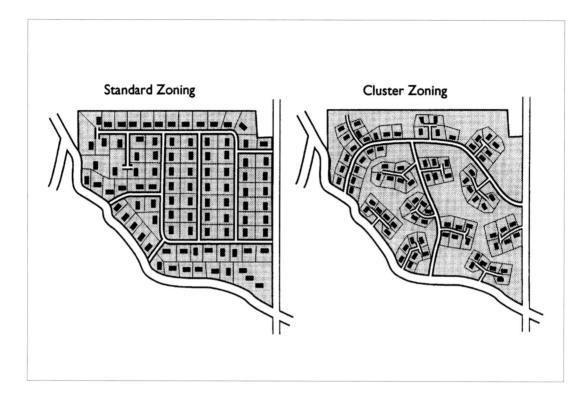

Figure 6.1 Standard Zoning and Cluster Zoning

confined to) commercial and industrial development. Instead of adhering to preset regulations, the PUD system gives developers the freedom to design developments which satisfy market demands. In place of elaborate lists of potential uses, a PUD simply sets out the criteria which have to be met (in relation, for example, to noise, vibration, smoke, odors, heat, glare, and traffic generation). Of course, this approach necessitates negotiations between the developer and the municipality: this is the mechanism by which flexibility is achieved. It is a far cry from traditional zoning.

One other feature of PUDs needs to be mentioned: the role of home owners' associations in managing commonly held property. Membership in such an association can be mandatory for the owners of dwellings in a PUD. Whether the result is a happy, democratic way of managing the local environment, or a financial, administrative, and political nightmare depends on the particular circumstances of the development – and the accidents of time, place, and neighbors.

NON-CONFORMING USES

The introduction of a zoning scheme presents obvious problems with regard to existing uses which thereby become non-conforming. It is impracticable to have these uses removed: indeed, any such threat would have been sufficient to kill off any idea of zoning. The general approach taken has been to hope that, in time, the non-conforming uses would pass away. This has typically proved not to be the case, and municipalities have striven to find ways to speed up the process. They have had little success.

The courts have been unsympathetic to municipalities which attempt to 'zone retroactively', though some of the early landmark decisions (such as *Hadacheck*, which is discussed in Chapter 4) apparently provided the constitutional basis where the offending use became a nuisance. The most common method of applying a control over non-conforming uses (limited though it is) forbids 'expansion' or 'alteration'. Sometimes, a restriction is imposed on rebuilding if a non-conforming use is 'destroyed', for example by fire. Moreover, a use which is 'abandoned' may be refused permission for resuscitation (after a certain number of years).

Unfortunately, all these terms have been, and continue to be, subject to intense debate and judicial differences. Perhaps the clearest case in which a non-conforming use may be eliminated is where a billboard is amortized over a number of years. Amortization is sometimes seen as the most painless way of ridding an area of an undesirable use, and the courts have been sympathetic, particularly where the non-conformer is given a reasonable amount of time. However, the political problems remain, as is dramatically illustrated by the success of the billboard lobby in preventing the use of amortization in connection with the federal highway advertising program. Not only was amortization prevented, but the Act actually requires the payment of compensation for the removal of billboards (see Chapter 8). There could be no clearer example of the force of politics in land use planning.

ZONING AMENDMENTS

A zoning amendment (or 'rezoning', or 'map amendment') is similar to a use variance in that it permits a use which is not allowed by the provisions of the zoning ordinance. However, while a use variance grants the owner an exemption (and leaves the ordinance intact), an amendment changes the ordinance itself. An amendment should be of greater consequence than a variance but practice does not always conform to theory, or even legality.

Amendments can be made to the zoning ordinance or to the map. The former deals with the written provisions of the ordinance; the latter with its detailed designation on a map of the area. The most

common is a map amendment which allows a more intensive use of a particular area. Such an 'upzoning' is usually to a more profitable use, and it is typically made in response to a request by the land owner. It is also regularly opposed by nearby residents: a more profitable use for an owner (for example, an increase in the permitted density of development) can arouse fears of unwelcome neighbors and a fall in property values – commonly expressed in terms of 'a change in the character of the area'. There are, however, circumstances in which a local government might 'upzone' on its own initiative, as for example, where it is seeking to attract development to its area.

SPECIAL DISTRICT ZONING

The term 'special district' is confusing since it has more than one meaning. Traditionally, special districts are governmental units established to perform specific functions which, for one reason or another, cannot be performed by the existing general-purpose local governments. Examples from the nineteenth century are the toll road and canal corporations. Today there are special districts for education, social services, sewerage, water supply, and natural resources. (Perhaps the most famous is the Port Authority of New York and New Jersey.) However, special districts that are so designated for zoning purposes are very different. These are areas to which an amendment of the zoning ordinance applies: they thereby become subject to 'special' zoning controls.

In an interesting and illuminating monograph *Special Districts: The Ultimate in Neighborhood Zoning*, Babcock and Larsen (1990) examine their contemporary use in a variety of contexts including, in New York, the Theater District (designed to preserve the area as such by forcibly bribing developers to build new theaters); the Special Fifth Avenue District (designed to stop the influx of banks and airline offices, and to encourage profitable residential uses above the stores); the Special Garment Center District (designed to prevent the conversion of manufacturing space to office uses and to safeguard the garment industry); and the ill-conceived Special

Little Italy District (designed to preserve the Italian character of the community, in disregard of the Chinese residents). San Francisco has sixteen special districts, several of which are Neighborhood Commercial Special Districts. This designation is applied to relatively small commercial corridors in residential areas, with the objective of 'preserving upper-floor residential units in commercial buildings, and keeping fast-food restaurants from taking over the street'. Chicago has a generic special district: the Planned Manufacturing District, designed to prevent the loss of industrial and manufacturing land to residential and commercial uses. This can be applied wherever it is needed, i.e. wherever the local electors pressurize their alderman for one.

In all cases, the intention is to shield the area from market forces. There is nothing 'special' in this: much zoning is essentially of this protectionist and exclusionary nature. The curiosity of special districts is that most of them have little that is special about them. The residents complain about unwelcome changes, or the threat of changes, and the zoning authority responds by giving them a special status. What appears to be special is the large degree of citizen involvement, not only in the designation of the area but also in enforcement.

It seems clear that special districts are being used in areas where they have no justification; and, once established, their popularity with the citizenry makes abolition extremely difficult.

CONCLUSION

There is no doubt that zoning is not the rigid, simple system of land use regulation that it is sometimes assumed to be. In fact, it is not rigid: it displays remarkable flexibility. It is not simple: it is increasingly complex. It is also, like any instrument of public policy, capable of good use and of misuse. Some local governments operate zoning in a highly responsible manner, with a careful balancing of private and public interests. At the other extreme are those who use it, sometimes blatantly, as an exclusionary technique. Of course, in a sense all zoning has inherent exclusionary features: the public policy problem arises when these are used, not as a means of regulating land, but of regulating people. This issue has been made clear in this chapter. We shall return to it in later chapters.

FURTHER READING

Any standard legal text discusses the court cases dealing with the various instruments of zoning, e.g. Callies *et al.*

BOX 6.3   WHAT IS SPECIAL ABOUT A SPECIAL DISTRICT?

Many in the Planning Commission believe that standard, generic zoning could be used to deal with local problems without the need to create a new special district for each neighborhood. The residents get a psychological lift from residing in an area that has a tag to it . . . They know the special regulations; some of them know the twists and bends of the provisions of their districts as well as the lawyers do and probably better than most of the administrators of the ordinance. They become, as Norman Marcus put it, 'zoning freaks'. Their zoning is the one part of the hopelessly complex myriad of municipal laws and policies that city residents believe they can understand . . . They can immediately spot a sign that violates the regulations of their special district or quickly detect a commercial establishment that operates in a way that is in violation of the labyrinthine district regulations.

Thus it appears that the professionals are losing a zoning conflict to the amateurs, a not unheard of event in the zoning arena.

Source: Babcock and Larsen 1990: 97

(1994) *Cases and Materials on Land Use*, and Mandelker (1993) *Land Use Law*. A concise summary is given in Wright and Wright (1985) *Land Use in a Nutshell*. There are remarkably few zoning books that are not legal in character, but Richard Babcock was a light-hearted lawyer with a knack of telling a good story. His books are insightful as well as enjoyable. See his early (1966) *The Zoning Game* and the later *The Zoning Game Revisited* (Babcock and Siemon 1985). A further volume is Weaver and Babcock (1979) *City Zoning: The Once and Future Frontier*. Bair (1984) *The Zoning Board Manual* is a practitioner's guide which discusses the day-to-day work of the zoning board.

For easy reference there are two APA publications: Meshenberg (1976) *The Language of Zoning*, and Burrows (1989) *A Survey of Zoning Definitions*.

Some issues relating to group homes are discussed in Steinman (1988) 'The impact of zoning on group homes for the mentally disabled: a national survey'; and Gordon and Gordon (1990) 'Neighborhood responses to stigmatized urban facilities'. See also Dear and Wolch (1987) *Landscapes of Despair: From Deinstitutionalization to Homelessness*.

On mobile homes, see Wallis (1991) *Wheel Estate: The Rise and Decline of Mobile Homes*.

For a statement of the law relating to variances, see Mandelker (1993) *Land Use Law*, pp. 640–52.

On cluster zoning and PUDs, see Tomioka and Tomioka (1984) *Planned Unit Developments: Design and Regional Impact*; and Moore and Siskin (1985) *PUDs in Practice*. Home owners' associations are discussed in Mckenzie (1994) *Privatopia: Home owner Associations and the Rise of Residential Private Government*.

Special districts are analyzed in depth (and in a particularly interesting manner) by Babcock and Larsen (1990) *Special Districts: The Ultimate in Neighborhood Zoning*.

QUESTIONS TO DISCUSS

1 **Is zoning inherently exclusionary?**

2 **Discuss the role of zoning in the protection of property values.**

3 **Why are 'variances' so called? Do you think that they should be subject to greater control, for example by a state review board?**

4 **What are the problems that arise over zoning for the single family home?**

5 **In what ways has flexibility been introduced into zoning?**

6 **'Zoning is more about law than policy.' Discuss.**

7

DEVELOPMENT CHARGES

These newcomers bring with them all their fondest hopes of the future. They bring dreams that are the same as ours – dreams of a better life and a better future. What they don't bring with them are the roads, the bridges, the schools, the hospitals, the libraries, the parks, the utilities, the sewers, the waterlines, and all the vast and varied human services that will be needed to realize our dreams.

Florida State Comprehensive Plan Committee, 1987

PAYING FOR THE COSTS OF DEVELOPMENT

The costs of development include not only the construction costs of buildings (houses, shops offices, etc.) but also the costs of the services and facilities which are needed to serve these. Sewage disposal, water supply, and other utilities are the most immediately obvious, but the full list ranges much more widely – highways, schools, day-care centers, hospitals and other social services, public transit, the provision of housing (or transportation) for low-income workers needed to service the development, and so on.

Who is to pay for these? and how? – the existing property owners through their property taxes, the developers through exactions, the new residents through special assessments? The possibilities are theoretically almost endless and, not surprisingly, the whole subject bristles with difficulty and controversy.

The story of development charges (or exactions or imposts – there is no standard terminology) in recent decades is one of an ever-expanding net, bringing more and more services within its grasp. The simplest, and oldest, is the development charge levied to pay for the provision of basic utilities on the site. These

charges arose in connection with subdivision control, and were legitimated in the Standard City Planning Enabling Act of 1928 which explicitly included a requirement for the provision of infrastructure internal to the development. Such services were normally limited to streets, sidewalks, street lighting, and local water and sewerage lines. Services external to the development were paid for by the appropriate suppliers.

This system worked satisfactorily until the housing boom of the post-World War II period, which placed a great strain on the budgets of the municipalities and school districts (and on the tolerance of property tax payers). Existing property owners were unhappy (sometimes vociferously so) at having to pay increased taxes for the benefit of newcomers and, increasingly, municipalities required developers to make contributions (dedications) of land for such purposes as schools and playgrounds (or, particularly in small developments, cash payments in lieu).

The next step was to extend these contributions to other services which are necessary to serve the development. Typically, these are off-site, such as sewerage and water supply systems, and arterial roads. These 'impact fees' have become increasingly popular for two reasons. First, the reluctance of existing property owners to pay for the servicing of new development grew substantially as federal aid to localities was reduced. At the extreme (as with California's Proposition 13) taxpayer 'revolts' brought matters to a head. More broadly, there has been the expansion of popular concern for the environment which has eroded the traditional belief in the benefits of never-ending growth: this culminated in an articulate and sometimes blinkered no-growth ethic.

In some areas, municipalities have for long required developers to provide or finance infrastructure which benefits not just a particular development but a wider area, or even the public at large. This has been particularly so in California where the state courts have taken an unusually relaxed view of the matter. As we shall see, this view was significantly affected by later Supreme Court decisions.

IMPACT FEES

An impact fee is a sophisticated mechanism for shifting from a municipality a part of the cost of the capital investment necessitated by new development. (The question of who bears the cost is discussed later in this chapter.) In tune with the spirit of the age, impact fees are much more complicated than the earlier charges. On the one hand, they can be far wider-ranging, extending to any municipal capital expenditure required to meet the needs of the inhabitants of the new development. On the other hand, they are subject to the restraints of a new calculus which attempts to calibrate the marginal impact of the new development upon a municipality. This is a field which has been extensively dealt with in the courts and in a newly developed area of planning expertise.

The crux of the matter is the determination of the 'rational nexus' – or, more simply (and therefore less appealing to lawyers) the 'connection' – between the charge levied upon a developer and the burden placed on the municipality by the development. Thus a new development might necessitate the building of a major arterial highway in an adjacent area, but it would be wrong to relate the whole cost of the road to the new development if, as is likely, the highway were required not merely for the new development, but for the area as a whole. There is a useful analogy in the last straw that broke the camel's back. The new development is the last straw: but the main burden on the camel is the straw that is already there.

The rational nexus has become increasingly popular as a broadly acceptable concept for debating the division of costs between the developer and the local authority. Basically, it uses cost accounting methods for calculating what share a new development has in

---

BOX 7.1   THE BASIS FOR CALCULATING IMPACT FEES

1 The cost of existing facilities;
2 the means by which existing facilities have been financed;
3 the extent to which new development has already contributed, through tax assessments, to the cost of providing existing excess capacity;
4 the extent to which new development will, in the future, contribute to the cost of constructing currently existing facilities used by everyone in the community or by people who do not occupy the new development (by paying taxes in the future to pay off bonds used to build those facilities in the past);
5 the extent to which new development should receive credit for providing common facilities that communities have provided in the past without charge to other developments in the service area;
6 extraordinary costs incurred in serving the new development;
7 the time–price differential in fair comparisons of amounts paid at different times.

Source: Nicholas *et al.* 1991: 91

creating the need for facilities. That proportionate share then becomes the basis for a charge.

THE RATIONAL NEXUS

The classic statement of the rational nexus concept was set out in the 1987 case of *Nollan* v. *California Coastal Commission*. This was the first exactions case heard by the Supreme Court, and its decision has been followed by an avalanche of writings reflecting a range of differing opinions (not all of which can be said to have clarified matters).

The case concerned an application by the Nollans to the California Coastal Commission for a permit to demolish their dilapidated beachfront bungalow and replace it with a new and larger house. The Commission, which has a policy of increasing access to and along the beach, gave permission conditional on the Nollans providing public access between the sea and their seawall. The dispute eventually found its way to the US Supreme Court where it was held that the Commission's requirement was an unconstitutional taking of property. The essence of the argument was that there was no nexus between the permit (which related to the building of a large house in replacement of a small bungalow) and the condition imposed (an easement for public access across part of the Nollan's land).

There have been many later cases, one of which was of particular importance. This (the *Dolan* case) related to the requirements imposed on the rebuilding of a commercial property in the commercial area of the Oregon town of Tigard. Some 10 per cent of the Dolan's land was required to be dedicated for storm drainage and a pedestrian and bicycle pathway. The US Supreme Court accepted the city's argument that the drainage requirement would mitigate the increase in stormwater runoff from the Dolan's property, and that the pathway would relieve additional traffic congestion on nearby streets. Thus there was a nexus; but were the conditions imposed by the city reasonable? The court thought not: they were in excess of a 'reasonable relationship' to the proposed development: a 'rough proportionality' was required

between the impact of the development and the conditions which could fairly be imposed.

Thus there are now two important tests: nexus and rough proportionality. No doubt later cases will develop (or at least debate) these concepts further: some fun should be had deciding whether there is any difference between the familiar 'reasonable relationship' and the new 'rough proportionality'. These tests give a logical framework for deciding what conditions can reasonably be imposed on developers. They also provide the basis for the design of impact fees.

THE INCIDENCE OF CHARGES

It is frequently assumed that charges imposed on developers are passed on to housing consumers. A classic case in support of this is the huge (100 per cent) increase in house prices in the Washington metropolitan area which took place in the early 1970s following the simultaneous takeover of the counties of Fairfax, Montgomery, and Prince Georges by antidevelopment politicians. However, a closer examination of this illustration shows its falsity. There are two important points. First, the *simultaneous* action of these three large counties created a regional land shortage. More typically, a single small municipality (or even a group of municipalities) around, say, Philadelphia or Chicago could not exert such a market influence: house-builders would simply move to more accommodating areas. (Suburbs are often essentially identical.)

The main issue is an elementary one: the incidence of a charge will be determined by market forces. These vary over time and space. The demand for housing in a particularly attractive area may be highly price-inelastic, and thus charges could readily be passed on to buyers. In an area with a plentiful supply of land of similar amenity, the tendency will be for charges to be passed 'backwards' to land owners: developers will pay less for land than they would have done in the absence of charges. At a time of rapid house price inflation, home buyers are tempted to pay high prices in the expectation that they will rise even further: here the developer should have no difficulty in simply passing on a charge to the buyers. On the other hand, high mortgage interest rates may make buyers resistant to prices which are increased on account of charges. In short, there is no single answer to the question: who pays?

EXISTING VS. NEW HOME OWNERS

One often hears the argument from residents of a growing community: 'Why should we pay for the expansion of public facilities? Let the developers or the newcomers pay for them. We have already paid our fair share.' Why indeed? But again on reflection, the obvious is not so. It is equally easy – and persuasive – to argue that since growth benefits the community as a whole (not just those involved in the new growth), no extra charges should be imposed on newcomers. In any case, is it not unfair for established households to change the rules for newcomers? Existing residents did not have to pay charges when they moved in: why should those who follow them be penalized?

These questions of 'intertemporal fairness' or 'inter-generational equity' are not easy to deal with, and there is no single answer which will fit the situation of widely differing municipalities. Much depends on the historical development of the particular community and the methods employed over time to finance infrastructure. For example, if annual growth is constant

and capital requirements incremental, the debt (both for replacement of facilities for existing residents and for expansion of facilities for newcomers) is spread over an ever-increasing population.

However, some of the investment is for future residents who will not pay their 'share' until further growth takes place. The net effect is problematic. Moreover, continued growth encounters 'thresholds' which require 'lumpy' investments (sewage treatment plants, for example – which cannot be expanded incrementally). The capital cost of these has to be carried for a lengthy period before the full complement of users (and therefore taxpayers) has arrived – by which time, of course, investment in the next 'lump' is necessary. Clearly, it is no easy matter to unravel all these (and similar) matters. The analyst's difficulties are compounded by the impact of inflation, which has a habit of destroying any notion of equity. Clearly there remains plenty of scope for argument at both the academic and the community levels.

LINKAGES

The inconstant way in which planning terms are employed is illustrated by the use of 'inclusionary zoning' to refer to housing (and other facilities) which developers of major downtown projects are required to make before permission to develop is

---

BOX 7.2   WHO PAYS FOR INFRASTRUCTURE?

With time, residents who were established in the community and those who arrived during that year will pay a decreasing share of the cost of facilities that were built for their use. Furthermore, the share of the cost they will bear declines even more as growth rates and financing periods increase.

On the other hand, growth requires new capital outlays for future residents, and established residents must help pay the debt for these capital expansions if they are to be publicly financed. Higher growth rates mean higher rates of facility expansion and, therefore, higher costs to established residents. This impact works as a counteracting balance against the dilution effect described above. The net effect of these forces is not easily determined. It depends upon the magnitude of the growth rate, the interest on borrowed funds, and the length of the financing period.

Source: Snyder and Stegman 1986: 42

given. This 'inclusionary housing downtown' is also, and better, termed 'linkage': it is 'included' in a scheme only in the sense that it is linked to it. However, there is some doubt as to how real this link is. Certainly, it is linked in the minds of the municipal officials who can point to an observed relationship between, say, major downtown development, increases in employment, and rises in housing prices. It is also argued that such development increases the demand for central city housing which, in turn, leads to gentrification and the displacement of poorer residents.

The case is far from universally accepted. Developers, as might be expected, see the matter in a different light. Among their many arguments is that downtown development *follows* and accommodates demand: it does not create it. As one critic nicely put it: 'additions to the supply of office space do not create office employment any more than cribs make babies'. Moreover, there is a danger that linkage fees will kill the golden goose of downtown development, particularly if they are set at a level which will finance significant amounts of housing.

There is merit in both sets of arguments and, as usual, which one is valid will depend upon the particular circumstances of the time and place. Linkage programs have typically been introduced during a major real-estate boom, and usually in connection with downtown commercial development (and mostly with density bonuses as an inducement). The end of the boom drastically changed the economics of the programs, and they were severely cut back. Most cities eschewed them: even if there was concern about the impact of large-scale development, there was a greater fear of scaring away private investment.

Linkage schemes are, in general, limited. There is less activity in the real world than the planning literature might suggest. (It is the exceptional which makes good news.) However, where these schemes carry an incentive or bonus, there is a danger that a municipality's desire to obtain contributions from the developer might overwhelm the requirements of good planning in the area. From this odd point of view it is an advantage that municipalities have so

little in the way of plans: their absence means that they cannot be sabotaged. But where there is an effective plan, bonusing can destroy it. Seattle provides a good example of what can emerge as a result of an assembly of bonuses. The Washington Mutual Tower gained twenty-eight of its fifty-five stories on account of the amenities offered by the developer. As of right, the developer was allowed twenty-seven stories. In addition to this, he obtained thirteen stories for a $2.5 million housing donation, one story for a transit tunnel entrance donation, two stories for a public plaza, two stories to compensate for mechanical space, a half-story for a public atrium, a half-story for a garden terrace open to the public, one story for a day-care facility, two stories for space lost to a sculptured top to the building, two stories for the provision of retail space, and two and a half stories for a public escalator to help pedestrians climb Seattle's hills. This was a truly remarkable example of private munificence! (See Plate 15)

Some of the initiatives of which this is a striking example are examined further in the following pages. Again, there is a difficulty over terminology. When is an incentive a linkage? When is a bonus a charge? When is an impost a development fee? There may be core elements in each of the terms which distinguish one from another, but these are obscured in the real world by local usage and also by the way in which the elements are combined to suit particular objectives. A good case in point is New York's 'incentive zoning' scheme. This is a tale with a moral.

INCENTIVE ZONING IN NEW YORK

New York's incentive zoning scheme started during one of the city's development booms, and was initially concerned with the provision of urban amenities such as plazas. In the words of William H. Whyte, in his highly enjoyable *City: Rediscovering the Center* (1988: 229):

> It seemed a splendid idea. Developers wanted to put up buildings as big as they could. Why not harness their avarice? Planners saw a way. First, they would down-zone. They would lower the limit on the amount of bulk

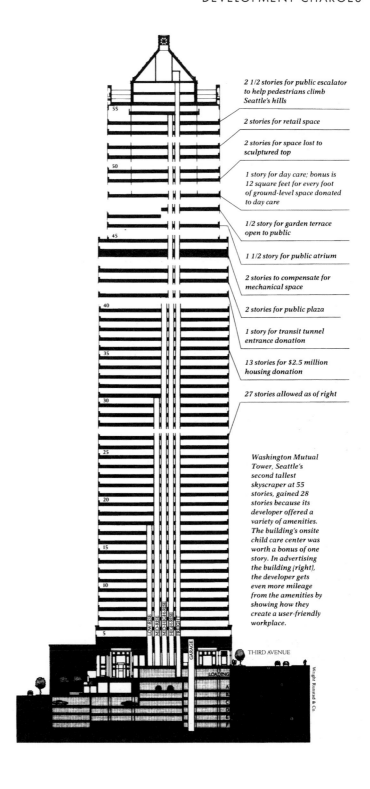

2 1/2 stories for public escalator to help pedestrians climb Seattle's hills

2 stories for retail space

2 stories for space lost to sculptured top

1 story for day care; bonus is 12 square feet for every foot of ground-level space donated to day care

1/2 story for garden terrace open to public

1 1/2 story for public atrium

2 stories to compensate for mechanical space

2 stories for public plaza

1 story for transit tunnel entrance donation

13 stories for $2.5 million housing donation

27 stories allowed as of right

Washington Mutual Tower, Seattle's second tallest skyscraper at 55 stories, gained 28 stories because its developer offered a variety of amenities. The building's onsite child care center was worth a bonus of one story. In advertising the building (right), the developer gets even more mileage from the amenities by showing how they create a user-friendly workplace.

THIRD AVENUE

Wright Runstad & Co.

Plate 15 Washington Mutual Tower, Seattle: 28 Extra Stories for Public Benefits
Courtesy Wright Runstad & Co.

a developer could put up. Then they would upzone, with strings. The builders could build over the limit *if* they provided a public plaza, or an arcade, or a comparable amenity.

At first, the scheme was across-the-board: bonuses were given as of right to developers who met the requirements set out in an ordinance. There was, therefore, no negotiation: thus for every square foot of plaza space provided the developer could claim an extra 10 square feet of office space. The scheme was a great success – in terms of the number of plazas provided. Indeed, it was really too successful: between 1961 and 1973, over a million square feet of new open space was created in this manner. The incentives greatly exacerbated the overbuilding boom of the 1960s which led to the high vacancy rates and lost real estate taxes in the 1970s. Moreover, developers found loopholes in the scheme which allowed them to make windfall profits. Even amenities provided did not escape some criticism. 'A lot of the places were awful: sterile, empty spaces not used for much of anything except walking across.'

THE NEGOTIATION SYNDROME

There were two ways of improving the situation. The first was to elaborate the guidelines. This was done, despite fears that it would be too inflexible and would unduly constrain architectural design. The new (1975) guidelines spelled out the rules of the game in considerable detail, and included the maximum height of the plaza, the amount of seating, the minimum number of trees, and so on. The second was to replace the mechanical as-of-right scheme with a special permit process (later designated by the ungainly term Uniform Land Use Review Process – unaffectionately known by its acronym ULURP). Essentially this was a negotiated agreement.

By 1980, it was clear that zoning in New York City was in real difficulty. Anticipation of bonuses fed back into higher land prices (though developers sought some measure of protection by signing contingency agreements with land owners, with the higher price to be paid only if the anticipated bonus was

granted), and buildings became larger and larger. Promised (negotiated) amenities were sometimes not provided. Citizen groups became increasingly loud in their complaints. Finally, in 1982, midtown zoning was subjected to a sweeping revision. Densities were reduced, and bonuses were largely dropped except for plazas and urban parks. Amenities which had formerly been obtained by way of bonusing now became mandatory. The cumulative effect of these provisions, it was anticipated, 'would go a long way toward eliminating negotiated zoning. They would permit development to proceed on a more predictable and as-of-right basis.'

THE DANGERS OF BONUSING

The lesson is clear: once introduced, incentive zoning is difficult to control. In the absence of any overall official plan policy framework, the process 'can engender considerable uncertainty respecting the city's intentions and can give the impression that the underlying basis of the plan is being subverted' (Toronto City 1988: 9). It can also be difficult to ensure that all land owners are being treated equitably and consistently. Moreover, the absence of basic ground rules results in a process which 'can be extremely time consuming (and costly) and require extensive professional involvement as each application is negotiated'.

A fundamental problem with individually negotiated bonusing is that it leads to a situation in which 'it becomes almost politically impossible for a municipality to approach a density increase without demanding the contribution of some amenity'. Indeed, since it would be difficult for a municipality to grant an increase in density to one owner on more favorable grounds than preceding owners, each grant of a bonus is likely to involve a demand for a higher contribution. 'In other words, from the municipality's perspective, each deal becomes the starting point for the next deal.' The argument does not have to be accepted in its entirety for its force to be felt.

A different argument contends that 'physical planning standards undermined by [incentive zoning]

Plate 16 New York
Courtesy Lee Carter, Viewfinder Colour Photo Library

are not the only interests important to communities. Other values, including those represented by social amenities, contribute to the quality of life, and a city might reasonably resolve that it will tolerate taller buildings and greater congestion in return for more low income housing or daycare facilities' (Kayden 1990: 101). A now classic case of the unhappy times on which New York City bonusing fell is that of the Columbus Circle project which eventually came before a New York trial court at the instigation of the Municipal Art Society of New York. The agreement reached in this case provided for the acquisition by the developer of the city-owned site, a 20 per cent increase in density, and the payment to the city of $455

BOX 7.3   A CRITIQUE OF DENSITY BONUSES

If our city planning theories about appropriate development, servicing and transportation have any validity, the extra density created in one place must either be denied somewhere else or paid for over decades in the expansion of services, utilities and transportation corridors. The costs are not as direct or quantifiable, but the taxpayers will bear them nonetheless.

. . . [what is created is] a circumstance in which one municipal goal (housing, for example) is traded off against another municipal goal (consistent planning) with no necessary relationship between them. If a municipal statement with regard to maximum densities is defensible by planning rationale why should the municipal need for a public swimming pool alter that rationale? Is there not a danger that the planning theory itself will come to be treated as arbitrary and unprincipled – simply one more chip to throw into the urban development poker game?

Source: Bucknall 1988

---

million, plus another $40 million for improvements to the nearby Columbus Center subway station. The city would also have realized about $100 million in taxes each year from the 2.7 million square feet development. Such riches were tempting indeed, and the city did not resist the temptation. But it fell foul of legal hurdles. The crucial point at issue was the fact that a substantial part of the payment to the city was to be for citywide purposes. Most damaging was the appearance of some $266 million of the proceeds in the city's 1988 budget, in advance of final approval of the sale. The court invalidated the sale on the grounds that the incentive provided by the city constituted improper 'zoning for sale'. However, though much of the debate was focused on these financial matters, the underlying issue was the huge size of the proposed development and the shadow it would cast over Central Park. (On 18 October 1987, the Municipal Art Society gathered more than 800 people with umbrellas to form a line from Columbus Circle to Fifth Avenue. On a given signal at 1.30 p.m. all the umbrellas were opened – thus demonstrating the shadow which the building would cause.) The city eventually redesigned the project on a smaller scale.

It would be quite wrong to regard this notorious case as the death-knell of incentive zoning: far from it, as can be seen by reference to Lassar's 1989 study, neatly entitled *Carrots and Sticks: New Zoning Downtown*. This documents in detail the wide range of incentive schemes which are being operated in many American cities. And, so far as the Columbus Circle case is concerned, New York transgressed because it 'upset the delicate balance between competing public and private interests . . . It made economic return the deciding factor, with scant attention to other public goals and land use considerations' (Lassar 1989: 38). In other words, incentive zoning is acceptable as long as it is kept within bounds and does not become a technique for raising additional municipal funds; but the temptation is difficult to resist.

PURPOSES OF BONUSING

All too often bonuses are seen to be self-evidently beneficial. That this is not always so is apparent from the previous discussion. Here we discuss some of the useful purposes which bonusing can achieve.

Ideally, a bonus should be an incentive for a developer to provide an amenity or facility which is of public benefit and which the developer would not provide voluntarily (such as a day-care center). There is an immediate difficulty with this: even if there is an agreement about which benefits are desirable, how can it be determined that they will be provided only if an incentive is offered? It has already been noted that New York, in 1982, abandoned many bonuses for mandatory requirements. San Francisco did likewise with matters of downtown design (Getzels and

---

BOX 7.4   SEATTLE'S RETAIL SHOPPING BONUS

The intent of the retail shopping bonus is to generate a high level of pedestrian activity on major downtown pedestrian routes and on bonused public open spaces. While retail shopping uses ensure that major pedestrian streets are active and vital, a limit to the amount eligible is set in each zone in order to maintain the dominance of the retail core as the center of downtown shopping activity.

Source: Getzels and Jaffe 1988: 3

Jaffe 1988: 2). However, these two cities are hardly representative: many cities are extremely anxious to attract downtown development and are therefore far more inclined to provide incentives rather than disincentive conditions.

A favorite objective of bonusing is the promotion of lively street-level retailing in downtown areas – in contrast to the dead blank walls which so seriously diminish the attractiveness of a street. A good statement of purpose is provided in the Seattle ordinance (see Box 7.4).

Seattle's downtown code provides brief 'statements of intent' for each bonusable amenity. Shopping corridors, for example, are 'intended to provide weather-protected through-block pedestrian connections and retail frontage where retail activity and pedestrian traffic are most concentrated downtown. Shopping corridors create additional 'streets' in the most intensive area of shopping activity, and are intended to complement streetfront retail activity.'

Lassar comments that bonus activities 'run the gamut' and can be clustered around several general categories: building amenities such as urban spaces and day-care centers; pedestrian amenities such as sidewalk canopies and landscaping; housing and human services such as job training; transportation improvements such as station access and parking; cultural amenities such as art galleries and live theaters; and preservation of historic structures. There is, it seems, no end to the ingenuity which can be employed in this area. (Lassar's book covers all these in useful detail.)

IN CONCLUSION

It is interesting to note that the trend towards greater private 'participation' in the financing of infrastructure is not restricted to the US. In a comparative study of the US and Britain, it was noted that in both countries 'the external costs of private land development have, over the past fifteen years, been increasingly borne by private land developers rather than public agencies'. The root cause is the inability of local governments to shoulder the increasing demands being made on them. They are therefore searching for new sources of revenue. Imposing levies on new development is a politically painless way of obtaining extra funds.

FURTHER READING

There is an enormous, and continually growing, literature on developer contributions to infrastructure and other public benefits. Two good sources (both of information and of references) are Nicholas *et al.* (1991) *A Practitioner's Guide to Development Impact Fees*, and Lassar (1989) *Carrots and Sticks: New Zoning Downtown*. See also Getzels and Jaffe (1988) *Zoning Bonuses in Central Cities*.

Changes in the law and practice of charges can be monitored in the quarterly *Urban Lawyer*. A number of papers previously published in this journal have been collected in Freilich and Bushek (1995) *Exactions, Impact Fees and Dedications*. A very full treatment of the situation in California is given in Abbott *et al.* (1993) *Public Needs and Private Dollars: A Guide to Dedications and Development Fees*; this also has a 1995 Supplement.

The 'rational nexus' rule is a cost accounting method which was first elaborated by Heyman and Gilhool (1964) 'The constitutionality of imposing increased community costs on new subdivision residents through subdivision exactions'. See also Ellickson (1977) 'Suburban growth controls: an economic and legal analysis'.

Intertemporal fairness is discussed in Beatley (1988) 'Ethical issues in the use of impact fees to finance community growth'. On 'intergenerational equity', see Snyder and Stegman (1986) *Paying for Growth: Using Development Fees to Finance Infrastructure*.

An interesting comparative study by Callies and Grant (1991) is 'Paying for growth and planning gain: an Anglo-American comparison of development conditions, impact fees, and development agreements'.

Whyte (1988) *City: Rediscovering the Center* gives a fascinating account of incentive zoning (and much else about planning) in New York City.

QUESTIONS TO DISCUSS

1 **List the costs which are involved in residential development. Discuss who should bear these.**

2 **Argue the case for and against impact fees.**

3 **Describe 'the rational nexus'. Do you think it is a useful concept?**

4 **'It is the purchaser who bears the cost of charges imposed on developers.' Discuss.**

5 **Discuss the uses and problems of incentive zoning and bonusing.**

6 **'Much of planning is in fact negotiation.' Discuss.**

PART III

QUALITY OF THE ENVIRONMENT

Most land controls deal with readily measured matters such as the height and bulk of building, the use which it is to serve, and perhaps (more problematically) the transportation implications. However, as the two chapters in this part show, there are increasing concerns for the quality of development. Chapter 8 discusses aesthetic controls. These began with billboards, for reasons which today seem quaint (they were considered to be 'hiding places and retreats for criminals and all classes of miscreants'). Nowadays, aesthetic controls are explicitly accepted (though not always without protest) in terms of design quality. Billboards still attract attention and litigation, but controls now extend to building design and the skyline. Lawyers and architects tend to take a different approach to the issues involved, and there is a wide range of opinion throughout the country on the ethics, law, and practicability of controls. Many areas, however, seem to get along nicely, either with an absence of controls or with a system for judging what is acceptable.

The concern for aesthetic values joins with an interest in history, architecture, and culture in a movement for historic preservation: this is the subject of Chapter 9. At first this was entirely voluntary but government has assumed increasing responsibilities. This was inevitable since the issues involved are often political and constitutional. Preserving things of value from the past can raise issues of property rights, of regulatory controls, and conflicts with other governmental policies (as, for example, when a new road is proposed through a historic area). Interest in the past has continued to increase and, as with natural areas, there is now a problem of dealing with very large numbers of visitors. At the same time, concerns for historic preservation have taken on a wider interest in the human heritage.

8

AESTHETICS

I think that I shall never see
A billboard lovely as a tree
Indeed, unless the billboards fall,
I'll never see a tree at all.
                    Ogden Nash

Ogden Nash may never have seen
A billboard he held dear
But neither did he see
A tree grossing 20 grand a year.
                    David Flint,
        Turner Advertising Company

REGULATING AESTHETICS

Despite the City Beautiful movement, aesthetic considerations have always been problematic in American land use planning. They involve questions of preference and taste on which opinions differ, as the following examples illustrate:

> The American Institute of Architects' choice of the best builder's house of 1950 was refused a mortgage by the Federal Housing Administration. Again, the Veterans Administration imposed a $1000 design penalty on an architect-designed house in Tulsa, Oklahoma, that *House and Home* had displayed on its 1954 cover. The Pruitt-Igoe public housing, which starred in a TV vehicle when HUD Secretary George Romney had it blown up, had won an architectural award in its day.
> (Haar and Wolf 1989: 533)

By contrast, designs once despised can become popular icons: the Eiffel Tower was once described in terms of 'the grotesque mercantile imaginings of a constructor of machines'. Now it is 'the beloved signature of the Parisian skyline and an officially designated monument to boot' (Costonis 1989: 64).

The difficulties of aesthetics are great at both the practical and the philosophical levels, yet, in simple terms, Americans like their neighborhoods to be pleasant and attractive, free of noxious intrusions. Fear of falling property values and unwelcome social groups play their role here too, but there remains a real, and increasing, concern for environmental quality.

This can be seen, for example, in the increased regulation of billboards; in the adoption of landscape ordinances, parking lot regulations, appearance codes, and design guidelines; and in the establishment of advisory or administrative design review boards. This chapter discusses a number of these planning mechanisms. In line with the historical developments, the first to be considered is the control over billboards.

BILLBOARDS

In one sense, all zoning involves aesthetic considerations even if they are as mundane as height and bulk; but other factors are also present, such as infrastructure, congestion, fire prevention, and so forth. Aesthetics first arose explicitly with billboards – and initially the overwhelming judicial view was that controls imposed for such reasons would not pass constitutional muster. A 1905 New Jersey case (*Passaic*) is illustrative:

> Aesthetic considerations are a matter of luxury and indulgence rather than of necessity, and it is necessity alone which justifies the exercise of the police power to take private property without compensation.

Similarly, a Denver ordinance of 1898 was held to be unconstitutional because it had specific requirements solely for billboards, including a ten-feet setback from the street line. The wording of the decision became quite lyrical (see Box 8.1).

Nevertheless, a minority of courts did hold that aesthetics was a legitimate consideration in the exercise of the police power, and by the 1930s it was generally accepted that aesthetic factors could be taken into account. This involved a legal fiction, namely that while aesthetic regulations were not acceptable in themselves, they could be justified on the grounds of associated evils. A 1932 New York decision stated the view nicely: 'Beauty may not be queen but she is not an outcast beyond the pale of protection or respect. She may at least shelter herself under the wing of safety, morality or decency.' A classic statement of this view occurs in a Missouri case of 1913 (see Box 8.2).

The majority of courts today hold that the police power can be used for aesthetic purposes, whether these have the ulterior purpose of promoting some other public good such as tourism or economic development, or for 'pure' aesthetic objectives. An important factor in this change was the 1954 Supreme Court case of *Berman* v. *Parker*. In his decision, Justice Douglas delivered the following *dictum* (that is, it was a gratuitous comment, not crucial to the case in question):

> The concept of the public welfare is broad and inclusive . . . The values it represents are spiritual as well as physical, aesthetic as well as monetary. It is within the power of the legislature to determine that the community should be beautiful as well as healthy, spacious as well as clean, well balanced as well as carefully patrolled.

A later case of some notoriety concerned a Mrs Stover who, for several years, hung clotheslines of rags in the front yard of her house in Rye, New York, as a protest against the high taxes imposed by the city. Each year an additional line was festooned with a remarkable range of materials: tattered clothing, old uniforms, underwear, rags and scarecrows. Neither the neighbors nor the city were amused, and after six years the city passed an ordinance prohibiting the erection and maintenance of clotheslines on a front or side yard abutting a street; exceptions could be granted where there were real practical difficulties in drying clothes elsewhere on the premises. Mrs Stover applied for an exemption but was refused, but she retained her

---

BOX 8.1   TASTE AND THE CONSTITUTION

The cut of the dress, the color of the garment worn, the style of the hat, the architecture of the building or its color, may be distasteful to the refined senses of some, but government can neither control nor regulate in such affairs. . . . Ours is a constitutional government based upon the individuality and intelligence of the citizen, and does not seek, nor has it the power, to control him, except in those matters where the rights of others are impaired.

Source: Curran Bill Posting Co. v. *City of Denver*, 1910

BOX 8.2   THE IMMORALITY OF BILLBOARDS

Billboards endanger the public health, promote immorality, constitute hiding places and retreats for criminals and all classes of miscreants. They are also inartistic and unsightly. In cases of fire they can often cause their spread and constitute barriers against their extinction; and in cases of high wind, their temporary nature, frail structure and broad surface, render them liable to be blown down and to fall upon and injure those who may happen to be in their vicinity. The evidence shows and common observation teaches us that the ground in the rear thereof is being constantly used as privies and dumping ground for all kinds of waste and deleterious matters, and thereby creating public nuisances and jeopardizing public health; the evidence also shows that behind these obstructions the lowest form of prostitution and other acts of immorality are frequently carried on, almost under public gaze; they offer shelter and concealment for the criminal while lying in wait for his victim; and last, but not least, they obstruct the light, sunshine, and air, which are so conducive to health and comfort.

Source: *St Louis Gunning Advertising Co.* v. *St Louis*, 1913

---

clotheslines. The case (*People* v. *Stover* 1963) went to court, and it was ruled that the city was justified in preventing Mrs Stover from her unusual form of protest: a form which was 'unnecessarily offensive to the visual sensibilities of the average person'.

Most courts now take the view that aesthetics alone is a legitimate public purpose and can be controlled by land use regulation. (The same logic is also applied to pornography.) It still remains, of course, for a municipality to ensure that the controls are properly applied.

As the *Berman* and *Stover* cases illustrate, some important court decisions on aesthetics are only indirectly concerned with signs. In the following pages, cases dealing specifically with the issues raised by signs (and billboards in particular) are discussed.

Signs can be of various kinds: directional, political, on-site business, freestanding adverts (billboards), and so on. The crucial distinction, however, is between 'informational' signs and billboards. On-premises signs (which, of course, can be as obnoxious as the worst billboard) are generally accepted in principle, though restrictions are common on their size and number. Billboards, on the other hand, arouse a great deal of controversy – fueled by two active lobbies: one promoted by the wealthy and powerful billboard industry, and the other by Scenic America (formerly the Coalition for Scenic Beauty), dedicated to 'curb an industry that . . . has run amok'.

No holds are barred in the open warfare on billboards. In his legal treatise, Norman Williams (1990: 118.02) refers to the billboard lobby as 'quite intransigent in demands and quite ruthless in tactics'. He comments:

> it has been common gossip among leading planners that the billboard industry maintains (or used to maintain) a blacklist. It is certainly true that on occasion segments of the industry have intervened to try to keep a planner known to be 'uncooperative' out of an important job.

Former New York Senator Thomas C. Desmond is quoted as saying that the billboard lobby 'shrewdly puts many legislators in its debt by giving them free sign space during election time, and it is savage against the legislator who dares oppose it' by favoring anti-billboard laws (Blake 1964: 11).

The billboard industry endeavors to enhance its public image by donating billboards to good causes, such as First Lady Barbara Bush's campaign to promote family literacy, and the boosting of morale in the San Francisco Bay area following the October 1989 earthquake. These public benefits are regularly reported in *Outlook: The Newsletter of the Outdoor Advertising Association of America*.

RURAL SIGNS

Billboards are the art gallery of the public
B.L. Robbins, President,
General Outdoor Advertising Company

With rural signs, the focus of the debate is on the location of billboards in the open countryside alongside major roads and, to a lesser extent, in commercial areas. (There is relatively little controversy about the undesirability of billboards in residential areas, though, as we shall see, there is a distinction to be drawn between on-site business signs and freestanding advertisements.) Billboards along highways and in rural areas have been objected to on aesthetic, safety, and other more ingenious grounds. Among the latter is the argument that regulation of billboards takes away only that value which is created by the building of the road from which the billboard can be seen. Thus the erection of a billboard takes for private gain the value of an opportunity created by public expenditure. In New York, the state erected a screen on public land to hide a dangerously sited billboard. In 1932, the court upheld this action, claiming that no owner had a vested right for his billboard to be seen from the road.

A few states, such as Vermont, Hawaii, Maine, and Alaska, have completely banned rural billboards. In some states, existing billboards can be amortized without compensation, but this policy has been affected by federal legislation concerning highways. Two years after the commencement of the building of the federal interstate highway system, the Federal-Aid Highway Act of 1968 (the 'Bonus Act') provided for a voluntary program under which states could enter into an agreement with the federal government on the control of outdoor advertising within 660 feet of the edge of interstate highways. The incentive was a bonus federal grant of ½ per cent of the construction cost of the highway project. The legislation provided for the prohibition of most off-premises signs, and some controls over on-premises signs. Later amendments exempted from control certain parts of the system: (a) areas that had been zoned or were in use for industrial or commercial purposes in September 1959, and (b) older rights of way which were incorporated into the interstate system.

Only half the states took advantage of this scheme. Three states used the power of eminent domain to eliminate non-conforming signs; seven used a combination of eminent domain and police power controls; and the remainder used police power controls alone. Six of the latter were challenged in court, but in only one case was the action declared unconstitutional: this was the highly conservative Georgia court (Floyd 1979b: 116).

A more elaborate system was introduced by the Highway Beautification Act of 1965 (sometimes known as the Lady Bird Johnson Act) which, in President Johnson's words, would bring about a new approach to highway planning (see Box 8.3). The reality bore little relation to the rhetoric. The lofty intentions of the Act were assailed by the billboard lobby and, instead of a system of effective control over roadside advertising signs – and also junkyards, a vast number of signs were in fact removed from control.

The Act was intended to make billboard control

BOX 8.3   HIGHWAY BEAUTIFICATION

In a nation of continental size, transportation is essential to the growth and prosperity of the national economy, but that economy, and the roads that serve it, are not ends in themselves. They are meant to serve the real needs of the people of this country. And those needs include the opportunity to touch nature and see beauty, as well as rising income and swifter travel. Therefore, we must make sure that the massive resources we now devote to roads also serve to improve and broaden the quality of American life.

Source: President Lyndon Johnson

mandatory in all the states and to extend the controls to major roads in addition to the interstates. The provisions of the Act were made mandatory (with a withdrawal of 10 per cent of federal highway funds from states which did not comply), but the provisions themselves were emasculated by the efforts of the billboard lobby. Though new off-site signs are limited to commercial and industrial areas, the actual controls in these areas (which include *unzoned* commercial and industrial areas) is minimal. The controls are agreed between the federal government and the individual states, but there are no *national* standards: the criteria for control are based on state law and 'customary use'. On-premises signs are totally exempted from control: hence the extremely high signs that are exhibited by gas stations close to the interstates.

The biggest victory for the billboard lobby, however, was the introduction of mandatory compensation for the removal of non-conforming signs. This precluded the elimination of billboards by amortization – a favorite technique among anti-billboard communities. The provision was extended in 1978 to require compensation for the removal of billboards under *any* legislation (not solely under the federal Act). This constitutes a boon to owners of obsolete and abandoned signs, who can off-load them on to the states and receive compensation.

A major problem here, as in the whole of this area, is that federal funds have been very small; as a practical result of this, many states have used all their funds for the acquisition of signs voluntarily surrendered by their owners. A report by the US Department of Transportation (1984: 8) on the operation of the Highway Beautification Program in Florida and Alabama notes that:

> These voluntary sales resulted in many spot purchases from areas where other signs remained. Federal Highway Administration officials generally believed that the only signs acquired under the program were those that were no longer economically beneficial to their owners. The remaining nonconforming signs, presumably of value to the owners, are still visible to the traveling public, and little or no benefit can be seen from the spot purchases.

The restriction of billboards to commercial and industrial areas is a much more limited provision than appears at first sight. Many municipalities (eager for the property tax on billboards – meager though it is) have zoned large areas along interstate and other major highways as commercial. Moreover, an area can be regarded as commercial or industrial even if it is unzoned: all that is necessary is some adjacent activity that could be regarded as falling into one of these two land use categories. Floyd has described the ingenuity of some advertising companies (see Box 8.4). There are many similar stories. The problems are exacerbated by the widespread practice (whether permitted or not) of vegetation cutting undertaken to extend the economic life of signs, misunderstandings (whether intentional or not) between the states and the federal government, and weaknesses in the enforcement of violations. Underlying these specific points, however, is the general lack of political support for the program. Despite the removal of a large number of non-conforming billboards, the legislation is a failure, and is more a testimony to the resourcefulness and power of the billboard industry than to effective controls.

BOX 8.4   INGENUITY IN EVADING BILLBOARD CONTROLS

In Georgia one property owner erected a small shed in a rural area and put up a sign designating it as a warehouse. A large billboard was erected next to this 'warehouse' and the outdoor advertising firm then applied for a permit based on the area's being an unzoned industrial area. In South Carolina, a large national advertising company helped set up a small radio repair shop in a residence that happened to be located near Interstate 95, and then used this 'business' as justification to erect several large billboards.

Source: Floyd 1979b: 119

URBAN SIGNS

Sign controls in urban areas present trickier problems than those in rural areas, where protection of the character of the landscape is usually more clearly evident. In residential areas, aesthetic issues more often relate to the 'harmony' or otherwise between new and existing developments. In commercial districts, the felt need to protect the view of a famous building, or mountain range, or vista can involve extensive controls, as can offensive satellite dishes. Some of these matters give rise to concerns about the infringement of the freedom of speech clause of the First Amendment.

On this, a distinction is frequently made between commercial and non-commercial free speech: commercial speech tends to receive less protection. The current situation (though by no means entirely clear) can be summarized simply. Most federal and state courts now reject free speech objections to sign ordinances; signs create visual problems that justify aesthetic controls. On-site signs advertising the business carried on there tend to be exempt from prohibitions though they may be banned from certain areas for aesthetic reasons. Signs which are not subject to a blanket prohibition can, nevertheless, be subject to controls over their placement and size.

ARCHITECTURAL DESIGN REVIEW

Good design is an elusive quality which cannot easily be defined (see Box 8.5). Yet, if it is to be regulated,
definitions – or at least guidelines – are essential. If an owner cannot understand what is or is not permitted under an ordinance, there is a basic unfairness. The municipality has too broad a discretion, and there is a likelihood of arbitrary action. On the other hand, aesthetic matters cannot be set out in the detail possible in, for instance, a building code. The problem is exacerbated by a lack of clarity as to what 'the underlying public purpose' actually is. A survey by Habe (1989: 199) concluded that while most communities with design control measures seemed to know why they wanted these, very few demonstrated a clear understanding of what was involved. There was little understanding of how controls could be translated into practice, how effective they might be in attaining objectives, and what their long-term implications could be.

One of the difficulties (as in many areas of public policy) is that there is typically more than one objective. Habe's survey of sixty-six American cities, showed that, in addition to aesthetic considerations, each city had at least two other objectives unrelated to aesthetic concerns. These included general 'economic' and 'public welfare', protection against urban problems such as crime, slums, and traffic congestion, 'psychological well-being, ecological concern, historic/cultural concern, facilitating the functional aspect of community life, accommodating user need, and maneuvering migration'. The vagueness of many of these objectives is noteworthy, and is common in this field.

A particularly frequent objective is the preservation of community character. This can, in practice, mean

---

BOX 8.5   THE ELUSIVENESS OF GOOD DESIGN

Short of requiring the builder to copy specific prototypes, it is impossible to legislate good design. No set of rules can anticipate all the situations and conflicts that will eventually surface, and there is a tendency that rules designed to prevent something bad will also prevent something good from happening. At best, we stack the odds against the worst and hope for the best. However cleverly the controls have been structured, designers have demonstrated an uncanny ability to technically meet every requirement and still evade the spirit of the underlying design objectives.

Source: Hedman and Jaszewski 1984: 136

anything from the perpetuation of an architectural style to the exclusion of different social groups. Perhaps the most popular design control is the 'no excessive difference' rule. This is typically expressed in terms such as 'new buildings must reflect the existing character of the area', or 'be sensitive to existing architecture':

> According to one community, harmony was defined as 'pleasant repetition of design elements to provide visual linkage, direction, orientation and connection of areas'. Often the concept is interpreted as similarity: 'cornice lines, openings and materials of new structure to be similar to those of adjacent buildings' (Concord, CA); or 'retaining and freestanding walls should be finished with brick, stone or concrete compatible with adjacent buildings' (Rochester, NY).
>
> (Habe 1989: 202)

Habe comments that 'such overemphasis on similarity of design encourages the trend towards specificity of standards, including setting specific architectural styles, rather than encouraging innovative solutions from designers'.

'No excessive difference' seems to be generally acceptable, but 'no excessive similarity' is more problematic. However, it is inappropriate to be dogmatic on this issue since remarkably few cases involving architectural review have come before the courts. Indeed, relative absence of litigation is a feature of aesthetic controls. The reasons for this, though speculative, are interesting. A major factor is that developers prefer to have community support for (or at least to avoid community opposition to) their proposals. They are therefore generally willing to negotiate: after all, the issue at stake is 'only' one of design, not one of significant cost. And who wants to build, or live in, a dwelling to which neighbors are hostile? If a developer (or a developer's client) wants a dwelling which is unusual, the obvious path of least resistance is to choose a site occupied by, or being developed for, similar deviants. The lower the density, the easier it is to be different in peace.

The negotiation of good design is a striking feature of a number of control schemes. For instance, the Lake Forest, Illinois, ordinance provides for review by a five-member board before a building permit will be issued:

> The board has not denied a permit in its twenty years of existence, choosing instead to negotiate with designers and developers over points of disagreement . . . The board's approach has been to seek improvement rather than censorship of design. The board is yet to be challenged through a lawsuit.
>
> (Poole 1987: 305)

Architectural design controls involve particular difficulties in the large cities affected by successive property booms. San Francisco can be quoted as an illustration. After a lengthy period of public controversy, the city enacted a series of design-related ordinances (an extract from which is given in Box 8.6). Whether the 'fancy tops' controls have proved effective in improving the skyline of San Francisco is debatable: they have certainly produced some very

---

BOX 8.6   SAN FRANCISCO DESIGN ORDINANCE

- The upper portion of any tall building be tapered and treated in a manner to create a visually distinctive roof or other termination of the building facade, thereby avoiding boxy high rise buildings and a 'benching' effect of the skyline.
- New or expanded structures abutting certain streets avoid penetration of a sun-access plane so that shadows are not cast at certain times of the day on sidewalks and city parks and plazas.
- Buildings be designed so the development will not cause excessive ground level wind currents in areas of substantial pedestrian use or public seating.
- The city consider the historical and aesthetic characteristics of the area along with the impact on tourism when issuing a building permit.
- Building heights downtown be reduced from 700 to 550 feet (from about 56 to 44 stories).

Plate 17 San Francisco
Courtesy Alex MacLean/Landslides

curious buildings (perhaps a nice case of beauty being in the eye of the beholder?). Seattle has similar, though less detailed restrictions in the downtown area: 'the requirements limit building heights, establish setbacks to maintain light and air, and ensure designs that reduce wind-tunneling and retain views of Elliott Bay' (Duerksen 1986: 14).

Boston has produced a volume of *Design Guidelines for Neighborhood Housing*. This is part of an ambitious project 'to transform all of the city's vacant buildable lots into attractive and affordable housing'. The guidelines emphasize existing neighborhood character and also cover such matters as the site, 'the organization of the residences' (by which is meant 'public and private territory and views, security and surveillance, and construction materials and maintenance'), and the residence itself.

Portland, Oregon, has received much publicity (deservedly so) for its urban planning and design. Of particular interest is the incorporation of design into the urban planning process. According to Abbott's analysis, this came about in three stages:

> During the 1960s, design issues were raised piecemeal in response to specific projects and problems. During the 1970s, design goals were incorporated into general planning policies. In the 1980s, design considerations have become an accepted part of the regulatory planning system.
>
> (Abbott 1991: 1)

Though the city of Portland has its own particular character (which Abbott describes as its 'orientation to a moralistic political style which accepts the possibility of disinterested civic decisions'), some other cities are moving toward a similar use of external

Plate 18 Horton Plaza, San Diego
Courtesy Rick Buettner, Viewfinder Colour Photo Library

standards and comparisons, and toward the integration of design review with other planning goals for the area.

It would, however, be wrong to give the impression that there is a widespread movement in this direction. Much of the US has no design control (or certainly none that is apparent). There is considerable controversy on the reasonableness and effectiveness (and, despite evidence to the contrary, constitutionality) of design controls, except in areas with highly distinguishing features such as historic districts. One compromise is to have informal guidelines, but these are no substitute for legal sanctions (even when, as in the case of Lake Forest, they are held in abeyance). A telling case in point is the city of Philadelphia, where an unofficial height limit was set at the top of William Penn's hat on the City Hall. This limit operated from 1894 to 1984 when it was finally breached.

Poole maintains that design controls directed at preventing the construction of excessively different buildings violate the First Amendment. Kolis (1979: 304) argues that 'the general public welfare will be better served by recognizing the First Amendment rights of architects and their clients so that they may achieve great architecture'. Habe complains that in attempting to ensure legality (and also to achieve maximum efficiency) design controls tend to emphasize details (which are easier to define) and adopt the use of generalized conditions from a standard list. S. F. Williams (1977: 33) suggests establishing criteria similar to those for obscenity (for example, that the proposed design is 'blatantly offensive' to community standards). Poole has also argued that 'architectural designs sufficiently distasteful to cause measurable harm to a neighborhood occur so rarely (if ever) that regulations to prevent them amount to making mountains out of molehills'. Municipalities should 'get out of the role imposing majoritarian notions of tastefulness on the community at large. Tastefulness by a committee assures nothing more or less than mediocrity.' The final word can rest with a view from the science of economics: Hough and Kratz (1983) assert, on the basis of an hedonic price equation for office space in downtown Chicago, that 'good' new architecture passes the market test: 'tenants are willing to pay a premium to be in *new* architecturally significant office buildings, but apparently see no benefits associated with *old* office buildings that express aesthetic excellence'. In short, the market can be left to look after new buildings; for historic buildings 'those who value them must devise feasible non-market mechanisms so that their preferences for these buildings are revealed and their dollars are contributed'. Lake Forest is hardly likely to be impressed.

FURTHER READING

Costonis (1989) *Icons and Aliens: Law, Aesthetics, and Environmental Change* is a lively, concise, illustrated account of the aspects of aesthetics listed in the subtitle. It nicely bridges the fields of law and design.

Duerksen (1986) *Aesthetics and Land Use Controls: Beyond Ecology and Economics* is a useful short monograph in the APA Planning Advisory Service series.

A collection of essays on a wide range of design review issues is edited by Scheer and Preiser (1994).

Floyd (1979a and b) provides good contemporary accounts of the passage of the billboard controls in two articles: 'Billboard control under the Highway Beautification Act', and 'Billboard control under the Highway Beautification Act – a failure of land use controls'. There is an extended discussion in (1979c) *Highway Beautification: The Environmental Movement's Greatest Failure*.

There are many books on urban design, e.g. Hedman and Jaszewski (1984) *Fundamentals of Urban Design*. The perspective of the practicing planner-politician is given by Barnett (1982) *An Introduction to Urban Design*. A useful article is Habe (1989) 'Public design control in American communities'.

QUESTIONS TO DISCUSS

1 What are the arguments for and against the regulation of aesthetic matters?

2 Do rural signs raise different issues for planners than urban signs ?

3 In what circumstances (if any) do you consider that billboards should be banned?

4 Is 'preservation of community character' a legitimate matter for land use regulation?

5 Do you think that good design can be measured by market prices?

9

HISTORIC PRESERVATION

If it is the role of the planner concerned with land use patterns to understand them in relationship to the dynamics of the contemporary land market and its interplay with social and cultural values, then it is the task of the historic preservation planner to understand the evolution of those patterns over time and to assess the significance of remaining fragments. Historic preservation planning is one of several perspectives on, and public interests in, land.

Ames *et al.* 1989

PRESERVATION AND PROFIT

Planning involves the resolution of conflicting claims on the use of land. This is particularly clear in the case of historic preservation since the nature of the conflict is so readily apparent. Typically, one party (often more than one) wants to preserve a historic structure for public enjoyment now and in the future. The other party (often one only) wants to use the site for a new use which produces a higher profit. The traditionalists use the language of culture and history; the redevelopers speak in terms of market trends and economic returns.

In the last century, the controversy was normally between public and private interests. This changed as it became evident that history could be molded to produce profits and (what may amount to the same thing) a good public image. For example, there was capital to be made out of a company's environmental concerns if these were manifest in the preservation of an historic building for modern use. Further profits were to be realizable from tourist attractions. Above all, changes in tax provisions transformed the attitudes of land owners and developers to preservation.

The new enthusiasm for historic preservation was

not to everyone's liking. As in other fields (national parks for instance – which in the US are in danger of becoming theme parks or zoos) too many people seeking to enjoy 'a piece of history' can overwhelm it and destroy the very experience which is sought. Moreover, both preservationist and developer interests have become much more sophisticated than in earlier times. The step from preserving a physical structure to preserving a community is not a large one (as New York experience with landmark preservation clearly shows). Community groups and preservation societies can be bought out by generous contributions to their good work from developers. Preservationists sit on the boards of development companies; and their development interests in turn are to be found on the managing bodies of voluntary bodies.

Thus the old lines of demarcation have become blurred. For the planner, the situation has become confused, and frequently an apparently simple clash of development and protectionist interests turns out to be something much more complex. In this chapter, a number of these issues are discussed, but the main focus is on the evolution from a simple approach to the historic preservation of landmarks toward a 'planning perspective' on cultural matters.

This perspective has now merged with a concern for urban design. On this, as Abbott (1991) has pointed out, 'design considerations have become an accepted part of the regulatory planning system'. Added to this there has been such an extraordinary expansion of the field of interest of what used to be simply called 'historic preservation' that the very term is now of vintage stock.

THE EARLY DAYS OF HERITAGE PRESERVATION

Historic preservation in the US grew from the grass roots in an unorganized way. Its early development is the story of a large number of predominantly private endeavors to save individual structures or sites. Many of these failed, like the attempt to save the so-called 'Old Indian House' in Deerfield, Massachusetts, which was the last home in the town to escape the famous massacre of 1704 (demolished in 1848 because it had 'no intrinsic value'). Similarly, the John Hancock House in Boston was destroyed in 1863. Others had a near miss, like Independence Hall, which the City of Philadelphia purchased for $70,000 in 1816. One of the most notable successes was Ann Pamela Cunningham's crusade to save Mount Vernon. The essentially indigenous character of the historic preservation movement in the United States was not changed by the occasional action of the federal government. This was restricted mainly to the acquisition of a small number of landmarks and individual park sites (such as Shiloh National Military Park in 1894, and Morristown Historical Park in 1933).

The national parks, of course, were already in the public domain and thus sites within these parks which needed public protection did not require acquisition. Public lands, in fact, were the scene of another development in historic preservation. This was the preservation of 'antiquities'. The Antiquities Act of 1906 provided for the designation as National Monuments of areas *in the public domain* which contained 'historic landmarks, historic and prehistoric structures, and objects of historic or scientific interest'.

This was broadened in 1935 with the introduction of the Historic Sites, Buildings and Antiquities Act, aimed at fostering 'a national policy to preserve for the public use historic sites, buildings and objects of national significance for the inspiration and benefit of the people of the United States'. It called upon federal agencies to take account of preservation needs in their programs and plans and, for the first time, promoted the surveying and identification of historic sites throughout the country. This program became the base for the National Register of Historic Places some thirty years later.

These early endeavors in preservation were essentially concerned with history and cultural values, as distinct from architectural quality (though the line was sometimes blurred, as with Monticello which had both historical and aesthetic features). At this time, buildings, structures, and sites were of appeal because of their associative and inspirational values. There was, however, an increasing concern for architectural values toward the end of the nineteenth century, neatly expressed in William Sumner Appleton's statement of purpose of the Society for the Preservation of New England Antiquities. He organized this in 1910 'to save for future generations structures of the seventeenth and eighteenth centuries, and the early years of the nineteenth, *which are architecturally beautiful or unique*, or have special historical significance. Such buildings once destroyed can never be replaced' (Hosmer 1965: 12). The added italics emphasize the primacy here accorded to architectural values. Of course, the historical and associative elements remain important today: in fact it is often difficult to disentangle them.

The Depression years were a lean time for historic preservation, although there were notable exceptions such as the creation of preservation commissions in Charleston, New Orleans (the Vieux Carré Commission), and San Antonio. The World War II and early post-war period was even leaner: indeed, urban renewal and highway projects destroyed many buildings which a few years later might have been preserved. It was this very destruction which (together with the reaction to the sterility of the International Style in new architecture) acted as a

Plate 19 Faneuil Hall and Market, Boston
Courtesy David Williams, Viewfinder Colour Photo Library

catalyst to an unprecedented burst of activity in the mid-1950s. The culmination of this was the publication in 1966 by the US Conference of Mayors and the National Trust for Historic Preservation of a powerful, eloquent manifesto *With Heritage So Rich*.

WITH HERITAGE SO RICH AND SUBSEQUENT LEGISLATION

The report *With Heritage So Rich* had the advantage, which many reports lack, of appearing at precisely the right time for a positive political response. It was cogently argued, dramatically illustrated, and persuasive. It consisted of a series of essays and a concluding set of recommendations. Some of these were immediately implemented by the 1966 National Historic Preservation Act (NHPA): for example, the introduction of a National Register of Historic Places, and the establishment of an Advisory Council on Historic Preservation (ACHP) (see Box 9.1). A remarkable change of policy in relation to highway construction was introduced in a provision of the Transportation Act of 1966 which requires the Secretary of State for transportation to refuse approval for projects which would involve damaging or demolishing historic sites unless there is 'no prudent and feasible alternative'.

A similar provision was included in the Model Cities Act 1966 in relation to urban renewal plans. Later amendments extended this policy to all federal departments. Changes in taxation were made by other legislation, such as the Tax Reform Act 1976 and the Economic Recovery Tax Act 1981, to encourage historic preservation (for example, by way of tax deductions for rehabilitation). A separate Act, the National Environmental Policy Act (NEPA), included additional provisions for preserving 'important

historic, cultural, and natural aspects of our national heritage'.

As this brief summary demonstrates, the fifteen years following the publication of *With Heritage So Rich* witnessed a veritable orgy of legislative activity. In the following pages, the more important features of this are discussed.

THE NATIONAL REGISTER OF HISTORIC PLACES

The National Register of Historic Places is maintained by the Keeper of the National Register in the National Park Service of the Department of the Interior. It lists districts, sites, buildings, structures, and objects which are significant on a national, state, or local level in American history, architecture, archeology, engineering, and culture: in short, America's cultural resources. These are provided with a degree of protection from the harmful effects of federal action. The federal government is committed, by law, to protect these resources: agencies are required to follow a statutory process of review and consultation with the ACHP in connection with any undertaking affecting properties included in the list. Additionally, and at first sight curiously, this requirement extends to properties which, though not listed, are eligible for listing. (The rationale for this is summarized in Box 9.2.) Though both listed and eligible properties are subject to the review process, only listed properties are qualified to receive grant aid or tax advantages (discussed below). The process (popularly known by its legal reference, as the 'section 106 process') is not a mere formality: all federal actions and federally funded projects are monitored or reviewed by preservationists. This usually occurs at the State Historic

---

BOX 9.1   ADVISORY COUNCIL ON HISTORIC PRESERVATION

The Advisory Council on Historic Preservation is composed of the heads of federal agencies whose departmental activities regularly affect historic properties; experts in historic preservation; a governor; a mayor; private citizens appointed by the President; and representatives of the National Trust and the Conference of State Historic Preservation Officers.

BOX 9.2   LISTING AND ELIGIBILITY

Before 1980, owners had no right of objection to listing. The rationale for this was:

1 the Register is a list of properties that meets an objective evaluation, which applies criteria and professional standards regardless of a current owner's opinion of the property;
2 the owner's opinion has no bearing on whether a property is historic; and
3 inclusion in the Register does not directly restrict a private owner's use of his property in any manner.

However, as a result of political pressures, the right to object was introduced in the 1980 amending act. As a compromise, the concept of 'eligibility' for inclusion in the Register was added.

Preservation Office level. Indeed, 'review and compliance', as it is called, now occupies a dominant position in the state programs. However, as always, much depends upon the quality of the local administration.

STATEWIDE COMPREHENSIVE HISTORIC PRESERVATION PLANNING

Though historic preservation is very much a local matter, it is more than this: as with all local plans, relationships with wider plans have to be forged. (The imperative is misleading since, in practice, as has already been stressed, there are so few *plans* – as distinct from zoning provisions.) Ideally, these would include such functional elements as transportation planning, economic development planning, and environmental planning. The most promising approach is where different planning agencies integrate (or at least cooperate in) their planning processes. In the words of the Delaware Comprehensive Historic Preservation Plan:

> It is very difficult, if not impossible, to integrate complete plans that can translate the recommendations of one plan or functional area into terms relevant to another. Plans must be integrated, or information exchanged, at the points in the planning process when problems and alternative goals are defined and analyzed and decisions made.
>
> (Ames *et al.* 1989: 9)

Moreover, without coordination, historic preservation policies may conflict with land use policies. In his

Handbook on Historic Preservation Law, Duerksen quotes the case where

> preservationists have struggled to enact an ordinance to control design details or forbid demolition by private developers in a historic neighborhood, only to discover that the real threat in the area is a city zoning policy encouraging high-rise development. In short, preservationists have focused on design issues and on saving threatened buildings when the key issue is more often how landmarks and their surrounding areas will be developed according to local zoning classifications and redevelopment programs.
>
> (Duerksen 1983: 44)

Coordination has other advantages, not the least being that it impresses courts that the municipality has a comprehensive plan and is working to this rather than making a series of *ad hoc* decisions. It also facilitates the use of sophisticated zoning techniques such as incentives, bonuses, and the transfer of development rights.

Coordination is also desirable between the policies of the municipality and the state. A well known example is Oregon's statewide planning goals, which are mandatory on municipalities. One of these goals includes the requirement that local programs shall be provided which will 'protect scenic and historic areas and natural resources for future generations, and promote healthy and visually attractive environments in harmony with the natural landscape character'. Inventories are required of historic areas, sites, structures, and objects; and cultural areas. An historic area is defined as 'lands with sites, structures, and objects

Plate 20 Nemours Gardens, Wilmington, Delaware
Courtesy Alex MacLean/Landslides

that have local, regional, statewide, or national historical significance'. A cultural area is 'an area characterized by evidence of ethnic, religious or social group with distinctive traits, beliefs, and social forms' (Rohse 1987: 261). Local comprehensive plans and land use regulations are required by statute to comply with these goals.

HIGHWAYS AND HISTORIC PRESERVATION

The ravages of highway construction constituted one of the major reasons for the swell of public opinion against 'the federal bulldozer'. It is therefore perhaps fitting that the strongest federal provision is to be found in a transportation act. The Department of Transportation Act declares that it is a matter of national policy that a 'special effort' shall be made to preserve and enhance the natural beauty of lands crossed by transportation lines (section 1653(f); commonly referred to (by a previous numbering) as section 4(f)).

The scope of the provisions of the Transportation Act are much broader than that of the NHPA: it gives protection to any site considered by officials to be of historic significance – not only those listed, or eligible to be listed, on the National Register. Moreover, whereas the NHPA merely gives the ACHP opportunity for 'comment' on harmful action, the Transportation Act's more stringent provisions permit harmful use only if two conditions are met: first, that no feasible and prudent alternative exists, and, second, that all possible planning is carried out to minimize harm. The courts have held that there must be 'truly unusual factors' of 'extraordinary magnitudes' for this high standard to be met.

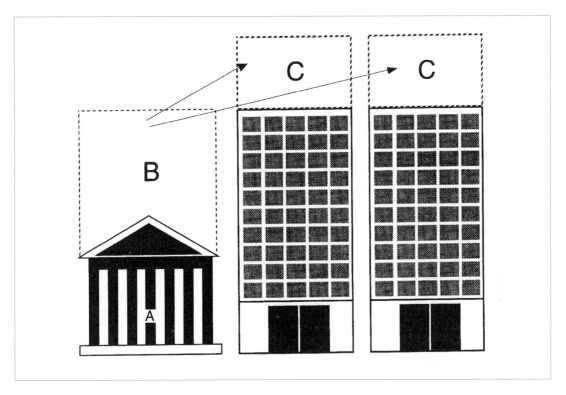

Figure 9.1 Transfer of Development Rights
To preserve the historic building (A), the development rights above it (B) are transferred to the nearby two buildings (C) which are thereby permitted an increase in development density. The owners of the latter purchase the development rights from the owners of the historic building, who thus obtain funding for preservation.

Section 4(f) has been used in relation to a wide variety of historic sites, buildings, and objects, from the French Quarter in New Orleans and the childhood home of Thomas Jefferson, to Hawaiian petroglyphic rocks, a truss steel bridge, Indian archeological sites, and many others. The section also applies to privately owned historic sites as well as those in public ownership. Indeed, most of the properties listed in the National Register of Historic Places are privately owned.

Other legislation dealing with specific modes of transportation have similar provisions, e.g. the Airport and Airway Development Act 1970, the Federal-Aid Highway Act 1968, and the Urban Mass Transit Act 1976.

THE NATIONAL ENVIRONMENTAL POLICY ACT

The National Environmental Policy Act of 1969 (NEPA) establishes a national policy of environmental protection. The historic preservation element of this refers to the preservation of 'important historic, cultural, and natural aspects of our national heritage' and the maintenance, wherever possible of 'an environment which supports diversity and variety of individual choice'. The legislation requires every 'major federal action' which 'significantly affects the environment' to be preceded by an environmental impact statement. This must contain a detailed analysis of the environmental impact of the proposed action, any adverse environmental effects that cannot be avoided if the proposal is implemented, and alternatives to the proposed action. (For further discussion See Chapter 15.)

There is some overlap between NHPA and NEPA (and the environmental protection acts passed by several states), and regulations have been issued by the Council on Environmental Quality in relation to coordination. The two Acts can be seen as reinforcing each other:

> The two laws reinforce each other and can be used effectively in tandem: if NHPA does not apply to a historic resource, NEPA might. While some courts may hold that agencies need not continue to comply with NHPA after a federal project has commenced, courts have generally agreed that NEPA does apply in such situations. If NHPA is weakened through funding cuts and revisions to the federal regulations to the ACHP, NEPA can still be used to compel agencies to consider historic properties.
>
> (Duerksen 1983: 305)

Nevertheless, the federal acts provide no guarantee that cultural resources will be protected: the only means which guarantees protection is acquisition.

ECONOMICS OF HISTORIC PRESERVATION

Taxation provisions often work against sectoral policies: they are typically designed to raise revenue, not to further public objectives. So it has been with historic preservation. Prior to 1976, the tax laws actually discouraged the preservation and rehabilitation of historic properties since tax deductions were allowable for the costs of demolition. The 1976 Tax Reform Act created a number of preservation incentives: tax credits for certain rehabilitation expenditures and (as a disincentive to the demolition of historic buildings) an increase in the 'tax cost' of demolition.

The use of preservation tax incentives increased enormously after the passing of the Economic Recovery Act of 1981 which introduced a new, and highly attractive, system of tax credits. By the mid-1980s, the program was running at an annual rate of 3,000 projects and an investment of $2 billion.

Reagan's Tax Reform Act of 1986 drastically cut these incentives but, even so, some 1,000 certified rehabilitations a year (involving $900 million of investment) were undertaken in the late 1980s. Between 1976 and 1989, a total of some 21,000 historic buildings were rehabilitated with the aid of tax incentives, representing private sector investment of almost $14 billion (Blumenthal and Siler 1990: 1).

In addition to the tax incentives program, the NHPA of 1966 provided matching grants to the states for historic preservation survey, planning, acquisition, and development. With the funding cuts made in the 1980s, little acquisition and development is being carried out, but most states continue to use grant funds for 'survey and planning' – which includes the preparation of nominations to the National Register, and developing technical preservation information.

States provide additional tax incentives. These vary considerably among the states, and take many forms. There are, however, basically six taxation methods used to encourage historic preservation: exemption, credit or abatement for rehabilitation, special assessment for property tax, income tax deductions, sales tax relief, and tax levies.

Subsidies of this nature recognize the public interests in – and benefit from – the preservation of historic buildings. Nevertheless, private owners of historic buildings may have to carry financial burdens. These may be in the form of maintenance costs which are not covered by the income from the property, or it may be in the form of the forgone higher profits from redevelopment. Generalization is difficult, but the constant battle to preserve buildings from demolition and redevelopment point to the frequency with which owners see redevelopment as being more profitable.

On the other hand, there are numerous examples of buildings which have been successfully converted into new uses. These range from modest adaptations for residential use to large schemes such as the Union Station in Washington and Faneuil Hall in Boston. Such schemes can be highly profitable as well as (and because they are) highly popular. But each site has problems and opportunities which are site-specific. There are many cases where a site cannot be restored without exorbitant cost and is also of insufficient community value to warrant public subsidy. There are also sites which are more economically valuable with a rehabilitated historic building than they would be after redevelopment.

STATE AND LOCAL PROGRAMS

As in so many areas of land use planning, it is at the local level that most of the real action takes place. All states have a State Historic Preservation Officer (SHPO). This is a federal requirement, and the Secretary for the Interior has the responsibility of approving state programs that provide for the designation of an SHPO, a state historic preservation review board, and a scheme for adequate public participation in the state program. Each SHPO is required to identify and inventory historic properties in the state; nominate eligible properties to the National Register; prepare and implement a statewide historic preservation plan; serve as a liaison with federal agencies on preservation matters; and provide public information, education, and technical assistance.

In 1956, New York State became the first to pass legislation enabling municipalities to enact an ordinance for individual landmark buildings (as distinct from historic areas). New York City was the first to take advantage of this (see Box 9.3). A New York City Landmarks Commission was established and empowered to designate properties of significant historic or aesthetic value. Designated properties cannot be demolished or altered without the approval of the commission. This is given only if the Commission decides that the proposed works will have no effect on the protected architectural features, is otherwise consistent with the purposes of the landmarks law, or is necessary to secure a reasonable return to the owner (assessed at 6 per cent).

Until the *Penn Central* case was settled by the Supreme Court in 1978, there was some doubt as to the constitutionality of such legislation. This case involved a proposal for the erection of a fifty-five story office building atop the city's *beaux-arts* masterpiece, the Penn Central Railroad's Grand Central Terminal: a building which the city had designated as a landmark. The New York City Landmarks Commission rejected the proposal, and the owners took the matter to court with two complaints. First, they argued, there had been, in effect, a taking of their property without just compensation. Second, by designating the terminal, the city had discriminated against the owners in requiring them to bear a financial burden which neighboring owners did not have to shoulder.

In a six-justice majority, the Supreme Court upheld the action of the Landmarks Commission. Though previous decisions had provided no clear rule for determining whether a taking has taken place, in this case it was determined that there was no taking: the owners had been left with a reasonable return, and the restrictions imposed were within the police powers of the city. It was explicitly stated that 'states

---

BOX 9.3   NEW YORK CITY LANDMARKS LAW

[The purposes of the Act are] to (a) effect and accomplish the protection, enhancement and perpetuation of such improvements and landscape features and of districts which represent or reflect elements of the city's cultural, social, economic, political, and architectural history; (b) safeguard the city's historic, aesthetic, and cultural heritage, as embodied and reflected in such improvements, landscape features and districts; (c) stabilize and improve property values in such districts; (d) foster civic pride in the beauty and noble accomplishments of the past; (e) protect and enhance the city's attractions to tourists and visitors and the support and stimulus to business and industry thereby provided; (f) strengthen the economy of the city; and (g) promote the use of historic districts, landmarks, interior landmarks, and scenic landmarks for the education, pleasure, and welfare of the people of the city.

and cities may enact land use restrictions or controls to enhance the quality of life by preserving the character and desirable aesthetic features of a city'.

The importance of the *Penn Central* case is underlined by Duerksen (1983: 19): '*Penn Central* made it clear that localities could forbid demolition or stop new construction for preservation purposes. Thus, a land owner who did not understand local preservation law could face serious economic consequences.' Together with the federal tax incentives introduced in 1976 (which provided land owners with significant benefits), historic preservation law suddenly emerged as a subject 'worth studying and practicing, just as environmental law had almost a decade earlier'.

The result was a major increase in historic preservation activity. Though there was some setback with the financial cuts imposed by the Reagan administration, historic preservation was clearly at the stage of becoming established as a significant land use control.

An additional note about the New York Landmarks Commission is appropriate. A major factor in its establishment was the widespread concern about the destruction of Pennsylvania Station in 1963. It was a curious twist of fate that made another railroad terminal (Penn Central) the subject of a case which confirmed the legitimacy of the Commission and its functions. There have, however, been constant rumblings about the way in which these functions have been carried out. In particular, the Commission has been accused of as acting 'as a kind of planning commission of last resort, stepping in to prevent or slow the pace of development in circumstances in which the planning commission had failed to act'. This difficulty is to be anticipated when land use planning and historic preservation planning are administered separately. But separate administration is normal; and whether the outcome is regrettable or desirable can be very much a matter of opinion.

HISTORIC PRESERVATION AND TOURISM

One of the objectives of historic preservation is often the promotion of tourism. Sadly, success here can bring its own problems. Too many people seeking a particular experience can result in its destruction. Fitch (1990) notes that many popular places – Kyoto, New Orleans, Paris, Leningrad – are facing threats to their actual physical fabric (see Box 9.4). More generally, Americans are in danger of loving their

---

BOX 9.4   HERITAGE TOURISM – TWO PERSPECTIVES

In many famous individual monuments, tourist traffic has reached its absolute limits: at Mount Vernon, George Washington's residence, stairs and floors have had to be reinforced to carry the weight of visitors; and the abrasion of flooring surfaces is so severe that protective membranes must be replaced in a matter of weeks. Faced with the noise, confusion, and downright squalor which such overcrowding often produces it would be all too easy to reject the whole concept of mass tourism and yearn for a return to the good old days of aristocratic travel.

Source: Fitch 1990: 78

Heritage tourism is just one way in which the preservation and maintenance of historic towns and urban areas may contribute to overall economic improvement. Innovative programs initiated at all levels, public and private, illustrate how preservation efforts can support and complement economic and social developments in urban areas.

Source: ACHP 1989: 35

national parks and historic sites to death. This problem is not, however, experienced in most places of historic interest. On the contrary, the economic development importance of historic resources is underlined by the use of the term 'heritage tourism'.

Under the heading 'economic revitalization' the 1989 ACHP *Report to the President and Congress* summarizes three initiatives in heritage (or 'cultural') tourism. In Lockport, Illinois, the Gaylord Building rehabilitation project is the first in the National Heritage Corridor, a 120-mile historic district designated in 1984, which stretches along inland waterways from Chicago to La Salle-Peru. This 'blend of natural and historic resources' is attracting tourist dollars to the Lockport area.

In the town of Port Townsend, Washington, tourism tripled between 1983 and 1988 as a result of promoting Victorian-era neighborhoods and downtown facade rehabilitation. In Georgia, the Antebellum Trail is being promoted (using hotel room taxes) as a tour of sites which Sherman missed on his march to the sea. Many other examples are given in the annual reports of the ACHP.

THE WIDENING SCOPE OF HISTORIC PRESERVATION

One of the most characteristic aspects of historic preservation today is that its domain is being constantly extended in two distinct ways. On the one hand, the *scale* of the artifact being considered as requiring preservation is being pushed upward to include very large ones (e.g. the entire island of Nantucket) as well as downward, to include very small ones (e.g. historic rooms or fragments thereof installed in art museums).

On the other hand, the domain is being enlarged by a radical increase in the *type* of artifacts being considered worthy of preservation. Thus in addition to monumental high-style architecture – traditionally the concern of the preservationist – whole new categories of structures are now being recognized as equally meritorious: vernacular, folkloristic, and industrial structures. In a parallel fashion, the time scale of historicity is being extended to include pre-Columbian settlements at one end and Art Deco skyscrapers at the other.

(Fitch 1990)

This lengthy quotation from Fitch (1990) nearly says

it all! The boundaries of 'historic preservation' are being stretched in such a way that the term is now a misnomer. One has only to peruse the volumes of *Perspectives in Vernacular Architecture* (e.g. Wells 1986; Carter and Herman 1989) to see the way in which interests have broadened. At the same time, new problems are arising and old problems are taking on new dimensions. Reference has been made above to the criticism of the New York City Landmarks Commission that it had, on occasion, exceeded its mandate. With heightened concern for 'historic' areas of the twentieth century, this may become more common. The issue is complicated by an overlapping concern to preserve low-income housing from redevelopment. In April 1990, for instance, the Landmarks Commission gave landmark status to a complex of fourteen buildings in the Yorkville section of Manhattan that were originally constructed as a privately financed experiment to provide housing for the poor. Here is a nice mixture of historic, architectural, social, and economic issues. Some idea of the flavor of the debate can be gleaned from the following quotation from the *New York Times* of 15 April 1990:

> Paul Selver of the law firm of Brown and Wood, which represents the owner, said there was 'nothing special' about the property to warrant landmarking . . .
>
> The commission, in its resolution, noted that the projects represented an attempt by a group of prominent New Yorkers 'to address the housing needs of the working poor'. Investors agreed to voluntarily limit their profits, and the apartments provided occupants with interior plumbing, more window space and more light and air than typical tenement apartments of the time.

The Commission also praised the 'distinction' of the architecture, and maintained that designation would help to protect an area which represents 'an important slice of history of the Upper West Side'.

Clearly, a host of different interests and values are at stake here. In such cases, the matter is settled (perhaps after recourse to the courts) by determining which interest – or interest group – is to prevail. And so we get into what Bishir (1989), in a stimulating and entertaining paper has termed 'the politics of culture'. She ends this by stressing that, since preservationists are participants in the politics of culture, it is necessary for them to be aware of the impact on their decisions of their value system. Whether self-knowledge is sufficient is an open question.

One final (and significant) point needs to be made. It was earlier stated that 'historic and associative' elements remain as important today as they were a century ago. It is, however, important to note the emergence (particularly in the western US) of scholarly studies of archeology which have profoundly affected our view of material history. The traditional approaches to the field varied. As Torma (1987) notes:

> At the onset of the national preservation program, the field was divided into at least two distinct camps – the archaeologists were on one side and the architectural historians and historians were on the other. While the orientation of the archaeologists was cultural, the orientation of the architectural historians and historians was traditional history and history of aesthetics. One group was trained in the social sciences (and some would say the sciences) and the other in the humanities. While the archaeologists looked at all aspects of the 'cultural picture' – economic base, diet, foodways, architecture and seasonal migration patterns (to name a few) – those working in the historic sites program were generally concerned with only two issues: is this structure aesthetically beautiful and/or does it have already demonstrated historic value?

The coming together of these different approaches has proved fruitful, and new perceptions of 'historic preservation' are emerging. In Stipe's words (1987: 274), the subject matter of historic preservation has become 'thoroughly democratized', and topics such as vernacular architecture and industrial and commercial archeology are now common and popular topics. The very term 'historic preservation' is being replaced by broader concepts of 'heritage'. It is an exciting time for students and practitioners in this field.

FURTHER READING

The best single book on historic preservation (and certainly the most enjoyable) is Costonis (1989) *Icons and Aliens: Law, Aesthetics, and Environmental Change*.

Major historical writers are Fitch (1990) *Historic Preservation: Curatorial Management of the Built World*, and Hosmer (1965) *Presence of the Past: A History of the Preservation Movement in the United States before Williamsburg*, and the same author's (1981) *Preservation Comes of Age: From Williamsburg to the National Trust, 1926–1949*.

A good and succinct account of the *Penn Central* case is to be found in Haar and Kayden (1989) *Landmark Justice: The Influence of William J. Brennan on America's Communities*.

There is a large and burgeoning literature on the developments taking place in this field. See, for example, Upton and Vlach (1986) *Common Places: Readings in American Vernacular Architecture; and Wells (1986)* Perspectives in Vernacular Architecture II.

For a discussion of the relationship between planning and historic preservation, see Birch and Roby (1984) 'The planner and the preservationist: an uneasy alliance'.

A useful reference book is *Landmark Yellow Pages*, edited by Dwight (1992).

QUESTIONS TO DISCUSS

1 **Is it worth preserving buildings that have historical associations, but no architectural merit? Give examples to support your argument.**

2 **In what ways can historic preservation be integrated with land use planning?**

3 **Discuss the ways in which historic preservation policies impact on the federal government.**

4 **On what grounds is it justifiable to give public subsidy for the preservation of old buildings?**

5 **How can the impact of 'excessive tourism' on popular sites of historic value be dealt with?**

PART IV

GROWTH MANAGEMENT

Growth management is one of the foremost issues in land use planning. At the local government level (discussed in Chapter 10), it has had a surprisingly long history, dating back to the late 1960s and early 1970s. (Involvement by the states – discussed in the following chapter – came later.) The most famous legal cases are *Ramapo* (New York) and *Petaluma* (California), which, in different ways, added the concept of timing to the two traditional planning dimensions of location and use. The idea is a simple, persuasive one: that development should proceed in parallel with the requisite infrastructure.

Local governments are severely limited in their ability to manage urban growth. The issues are essentially regional in character. Restraints in one area may simply result in development pressures moving elsewhere in the region. This is the rationale for state action. A number of states have assumed responsibilities for growth management, though there are marked differences in the extent to which they have been willing and able to shoulder these. There are also significant variations in the techniques adopted. In Chapter 11, the policies of six states are examined. Hawaii introduced state-wide zoning; Oregon set up a comprehensive system of local planning which had to conform to a long list of state goals; Vermont set up a system of citizen district commissions to administer a development plan system; Florida introduced state controls in 'areas of critical concern' and 'developments of regional impact'; California established a coastal planning system; and New Jersey battled with the introduction of a state plan intended to guide growth and conservation throughout the state. The effectiveness of these and similar policies is a matter of considerable debate, but it is clear that they have proved to be very difficult to implement.

10

GROWTH MANAGEMENT AND LOCAL GOVERNMENT

The slow-growth movement has proved that it can win elections. What it has not proved, however, is that it can stop growth.

Fulton 1990

ATTITUDES TO GROWTH

American zoning proceeds largely on the basis of decisions regarding individual lots. What is typically ignored is the cumulative effect of an enormous number of 'lot decisions'. This is partly because the zoning machine usually operates without the advantage of a guiding plan; partly because zoning has traditionally been unconcerned with the timing of development (or its relationship to the provision of infrastructure); and partly because the normal presumption of municipalities is in favor of development – the more, the better. The last point goes deep: instead of asking 'Is the proposed development desirable in the public interest at this place at this point in time?', the typical municipality starts from the presumption that any development is good and, in any case, it is unfair to penalize a particular owner with a refusal: if one farmer's land has been approved for development, why shouldn't his neighbor get equal treatment?

This traditionally positive attitude to growth is now reversed in some areas, particularly where development has been rapid. Suburban localities have for long put up formidable barriers to development which might attract low-income households, but this anti-growth stance has more recently spread, in some areas, to any development which might increase tax burdens or add to levels of traffic congestion which are already considered to be severe. As a result, restrictions have increased; but, as long as the siting of a development does not detract from the amenities enjoyed by an articulate minority, it is likely to go ahead.

Whether a municipality is for or against growth, however, it is unusual for it to embody its ideas in a formal land use plan. Zoning has generally remained the standard system of land use control. Yet zoning usually operates without concern for wider questions of planning: it is essentially reactive and timeless. The difficulties to which this may be expected to give rise are exacerbated by the fact that zoning maps usually have a similar timeless quality. They show the use to which individual lots of land may – in isolation – reasonably be put, but they do not take into account the effect of the timing of development applications or the effect of a number (and certainly not all) of the proposals emerging at a particular time. The availability of public services (from sewers to roads to schools) does not enter into the political calculus. Development patterns can therefore be haphazard, inefficient and wasteful, costly to service, and cumulatively disastrous – with inadequate public services, gridlock and the like.

Added to the political predispositions are a number of other complicating factors. The dictates of the Constitution are one – particularly the requirement for equal treatment (how does a political body, normally consisting of a very small number of members, defend unequal treatment to land owners on some fuzzy basis of the public interest?). Another is the division of responsibility between different agencies. Transportation planning is frequently the responsibility of an agency different from the one concerned with zoning; schools always are. As a result, zoning is the major discretionary function of municipalities – and sometimes the dominating issue at local elections.

By a curious twist of the tale, action to promote coordinated planning may be interpreted (not always unjustly) as an underhand means of excluding minority groups from an area – what Bosselman (1973: 249) has characterized as 'the wolf of exclusionary zoning under the environmental sheepskin worn by the stop-growth movement'. All these considerations help to explain the widespread popularity of large lot zoning: it results in development which makes the minimum demands on public services (and on the demand for an expansion of them); it pays for itself in the narrow terms of a municipal budget; and it excludes minorities from the area. It also enables a municipality to operate a primitive form of growth management by holding back development pressures. But it can be very inefficient. Large-lot zoning can lead to development which is scattered and expensive to service.

These ideas are explored in more depth in other chapters: they are mentioned here to provide a reference point for the ensuing discussion of the interesting, and largely unsuccessful, attempts to plan the location and timing of urbanization.

Considerable ingenuity has been displayed in devising techniques of growth management. They range from restrictive subdivision and zoning regulations, to permits to begin development, caps on the number of new dwellings (either annually or over a period of years), phasing development along with the provision of infrastructure, urban growth limit lines, and the preservation of land for agricultural or other highly restricted uses. A complete list of all the possible measures would be a very long one. Indeed, most planning techniques can be utilized for growth management purposes. The discussion of the subject

is therefore scattered among the various chapters of this book.

Any account of this subject must include two machinations in mathematical probity which assumed fame in the early 1970s – the growth control programs of *Ramapo* and *Petaluma*.

THE RAMAPO GROWTH CONTROL PROGRAM

Ramapo is a town in Rockland County, New York, about thirty-five miles from downtown Manhattan. At the end of the 1960s, it had a population of around 76,000, and was growing rapidly. As a result, there was an increasing strain on public services and infrastructure. A master plan had been adopted in 1966, followed by a comprehensive zoning ordinance, and then by a capital improvements program and a phased growth plan. The latter provided for the control of residential development in phase with the provision of adequate municipal facilities and services. The various plans covered a period of eighteen years.

The timed growth plan did not rezone any land: the restraint on property use was regarded as being of a temporary nature. This restraint took the form of a requirement that a special permit be obtained for residential development. Thus, where the required municipal services were readily available a special permit would be granted, but where a proposed development was located further away, development could not begin until the programmed services reached the location – unless the developer installed the services.

The court held that

> where it is clear that the existing physical and financial resources of the community are inadequate to furnish the essential services and facilities which a substantial increase in population requires, there is a rational basis for phased growth and, hence, the challenged ordinance is not violative of the federal and state constitutions.

This decision is of great significance: for the first time land use regulation became legally 'three-dimensional'. To the traditional dimensions of location and use was added the new one of time.

BOX 10.1   THE RAMAPO TIMED GROWTH PLAN

The standards for the issuance of special permits are framed in terms of the availability to the proposed subdivision plat of five essential facilities or services: specifically (a) public sanitary sewers or approved substitutes; (b) drainage facilities; (c) improved public parks or recreation facilities, including public schools; (d) state, county or town roads – major, secondary or collector; and (e) firehouses. No special permit shall issue unless the proposed residential development has accumulated fifteen development points, to be computed on a sliding scale of values assigned to the specified improvements under the statute.

THE PETALUMA QUOTA PLAN

The Ramapo plan implied a limit to the annual number of dwellings that could be built in the area. This was not a predetermined figure: the actual number of dwellings built was dependent upon capital improvements and the ability of developers to acquire points under the point system. The Petaluma plan, by contrast, operated by way of a fixed quota.

Petaluma lies some forty miles north of San Francisco. In the 1950s and 1960s, it experienced a steady population growth, from 10,000 in 1950 to 25,000 in 1970. By the latter date, however, this self-sufficient town had been drawn into the Bay Area metropolitan housing market, and development boomed. Whereas only 358 dwellings had been built in 1969, the number rose to 591 in 1970 and 891 in 1971. Alarmed at this rate of growth, the city introduced a temporary freeze on development. This provided a breathing space during which a growth management plan could be prepared. The plan, adopted in 1972, fixed development at a maximum rate of 500 dwellings a year (excluding projects of four or less units). To give effect to this control mechanism it was necessary to have a system which would choose between competing claimants. The instrument devised for this purpose was an annual competition among rival plans in which points were awarded for access to existing services which had spare capacity, for excellence of design, for the provision of open space, for the inclusion of low-cost housing, and for the provision of needed public services. (The policy allocated between 8 and 12 per cent of the annual quota to low- and moderate-income housing.)

Not surprisingly, the development interests in the area were highly alarmed, and a case was brought against the city. The district court declared the plan to be unconstitutional, but the Court of Appeals reversed. Though it accepted the view that the plan was to some extent exclusionary, it noted that 'practically all zoning restrictions have as a purpose and effect the exclusion of some activity or type of structure or a certain density of inhabitants'. The court's review did not cease upon a finding that there was an exclusionary purpose: what was important was to determine whether the exclusion bore any relationship to a legitimate state interest. The court held that the Petaluma plan did in fact serve such an interest. Moreover, the plan was certainly not exclusionary in the sense of keeping out low-income households. On the contrary, it was 'inclusionary to the extent that it offers new opportunities, previously unavailable, to minorities and low and moderate income persons'.

OTHER GROWTH CONTROL PROGRAMS

Ramapo and Petaluma are only two of many growth management schemes which have been introduced since the late 1960s. They are particularly notable because of the blessing bestowed upon them by the courts (and their prominence in standard texts). However, not all schemes have been approved by the courts. For instance, the attempt by Boca Raton, Florida, to place a cap (of 40,000) on the number of dwellings ultimately to be built in the city (agreed by a public referendum after a superficial review of

Plate 21 Housing Snake, Palm City, Florida
Courtesy Alex MacLean/Landslides

urbanization trends) was declared unconstitutional. Perhaps Boca Raton was just unlucky, though its action was not backed up by the supportive planning studies which courts like to see. Yet – a point which needs to be constantly borne in mind – most schemes are not in fact challenged. For example, there was no challenge to the Californian City of Napa's *residential urban limit line* which was intended to limit the city's population to 75,000. This *line* represented the boundary beyond which essential public services would not be provided. It was accompanied by the Napa Residential Development Management Plan, which imposed an annual ceiling on new residential construction.

The reasons for introducing growth management policies vary. Thus, while Napa's *residential urban limit line* was intended to cap the population growth of the area, nearby Santa Rosa had an urban boundary designed to permit all the development which was envisaged for the foreseeable future. If it transpired that further land was required, the boundary could be extended: the objective was not to prevent growth but to ensure that it took place in a desirable manner. It was aimed at the problems of 'scatteration' and the 'unnecessary use' of agricultural land. It also attempted to preserve environmental quality and enhance the aesthetic quality of new housing. Other California cities (such as San Rafael and Novato) have also been concerned essentially with the preservation of open space and the establishment of green belts. Similarly,

the comprehensive planning program in Montgomery County, Maryland, was 'intended to accommodate growth, and to manage it only to the extent needed to moderate its ill effect' (Porter 1986: 82).

GROWTH MANAGEMENT AND INFRASTRUCTURE

The use of infrastructure planning as a major element in land use controls has become popular. A common method of coordinating urban growth and the provision of infrastructure is to require developers to hold back until the necessary provision can be made (as in Petaluma). This also allows developers to proceed if they themselves provide the infrastructure, or the finance for it. (Impact fees, discussed in Chapter 7 may form an element of such a scheme, whether or not the overall intention is the limitation or the management of growth.)

Another permutation of this school of controls is *impact zoning*. This has been particularly popular in the towns of Massachusetts. 'Drawing on NEPA and the lawyer's continuing faith in procedural solutions, these towns have amended their bylaws to require a statement of the impact of proposed subdivisions on town services and the local environment.' The form such an amendment to the zoning bylaw may take is illustrated by the impact zoning scheme set out in Box 10.2.

BOX 10.2   IMPACT ZONING

In order to evaluate the impact of the proposed development on Town services and the welfare of the community, there shall be submitted an Impact Statement which describes the impact of the proposed development on (1) all applicable town services, including but not limited to schools, sewer system, protection; (2) the projected generation of traffic on the roads of and in the vicinity of the proposed development; (3) the subterranean water table, including the effect of proposed septic systems; and (4) the ecology of the vicinity of the proposed development. The Impact Statement shall also indicate the means by which Town or private services required by the proposed development will be provided, such as by private contract, extension of municipal services by a warrant approved at Town Meeting, recorded covenant, or by contract with home owner's association.

Source: Haar and Wolf 1989: 592

INITIATIVES IN BOULDER

A further example of the use of infrastructure planning in land use controls is provided by the experience of Boulder, Colorado. Boulder has had a long history of planning to preserve its dramatic natural surroundings. Frederick Law Olmsted Jr produced a report in 1910 on *The Improvement of Boulder, Colorado*, and in 1928 the city became one of the first western cities to introduce a zoning ordinance. Not surprisingly, pressures continued on the peripheral areas and in particular on the mountain foothills. To stem this, a 1959 charter amendment established an elevation of 5,750 feet along the mountainsides beyond which utility service could not be extended (Porter 1986: 35). In the 1960s, two-fifths of a one-cent city sales tax was earmarked for open space acquisition (the balance went to road improvements). By the end of the 1980s, the city had spent $53 million on the acquisition of 17,500 acres of open space most of which lay outside the city limits.

The city was able to exercise some control over development beyond its boundaries by virtue of its utility functions: in a part of the country where water is in short supply, Boulder had virtually complete control of the water in its area (Godschalk *et al.* 1979: 258); but it received a setback in 1976 when the courts ruled that it was unconstitutional to use the powers of a public utility for planning purposes. However, by cooperating with the surrounding Boulder County, a comprehensive plan for the larger area was agreed, and this enabled the two governments to coordinate planning and annexations. Boulder uses a system of phased-in development in which the area is divided into three sections. The first consists of the 19 square miles now within the city limits, and has a full range of public services. The second, 7.5 square miles under county jurisdiction, is targeted to be annexed and serviced within three to fifteen years. The third, some 59 square miles, is not projected for servicing until after fifteen years – if ever.

An interesting feature of Boulder's planning policy is its purchase of development rights, which keeps land in agriculturally productive use but prevents development upon it. Boulder is by no means alone in its concern for protecting agricultural land from urbanization, as the following discussion demonstrates.

SAFEGUARDING AGRICULTURAL LAND

The safeguarding of agricultural land has a strangely captivating and persistent appeal. It is uncritically accepted that food-producing land is 'under threat', that its loss is irreversible, and that it is folly to reduce national self-sufficiency in food supplies. This seems singularly inappropriate in a country with the vastness of the United States. Nevertheless, there is considerable controversy on precisely this point. The issue achieved salience in the 1960s with concern about environmental degradation, urban sprawl, and the pressure for national land use policies. The federal reaction was to mount the National Agricultural Lands Study (US Department of Agriculture 1981). The accuracy of the data presented in this study has been subject to intense debate, but little consensus has appeared. Action at the federal level has been minimal. The most significant legislation has been the Farmland Protection Policy Act of 1981 which requires federal agencies to have regard to the effect of their programs on the loss of agricultural land (see Box 10.3).

The regulations for implementing the Act appeared in draft form in 1982, but became the subject of much controversy. When the final version of the regulations appeared in 1984, they did little more than require that federal agencies consider the impact of their activities on the conversion of farmland: there was no requirement that the activities should be changed as a result of the impacts.

While federal action has been less than dramatic, there has been much action at state and local levels. Indeed, the position has not changed since the National Agricultural Lands Study noted that state and local governments are the prime instigators of agricultural preservation.

BOX 10.3   FARMLAND PROTECTION POLICY ACT 1981

The act requires federal agencies . . . to develop criteria for identifying the effects of federal programs on the conversion of land to nonagricultural uses, and to identify and take into account the adverse effects of federal programs on the preservation of farmland; consider alternative actions, as appropriate, that could lessen such adverse effects; and assure that such federal programs to the extent practicable are compatible with state, units of local government, and private programs and policies to protect farmland.

At the state level, the most common program is some type of favorable tax treatment, such as assessment at existing use (farming) value rather than market value (which may include potential development value). This can apply to both property and inheritance taxes. However, it is doubtful whether such tax benefits are very effective on their own: owners may simply enjoy reduced taxes until the time comes when they want to sell. They are, of course, popular with farmers, and therefore they enter into the arena of state politics.

Also popular are 'right to farm' laws: these protect farmers from local ordinances (and private nuisance suits) that restrict normal farming operations. There is little analysis of the effectiveness of such laws, though one study concluded that, while they reduced the number of private actions for farm-related nuisance, they had no effect on the loss of farmland to other uses, especially in peri-urban areas (Lapping and Leutwiler 1987). They may be of greater help to farmers, at least in the short run, than the strategy employed by one Delaware farmer who placed a huge notice on the boundary between his mushroom farm and a new housing development warning prospective buyers of the unpleasant environment into which they were being enticed to move. (The reader may wish to be reminded of the *Hadacheck* case, summarized in Chapter 4.)

State programs tend to be rather blunt instruments: the real action is at the local level (though in fact it is rather modest). Here, there are three main approaches: agricultural zoning, the purchase of development rights, and the transfer of development rights. Agricultural zoning, as its name suggests, restricts use in the defined zone to agriculture. It is a simple technique, but it is open to constitutional challenge and, not surprisingly, it faces strong political opposition from farmers. However, it can be useful when coupled with other measures such as tax incentives or the transfer of development rights.

The most effective way of safeguarding agricultural land (other than outright purchase at market value) is by the acquisition of the development rights. All the states have passed legislation enabling such acquisitions (usually at local level), but the costs are so high that few local governments can contemplate a program on any significant scale. There are, however, various devices for overcoming this difficulty by the transfer of development rights (TDR). This is a relative newcomer to the armory of planning techniques. It is simple in concept but complex in its details. In essence, it separates the development value of land from its existing use, and 'transfers' that development value to another site. The owners of the land in the area to be preserved can sell their development rights to developers in designated 'receiving' areas that are thereby allowed to build at an increased density reflecting the value of the transferred rights. Unlike traditional zoning techniques, TDR gives farmers an incentive to retain their land in agricultural use.

Few TDR programs have been implemented, though they have attracted considerable interest. The program in Montgomery County, Maryland, is one of the best known, and perhaps the most ambitious. This designates *preservation areas* where downzoning has reduced development density on about a third of the 500-square-mile county to one house per twenty-five acres. In addition, there is a transferable development right of one house per five acres – the density which the land had before designation. This can be sold to developers in receiving areas.

BOX 10.4   TDR IN MONTGOMERY COUNTY, MARYLAND

Receiving areas are the designated sites to which development rights can be transferred. They must be specifically described in an approved and adopted master plan, a process by which areas are screened to assure the adequacy of public facilities to serve them and to assure compatibility with surrounding development. Each receiving area is assigned a base density. Developers can build to this density as a matter of right. To achieve the greater density permitted under the TDR option, the developer must purchase development rights. No rezoning is necessary, but a preliminary subdivision plan, site plan, and record plat must be approved by the Montgomery County Planning Board.

Source: Banach and Canavan 1987: 259

All the evidence shows that the preservation of agricultural land can involve very large costs. In assessing whether these are justifiable, it is necessary to ask not only what the objectives are, but also who actually benefits. In a review of agricultural land protection policies in New England, it was concluded that there are four interrelated concerns behind the adoption and implementation of such policies in this region. These are the difficult-to-define but easily recognizable quality of the rural landscape; environmental degradation (pollution in all its forms); the quality and regional availability of food products; and the various economic benefits of agriculture (such as its beneficial impact on the economy, and the avoidance of the problems of land speculation and rising land prices). Interestingly, it is the 'aesthetic' concerns that predominate. The term is used here in a very wide sense to mean the general quality of the environment (Schnidman *et al.* 1990).

It is noteworthy that these issues are not only interrelated but also somewhat elusive; but clearly the

major issue is a vague unease and concern about the way in which a familiar and friendly environment is changing. This, of course, is a common feature of the operation of the planning and zoning system. Fischel (1982: 257) argues that 'the real beneficiaries . . . and the real force behind the farmland preservation movement, are local antidevelopment interests'. By contrast, the American Farmland Trust (1988), in its survey of schemes for the purchase of development rights in Massachusetts and Connecticut, underlines the benefits to individual farmers and to the farming industry generally. A more balanced viewpoint is expressed in a monograph emanating from the long-term research program on farmland protection carried out by the Florida Joint Center for Environmental and Urban Problems (Hiemstra and Bushwick 1989: xi). As with all good research projects, the conclusions raise as many questions as are answered:

> the Center has concluded that urban conversion of agricultural lands, while not posing an immediate threat to America's food supply or its strategic position in

---

BOX 10.5   FARMLAND PROTECTION IN NEW ENGLAND

State-by-state review of New England farmland protection efforts reveals that in every state one of the most important concerns was the desire to preserve certain aesthetic qualities which agricultural lands provide. The determination to protect open space and local community character evolved primarily from intangible motivations such as the value of farming as a lifestyle that is pleasing to the eye. The traditional Yankee farm, with its small fields surrounded by stone walls, woodlands, and rural architecture, has given the landscape a unique visual character that a majority of New Englanders, both urban and rural, want to protect.

Source: Schnidman *et al.* 1990: 322

international affairs, does warrant concern on other grounds. Certainly, other things being equal, it makes little sense for a society to shift agriculture from better to worse lands if planning and management would allow more efficient uses of land resources. Nor is it prudent to convert agricultural land to urban uses if the urban development in question is itself wasteful and socially expensive. The challenge is to develop farmland protection programs that distinguish inefficient from efficient land uses and promote objectives more complicated than simply indiscriminately saving all agricultural land.

CONCLUSION

In this chapter a brief account has been given of the urban growth controls operated by a number of local governments, together with a summary of some policies relating to the safeguarding of agricultural land. The policies discussed are interesting, and they are certainly popular with local residents – who, of course, have not themselves been prevented by the controls from living in the area. There is no lack of critics, some of whom are very sure of themselves. Ellickson (1977), for example, describes growth controls as a type of 'home owner cartel', while Frieden (1979) slates 'the defense of privilege'. However, the evidence is variable in reliability, and equivocal or contradictory in its results. Thus, while one study concluded that the Petaluma policy resulted in higher house prices, another found that (in both Petaluma and Ramapo) there was little demonstrable effect on subsequent development.

However, it may sometimes be that, though policy is expressed in terms of urban growth management, the real purpose is to secure leverage in the planning process to obtain benefits for the locality. Stiff restrictions on development may prompt developers to offer 'amenities' on a scale or of such a character as could not be legally required by the local government.

Moreover, what happens to growth pressures which are successfully stemmed in one area? Do they necessarily move to another area where development is in the public interest? How can any rational assessment be made of such matters without a proper land use plan? Interestingly, Judge Choy made a similar point in the *Petaluma* case where he noted that:

If the present system of delegated zoning power does not effectively serve the state interest in furthering the general welfare of the region or entire state, it is the state legislature's and not the federal courts' role to intervene and adjust the system . . . the federal court is not a super zoning board and should not be called on to mark the point at which legitimate local interests in promoting the welfare of the community are outweighed by legitimate regional interests . . .

In short, policies relating to growth management cannot be adequately designed and implemented on a local basis: a regional or state outlook is required. A few states have realized this, and are making attempts to create a new intergovernmental system of land use control. This is the subject of the next chapter.

FURTHER READING

By far the greater amount of writing on growth management is concerned with state (not local government) policies. The legal texts on land use typically have a substantial discussion of the main cases. Schiffman (1990) *Alternative Techniques for Managing Growth* is a modest but very useful book which outlines the various growth control measures available to local government. Porter (1986) *Growth Management: Keeping on Target?* details many of the measures in operation in the early 1960s, as does Godschalk *et al.* (1979) *Constitutional Issues of Growth Management*.

Critiques of growth management policies abound. See, for example, Fischel (1990) *Do Growth Controls Matter: A Review of Empirical Evidence*, and Landis (1992) 'Do growth controls work?'. See also Stanilov *et al.* (1993) *A Literature Review of Community Impacts and Costs of Urban Sprawl*.

There is a useful set of articles reviewing growth management programs and their achievements in the autumn 1992 issue of the *Journal of the American Planning Association* (58: 425–508). Also well worth studying is Chinitz (1990) 'Growth management: good for the town, bad for the nation?'.

On agricultural land, see Brower and Carol (1987) *Managing Land Use Conflicts: Case Studies in Special Area Management*, especially the paper by Banach and Canavan on the Montgomery County program (and its TDR scheme); and Schnidman *et al.* (1990) *Retention of Land for Agriculture: Policy, Practice and Potential in New England*. An article which combines theory, a literature review, and some empirical data from Oregon is Nelson (1992) 'Preserving prime farmland in the face of urbanization'. See also Daniels (1991) 'The purchase of development rights: preserving agricultural land and open space'.

There is, of course, the irony that while some public policies are geared to keeping land in agricultural use, others (particularly federal agricultural subsidies) are aimed at limiting output. For an absorbing account of *How the US Got Into Agriculture: and Why it Can't Get Out*, see Rapp (1988).

QUESTIONS TO DISCUSS

1 **Describe the various methods of managing local urban growth. Which do you think is the most effective?**

2 **Discuss the case for and against growth management policies.**

3 **What is involved in the Transfer of Development Rights (TDR)? For what planning purposes can this be used?**

4 **Evaluate the reasons given for farmland protection.**

5 **Is there a regional dimension missing from municipal growth management plans?**

6 **Consider the arguments of Chinitz (1990) on whether growth management is 'good for the town, bad for the nation' (*Journal of the American Planning Association* 56: 3–8); and the response by Fischel and Nueman (57: 341–8).**

11

URBAN GROWTH MANAGEMENT AND THE STATES

This country is in the midst of a revolution in the way we regulate the use of our land. It is a peaceful revolution, conducted entirely within the law. It is a quiet revolution, and its supporters include both conservatives and liberals. It is a disorganized revolution, with no central cadre of leaders, but it is a revolution nonetheless.

The tools of the revolution are new laws taking a wide variety of forms but each sharing a common theme – the need to provide some degree of state or regional participation in the major decisions that affect the use of our increasingly limited supply of land.

Bosselman and Callies 1972

URBAN GROWTH PROBLEMS

This now famous quotation from Bosselman and Callies was written in the heady days of the 1970s, and its promise has not been fulfilled. Only a few states have become involved in land use planning (as distinct from environmental planning); there are marked differences among them in the purpose and scope of their involvement; successes have been limited, and typically modest.

In this chapter, six illustrative types of state land use planning are discussed: Hawaii, Oregon, Vermont, Florida, California, and New Jersey. Each of these states faced one or more problems which was seen as requiring state action. Hawaii was troubled by the rapid urbanization of its valuable agricultural land; Oregon experienced growth pressures along its coastline, and also problems of urban development and speculation; Vermont faced a sudden large increase in development pressures; Florida experienced phenomenal growth; California was witnessing a large loss in public access to the coast; and New Jersey (the most urbanized state in the nation) was facing massive urbanization. There are clearly some similarities among these states: all have had to devise ways of dealing with growth. But, as will become apparent in the following account, each has its distinctive set of problems, goals, constraints, and plans. Thus Hawaii introduced state-wide zoning; Oregon set up a comprehensive system of local planning which had to conform to a long list of state goals; Vermont set up a system of citizen district commissions to administer a development plan system; Florida introduced state controls in 'areas of critical concern' and 'developments of regional impact'; California established a coastal planning system; and New Jersey battled with the introduction of a state plan intended to guide growth and conservation throughout the state. Many of the initial provisions have been revised, for a variety of reasons ranging from a recognition that the early provisions were inadequate to the impact of changes in political control. The story continues to unfold, of course.

Plate 22 Highway Infrastructure, Detroit
Courtesy Alex MacLean/Landslides

HAWAII

It all began in Hawaii.

Bosselman and Callies 1972

Hawaii's approach to land use control is, as elsewhere, a product of its history; but this history is very different from that of the other forty-nine states. Indeed, the land use planning system which has emerged is unique but, since it was the first, it is of some significance, as well as being of intrinsic interest. The discussion is, however, quite brief.

The particular history of Hawaii led to a concentration of land ownership which lasted until a combination of events brought into being the 1961 Land Use Law. The concern was not simply with land ownership, but with the effects of the policies being operated by the land owners. In short,

an increasing pace of development (particularly by way of premature subdivisions) threatened Hawaii's agriculture-based prosperity. Thus, unlike the situation which is typical of other states, controls were introduced, not to retard or control growth but to safeguard and promote economic development.

Another distinctive feature of Hawaii is its highly centralized governmental structure, with the state having responsibility for major services such as education, welfare, and housing. There is no tradition of autonomous local government: the islands are governed by four counties, and there are no lower levels of local government.

The 1961 Act established a Land Use Commission which was charged with designating all land in four 'districts': urban (5 per cent), agriculture (47 per cent), conservation (47 per cent), and 'rural' (1 per cent) (Callies 1984: 7). The urban districts cover land which is in urban use or which will be required for urban purposes in the foreseeable future. The administration of zoning in these districts is the responsibility of the counties. The designation provides no rights to urban development: it merely signifies that the county may zone the land for urban development under its zoning code. Thus the counties can impose more restrictive conditions, but they cannot relax the commission's regulations.

Agricultural districts cover land used not only for agricultural purposes but also for a range of other uses including 'open area recreational facilities'. In establishing these, the commission is required to give the 'greatest possible protection' to land which has a high capacity for intensive cultivation. Conservation districts are primarily forest and water reserve zones, but also include historic sites, mountains, and offshore outlying islands. The administration of planning in the conservation districts lies directly with the state government, operating through the Land Board of the Department of Land and Natural Resources. The final small category, of 'rural districts', was added to permit low-density residential lots. These are principally small farms and rural subdivisions which are inappropriate for either the agricultural or urban designations. Administration lies with the Land Commission operating within the framework of the state plan, which was approved by the Legislature in 1978.

Hawaii was the first state to enact a comprehensive plan. It is a short document which sets out a series of 'themes' and policies for the state covering a wide range of issues including health, culture, education, and public safety, as well as land use, population, and the environment. The provisions of the state plan relate to a number of important matters concerning population growth and distribution. These include the carrying capacity of each geographical area; the direction of urban growth primarily to existing urban areas; and the preservation of green belts and critical environmental areas.

Hawaii is the only state to operate a centralized statewide system of land use controls. As already explained, the particular history and governmental

system of Hawaii accounts for this. In 1978, Hawaii's four counties had only rudimentary land use and zoning schemes. However, matters have changed greatly since then, and they now have well-developed planning capabilities. This raises the question as to whether the statewide system of controls is still appropriate. There are also wider questions now being raised about the adequacy of the bureaucratic Hawaiian system to meet the needs of the island (Callies 1994).

OREGON

> The form of planning in Oregon is not so much different from that in other states, but the substance is. In most states, the cities and counties may plan and zone; in Oregon they must. In most states, standards for local planning are not uniform from one jurisdiction to another, are not particularly high, and are not enforced by any state agency; in Oregon, general planning standards (the goals) are the same throughout the state, they are high, and they are administered by an agency with clout.
>
> Rohse 1987

Oregon has had a good lengthy track record for state planning initiatives. For example, between 1969 and 1971 five laws were passed (the so-called 'B' laws) which provided for public access to the beaches; issued bonds for pollution abatement; banned billboards; earmarked funds for bicycle paths; and mandated returnable bottles. Other laws established the Oregon Coastal Conservation and Development Commission, and mandated local governments to prepare comprehensive land use plans and develop land use controls. In 1973 came the Land Conservation and Development Act which greatly increased the powers and responsibilities of (mandatory) local planning, and provided for a set of state planning guidelines which local plans are required to follow.

As usual, there is no single factor which explains why Oregon acted as and when it did. Certainly, a catalyst was political – in the form of Governor Tom

McCall and Senator Hector Macpherson who, in promoting new legislation in 1971, started what DeGrove (1984) has called 'a kind of blitz, in which powerful forces allied themselves on both sides of the issue, and a hard-fought series of compromises had to be worked out to obtain the bill's ultimate passage'. But there was a popular base on which this blitz was waged. Oregonians have a particular pride in the beauty of their state: they see it as a precious heritage which demands to be preserved. Bolstering this is a strong and vocal conviction that 'Oregon must not become another California'. (This conviction has been fueled in recent years by the influx of Californians seeking an environment similar to that of California, but with much lower house prices.) The concern emanates from visible pressures on the land: urban encroachment in the Willamette Valley, land speculation in the fragile landscape of the eastern part of the state, and degradation of the marvelous shoreline.

Oregon's land use law is comprehensive only in the sense that the planning guidelines apply to the whole of the state. The actual preparation and implementation of local plans is the responsibility of the 241 cities and 36 counties. Thus there is in no real sense a 'state plan': there are 277 local plans that have been developed in accordance with state standards and have been reviewed and approved by the state. Nevertheless, the importance of the state requirements should not be underestimated: plans have to conform to a range of specific state land use policies. These include the containment of urban growth: each municipality's plan has to delineate an *urban growth boundary* which defines the limit of urban development and its separation from rural land.

A state planning agency, the Land Conservation and Development Commission, was established to ensure that state policy is implemented. Its seven members are appointed by the Governor, and confirmed by the senate. Its first tasks were to adopt the statewide planning goals and to review and approve (technically termed the 'acknowledgment' of) local plans. The review of amendments (of which there are several thousand every year) is dealt with by the Commission's administrative arm, the Department of Land Conservation and Development. However, the

Department has no power to prevent a municipality from adopting an amendment to which it objects. In such cases, it would normally appeal to a body known as the Land Use Board of Appeals. This Board was established to provide a simple means for settling land use disputes without the need to go through the state circuit courts. The important lubricant in this system is the wide provision for citizen involvement. This is in the political tradition of Oregon: it is common for local governments to establish citizen advisory committees (CACs) for every city neighborhood and county district:

> The groups typically meet monthly. Their advice and concerns are given to the planning commission or governing body. CAC meetings are quite informal, and are open to all, without dues or formalities of membership. CAC meetings are often attended by members of the planning department, who can answer technical questions or keep a record of comments.
>
> (Rohse 1987: 57)

Local governments are also required to establish and support 'an officially recognized citizen advisory committee or committees broadly representative of geographic areas and interests related to land use and land use decisions'. At the state level, there is the Citizen Involvement Advisory Committee, which has several functions: to advise the Commission on matters of citizen involvement; to promote 'public participation in the adoption and amendment of the goals and guidelines'; and 'to assure widespread citizen involvement in all phases of the planning process'.

Intergovernmental relations are often characterized by a heavy measure of bluff. A state law may mandate a local government to do something, but if there is no machinery for ensuring compliance or no financial incentive, nothing may happen. The Oregon system has teeth in it. It provides tangible incentives to local governments to prepare plans and obtain state 'acknowledgment' of them. Certain state contributions to local budgets are dependent on this. In addition to such financial considerations is an incentive which is even more effective: once a local government's plan and land use regulations have been approved by the Commission, the state's role in local planning is greatly reduced. There is no longer any need for it: the goals have been incorporated in the plan.

Urban growth boundaries (UGBs) are, understandably, a source of contention: in more senses than one they are the cutting edge of planning implementation; and wherever they are drawn someone will be upset. Nevertheless, they have worked out much better than might have been expected. Each of Oregon's 241 cities has adopted an urban growth boundary which has been reviewed and approved by the Commission. Difficulties have generally been overcome by a judicious degree of flexibility: for instance, some of the areas just beyond the urban growth boundaries have been designated for eventual urban expansion on a comprehensively planned basis. The UGB system has been effective in controlling urban growth in the Willamette Valley. This is a significant achievement; but development pressures continue to mount. The plans were drawn up for the period ending in the year 2000, and need review for the rapidly approaching new century. In the judgment of Arthur C. Nelson:

> The challenge facing Oregon now is how to properly recognize the urban form it has created through UGB policies, and its implications, in what manner it should be reassessed, and how best to consciously facilitate that urban form.
>
> (Abbott *et al.* 1994: 45).

VERMONT

> It is the traditional settlement pattern (village, town, and countryside) that reflects the essence of Vermont. In order to maintain the essential character and ethic of Vermont's built environment, there should be a clear delineation between town and countryside through effective planning and supportive land development.
>
> Report of the Commission on
> Vermont's Future 1988

Vermont is a largely rural state in which the pressures for development come mainly from outside. In terms of population (some 563,000) it ranked forty-eighth in the 1990 census. It has a highly decentralized local government system of small New England communities with nine cities, 237 'organized towns', five unorganized towns (in sparsely populated areas) and fifty-seven incorporated villages which are urban in character. Counties largely exist only on paper, and there is therefore a void between state and local government which has been filled by regional state administration. Despite this, Vermont had no tradition of state planning; yet in 1970 it passed a growth management measure which introduced a state-wide planning system.

The major reason for this was a transformation of Vermont from a state whose young people traditionally left for better opportunities elsewhere into a state beset by the problems of unprecedented growth. This was caused by a number of factors. The extension of the interstate highway system brought Vermont within easy travel distance of the 40 million inhabitants of the urbanized areas to the south. Several economic changes also took place within the state, some of which were related to this new accessibility, including the expansion of the ski industry and the growth of second homes. Almost suddenly, Vermont changed from a remote area to an easily accessible vacation, second-home, and commuters' haven. It is, of course, a beautiful state.

The resultant growth in population led to development pressures and increased land costs (and taxes) which were alarming to the conservative Vermonters. A 14 per cent growth in population during the 1960s, though modest by the standards of California or Florida, was greater than the increase over the previous half-century. It was, moreover, concentrated in particular areas, and therefore its impact was striking. The local governments of Vermont were quite unable to deal with this unprecedented situation. None of the towns had a capital budget program, and few had a zoning ordinance. They were literally at the mercy of developers.

In a remarkably short space of time, the state government acted, and legislation (Act 250) was passed in 1970. This introduced a development permit system administered by an appointed environmental board and district environmental commissions. It also provided for the preparation of three state-wide plans: an interim land capability plan (an inventory of physical data); a land capability plan (to guide 'a coordinated, efficient, and economic development of the state'); and a final land use plan. The development permit system has worked reasonably well, but the plans have given rise a number of difficulties, and the land use plan never emerged.

Establishing new agencies of government is always problematic. In particular, there is the perennial issue of decentralized versus centralized control. In Vermont, it was clear that the local government system could not administer the development permit scheme, but there was little enthusiasm for giving more power to the state. The solution adopted placed the major responsibility for administering the development permit system on lay citizen district commissions – with the right of appeal to a lay state board. Thus the process is decentralized, but in a way which bypasses the established local governments. Ideally, of course, a plan should have preceded the introduction of this system, but there was insufficient time: the development pressures were too intense.

In the absence of a plan, Act 250 provided a list of criteria against which development applications are to be judged. These include a wide range of environmental, aesthetic, and land use issues. For example, development proposals are not to be approved if they would cause 'unreasonable congestion or unsafe conditions on highways', or create an 'unreasonable burden' on the ability of the municipality to provide services, or have an 'adverse effect' on the scenic beauty of the area. Such criteria clearly give a considerable range of discretion. However, they apply explicitly only to large developments (see Box 11.1),

and decisions can be appealed to the Vermont Environmental Board.

Vermonters were in favor of controlling unwanted development, but they were dubious about plans, particularly if these were to be drawn up by state bureaucrats. A land capability plan was accepted, but plans which would limit local discretion were strongly opposed. In the words of a local planning consultant, 'The idea had never been to limit the options for local folks, but rather to stem the destructive tide of "flatlanders" bent on citifying Vermont' (Squires 1992: 14). The anxieties here were so strong that the statutory provision requiring a state land use plan was repealed.

Of course, plans typically have multiple objectives, some of which may be difficult to harmonize, while others may be contradictory. The Vermont planning system was directed to improving the quality of large-scale developments; there is in reality little of a 'plan' in it. The approach is essentially 'reactive': it evaluates planning proposals which are submitted for approval; it does not direct growth to areas which are considered by planners to be suitable for growth. In short, as a growth management system, a lot remains to be desired. A 1988 report by a commission appointed by the Governor underlined the perceived weaknesses, and stressed the fact that 'a consequence of the failure to adopt comprehensive local and regional plans is that basic planning decisions are left to the regulatory process'. As a result of the local nature of most of the land use controls, suburban and resort developments were continuing at a rapid rate: there was an urgent need 'to introduce planning into the regulatory process'.

BOX 11.1   DEVELOPMENTS REQUIRING A PERMIT IN VERMONT

1 Housing developments of ten or more units by the same applicant within a five-mile radius.
2 Developments involving the construction of improvements for commercial or industrial purposes on a tract of more than one acre in towns without permanent zoning and subdivision bylaws and on a tract of more than ten acres in towns with such controls.
3 Developments involving the construction of improvements for state or municipal purposes of a size of more than ten acres.
4 All developments above an elevation of 2,500 feet.

Legislation was passed in the same year. This (Act 200) specified the minimum contents of local and regional plans (including land use, housing, transportation, utilities, education, and natural resources). It authorized impact fee ordinances for local governments that had adopted plans and capital improvement programs. It retained the existing regional planning commissions with wider powers and subject to a requirement that they cooperate with other agencies and levels of government. All regional commissions and state agencies are required to ensure that their planning is consistent with twelve broad state planning goals. In this revised system, the regional commissions become the vital force in growth management: they are assigned the responsibility for reviewing and approving local plans, and for commenting on state agency plans.

A new agency, the Council of Regional Commissions, was created to review regional and state agency plans for compatibility with state goals. Additionally, a Municipal and Regional Planning Fund was established to assist municipal and regional planning commissions. A geographic information system (to which all commissions and agencies contribute data) is financed from this fund.

Passed at a time of economic prosperity (and major governmental initiatives in environmental planning), the new legislation seemed to promise a major improvement in Vermont planning. But there was much that it did not do (partly because of opposition during its passage), and further opposition quickly followed. This increased as the economy deteriorated, but its origins were deep. A well-funded and organized Citizens for Property Rights group has attracted much support, and the controversy continues. The outlook is uncertain.

The requirements of Act 200 for consistency with the state's planning goals provide a substantive framework for plan preparation and implementation. (Plans have to demonstrate consistency with the goals, or good reasons for departing from them.) The goals are therefore a unifying element in the planning system. This appears to have had good effect in coordinating the plans of state agencies. The Act does not, however, mandate local planning; and municipalities can elect not to submit a plan for review by the regional planning commissions. Nevertheless, more planning than ever before is being undertaken.

The revised Vermont system is a neat balancing act between the requirements of area-wide planning and the strong proclivity of Vermonters for local control. But in essence the planning process works from the 'bottom up', though within the framework of state policies. There has thus been little change in the Vermont allegiance to local control.

FLORIDA

Deep-rooted love affairs are always difficult to terminate, and Florida's love affair with growth has been no exception.

DeGrove 1984

Florida's growth in post-World War II years has been phenomenal – a result of its attractive environment, its warm climate, and its low taxes. In 1950 the state had a population of 2.8 million. This increased to 5.0 million in 1960, 6.8 million in 1970, 9.7 million in 1980, and 12.9 million in 1990. It is now the fourth largest state in the Union. Such a growth would have presented problems in any state, but the problems in Florida are compounded by its unique, fragile, and complex natural environment. These are the most difficult in precisely the areas of the greatest growth – the southern part of the state. If ever a situation cried out for strong planning measures, this is it.

It took some time for Floridians to appreciate and acknowledge this, but concerns about rapid growth finally resulted in a legislative response after the serious drought of 1971. Of particular importance was the Environmental Land and Water Management Act of 1972 which provided for the designation of *areas of critical state concern* (ACSC) and for special measures for dealing with *developments of regional impact* (DRI). The appeal of these instruments was that they furnished a nice balance between state, regional, and local interests. Development normally remains the responsibility of the municipalities, but in the case of an ACSC or DRI, higher levels of government are involved.

Areas of critical state concern are recommended by the state planning agency. (Four areas have been designated: the Big Cypress Area, the Green Swamp, the Florida Keys, and the Apalachicola Bay Area.) While areas of critical state concern are designated by the state, developments of regional impact are a matter for local governments, subject to review by the regional planning council and the state. A DRI is designated only when a development is proposed. The system is therefore a reactive one, and it was made more difficult initially because of the absence of a comprehensive state plan. It was, however, an improvement on the previous system in that it brought into the development approval procedure the regional level of government. All of the state is now covered by eleven regional planning agencies (which are essentially multi-county councils of government).

The system was characterized by persuasion: persuasion of one level of government by another, and persuasion of developers by the municipalities. As a result, the great majority of developments were approved (though with conditions attached). The obvious weaknesses in the system (particularly the absence of a state plan, lack of funding for local planning, and the inadequacy of state review of plans) led eventually to the introduction of major changes in the planning system.

The turning point in Florida planning came in the mid-1980s, when the state overhauled its planning system at state, regional, and local levels. The revised system is in essence one of growth management. A hierarchy of plans features a comprehensive state plan with which the plans of state agencies ('functional plans') and regions (regional plans) must be consistent; similarly with local plans. The state plan adopts twenty-five goals and policies, ranging from education to housing, from health to natural resources, and from air quality to property rights and plan implementation. (See Box 11.2 for the land use goal and its associated policies.)

Florida planning has thus been transformed. In place of the 'bottom up' character of the earlier legislation, it is now unequivocally 'top down'. All municipalities and counties are obliged to prepare and adopt comprehensive plans. The requirement that

---

BOX 11.2   FLORIDA LAND USE GOAL AND POLICIES

Goal

In recognition of the importance of preserving the natural resources and enhancing the quality of life of the state, development shall be directed to those areas which have in place, or have agreements to provide, the land and water resources, fiscal abilities, and the service capacity to accommodate growth in an environmentally acceptable manner.

Policies

1 Promote state programs, investments, and development and redevelopment activities which encourage efficient development and occur in areas which will have the capacity to service new population and commerce.
2 Develop a system of incentives and disincentives which encourages a separation of urban and rural land uses while protecting water supplies, resources development, and fish and wildlife habitats.
3 Enhance the livability and character of urban areas through the encouragement of an attractive and functional mix of living, working, shopping, and recreational activities.
4 Develop a system of intergovernmental negotiation for siting locally unpopular public and private land uses which considers the area of population served, the impact on development patterns or important natural resources, and the cost-effectiveness of service delivery.

Plate 23 Suburban Housing, North of Tampa, Florida
Courtesy Alex MacLean/Landslides

these be consistent with the state plan is not mere rhetoric. If the plan is not 'compatible with the goals' of the state plan, the state can impose some severe sanctions, particularly the withholding of funds. A remarkable provision requires local governments to coordinate the provision of infrastructure with urban growth. Development can be permitted only to the extent that the infrastructure can support it: 'public facilities and services needed to support development shall be available concurrent with the impact of the such development'. Local governments are required to design adequate and realistic 'level of service' (LOS) standards for roads, sewers, drainage, water, recreation, and (if applicable) mass transit. Development which would fail to maintain LOS standards cannot be permitted unless the deficiency will be made good by the provisions of the capital investment plan.

These *concurrency* provisions, together with a strong emphasis on compact urban development, has given Florida a powerful tool of growth management. Implementation has been made feasible by the preparation of mutually consistent plans. (Local and regional plans are statutorily required to conform to the state plan.) It has, however, been weakened by inadequate state funding of infrastructure. This problem has been dealt with, to a limited extent, by increased local taxation and by the imposition of impact fees on developers. Nevertheless, the long-term viability of the system is dependent upon a stable solution to the infrastructure financing issue. Such a solution is not yet in sight. Until this intransigent hurdle is overcome, Florida's impressive planning system will have more unfilled promise than achievement.

CALIFORNIA

> Probably no state employs more planners or produces more plans, and probably nowhere else in the country does planning and development engender more discussion at the community level. But, for all that, California has proven, over the past decade, incapable of managing its growth.
>
> W. Fulton

California has had a long and checkered history of planning endeavors. It has seen a prodigious number of plans, which continue to flow – though their destination is more often planning libraries than implementation. There is, however, no machinery for state intervention in local land use planning. At first sight this is curious, since California is the leading state in environmental planning, and its elaborate system of environmental review has had a profound impact on environmental considerations in local planning. Nevertheless, no state agency exists to review or approve land use plans.

Instead of state intervention in local planning, there is a highly popular system of citizen enforcement. It is the citizenry who take responsibility for the enforcement of planning controls, who monitor the planning process, and who litigate against unacceptable actions by local planning agencies. But further, much planning legislation is initiated through the well-established ballot box system of participatory democracy (Caves 1992). The power of *initiative* and *referendum* is by no means unique to California, but nowhere else is it used to such a great extent. Between 1971 and 1989, there were 357 ballot-box measures concerned with land use planning: on average over two-thirds of these succeed. (Perhaps the most famous of California's ballot-box measures was concerned, not with land use, but with taxes: Proposition 13, passed in 1978, reduced property taxes and limited future increases. In fact, taxation and land use control are closely interrelated; and in California the connection is so close as to give rise to what is termed the 'fiscalization of land use'.)

Ballot-box planning results in a complex, diverse, locally controlled mosaic of planning policies. Policies which have been introduced in this way include caps on the amount of residential development (often based on the Petaluma model discussed in the previous chapter); density restrictions; infrastructure limits; minimum lot sizes for new residential building; moratoria on development; and designation of areas for conservation or development. Of increased popularity in recent years has been the reservation of certain planning decisions for future voter approval (such as any change to the local plan). For instance, an initiative of the city of Lodi effectively prohibited further development in the city's peripheral areas without specific approval of the voters – which was repeatedly denied (Fulton 1991: 145).

It is against this background that state involvement in land use planning is restricted. Curiously, however, there is a major exception which itself came about as a result of an initiative. After failing to pass the state legislature, an initiative promoted by environmental groups led to the enactment of the Coastal Act in 1972 (despite a well-funded aggressive counter-campaign by developers, oil companies and the like – and the opposition of Governor Ronald Reagan). Thus California has a full-scale coastal planning program, and this has survived several hundred bills to kill or cripple it (Fischer 1985). It is this program which is of relevance to this chapter.

Strictly speaking, California's coastal program is not a statewide comprehensive planning endeavor: as its name suggests, it is concerned only with the coast. But that coast is 1,100 miles long (and the coastal planning area is up to five miles wide in rural areas). It is therefore very much akin to a statewide planning area. It is administered by the California Coastal Commission.

The coastal plan is not concerned solely with environmental protection: it seeks to ensure that the coastline is used intelligently and sensitively, with due regard to both the environment and the needs of coastal-related development. However, the plan is highly restrictive in respect to the preservation of wetlands, historic, scenic, agricultural, and forest lands. The basic goals set out in the legislation are listed in Box 11.3.

BOX 11.3 CALIFORNIA COASTAL PLAN GOALS

1 To protect, maintain, and, where feasible, enhance and restore the overall quality of the coastal zone environment and its natural and artificial resources.
2 To assure orderly, balanced utilization and conservation of coastal zone resources taking into account the social and economic needs of the people of the state.
3 To maximize public access to and along the coast and maximize public recreational opportunities in the coastal zone consistent with sound resource conservation principles and constitutionally protected rights of private property owners.
4 To assure priority for coastal-dependent and coastal-related development over other development on the coast.
5 To encourage state and local initiatives and cooperation in preparing procedures to implement coordinated planning and development for mutually beneficial uses, including educational uses, in the coastal zone.

In the early years of the program, administration was by interim planning commissions. These were independent of local government, and they operated that way: they made very little effort to develop any collaborative relationships with their constituent municipalities. This changed dramatically after 1976 when, subject to conditions, plan-making and regulatory responsibility was returned to local government. The conditions were several. The statute mandated development of local coastal plans, with regulatory authority over most development to be transferred back to local government only after the Commission had certified that the plan was in conformity with the policies of the Coastal Act. Further, the Commission retains some important planning and regulatory responsibilities, including permanent jurisdiction in some areas such as tidelands, submerged lands and trust lands; reviewing and acting upon appeals from local permit decisions; reviewing and authorizing amendments to plans; implementing public access programs; and periodically reviewing the implementation of certified plans to determine if they are being implemented in conformity with provisions of the Coastal Act; and making recommendations to local governments or the legislature. Thus the legislation clearly establishes a shared responsibility between the Commission and local governments.

The Coastal Commission is a regulatory body. To complement its operations, a State Coastal Conservancy was established. Among its many functions, this helps to carry out coastal improvement and restoration projects to implement policy established through the plans and regulations of the Commission and local governments. It is empowered to buy land and restore or resubdivide it, or sell or transfer it to others (whether at a profit or a loss). It carries out a wide range of functions in furtherance of the policies and regulations of the Commission and the local governments.

One priority is the maximization of public access to and along the shoreline. The Commission requires, as a condition for the granting of a permit, that public access be provided. (This is a condition which achieved national publicity in the planning world with the 1987 case of *Nollan* v. *California Coastal Commission* which is briefly discussed in Chapter 7.) This involves the dedication of an easement to a public agency that is willing and able to accept responsibility for maintenance and liability. Since huge numbers of conditional permits have been issued (1,800 in the twelve years up to 1985: a potential of more than fifty miles of additional shoreline access), this is no small task; and it is one which financially hard-pressed local governments are none too happy to accept. The future of the coastal program is uncertain. It has aroused a great deal of continuing opposition, and its budget is under constant attack. More directly concerned with growth management are the attempts at regional planning which have been made in some of the major urban areas of the state. Here we look at two metropolitan agencies (for the San Francisco Bay Area and for the Los Angeles and Southern California region), both of which are struggling with widely conflicting views of the future regional planning of their areas.

The Bay Area 'Bay Vision Commission' has stressed the diversity of views which were represented on the Commission and the difficulty of reaching agreement for its *Bay Vision 2020* report (see Box 11.4). It is not clear which view will prevail, though it seems certain that whatever policy is adopted will allow continued growth. Hopefully, it was proposed that a regional commission be set up that would combine, for a start, the functions of the Bay Area Air Quality Management District, the Metropolitan Transportation Commission, and the Association of Bay Area Governments (and later the Regional Water Quality Control Board and the Conservation and Development Commission), but this still leaves sixty-two other agencies! There is, moreover, disagreement on the constitution of the proposed regional commission, its role in equalizing tax burdens, and its power to control developments of regional importance. But 'we strongly believe in maintaining the integrity of existing local governments and their autonomy over local decision-making' (ibid.: 38). As Joseph E. Bodovitz, the Commission's project manager, has commented: 'there is no ground swell of readiness to plug into a regional political system. Indeed, there is great antipathy' (Stanfield 1991: 2330).

There is a clear parallel with the 1990 report on Los Angeles and Southern California by 'The 2000 Partnership'. Though existing governmental agencies 'cannot adequately plan for and manage growth on a regional level', no new planning authority is proposed; instead a new council would consolidate the current planning powers of existing agencies, and subregional councils could be formed on the basis of cooperation between local governments. It is suggested that the new regional council 'would have the authority to make and implement policies when a city, county, or special district was determined to have failed to meet regional objectives within a specified time limit'. This sounds as if the regional council would have some teeth, but these are quickly drawn: regional and subregional plans would be subject to the agreement of the constituent authorities who would have ample opportunity for 'consultation' and 'bargaining'. At the most, the proposal amounts only to control over 'limited areas of regional impact'. It remains to be seen what the effect will be of the federal planning requirements for transportation planning introduced by the Intermodal Surface Transportation Efficiency Act. This is discussed in the following chapter.

NEW JERSEY

Statewide comprehensive planning is no longer simply desirable, it is a necessity.

Mount Laurel I 1975

BOX 11.4   BAY AREA REGIONAL PLANNING DEADLOCK

We have noted that current forecasts predict an increase in the Bay Area's population from the current six million to well over seven million by the year 2000. Some of us have concluded that there is a point beyond which the Bay Area's population must not be allowed to grow if the natural resources of the Bay Area are to be protected adequately. Others of us believe that such a population limit is neither desirable nor possible to achieve. Still others believe that the issue is not population growth itself, but the need to manage development so that natural resources are not degraded as population increases. All of us agree, however, that the environmental impacts of an increasing population and an expanding economy will require a new, more comprehensive ability to plan and make regional decisions for the Bay Area.

Source: Bay Vision 2020 Commission (1991): 4

New Jersey has had a series of important regional planning initiatives. The most famous is the Pinelands Commission, established in 1979, which is responsible for the planning of some one million acres in the southern part of the state. Earlier, the Coastal Area Facility Review Act of 1973 created a regional commission to regulate large developments in the coastal area.

In 1980, the Democratic administration of Governor Brendan Byrne created a State Development Guide Plan. Though this was short-lived (Republican Governor Thomas Kean abolished it in the following year) the courts continued to use it, in the implementation of the Mount Laurel policy, to identify areas where municipalities were required to set aside some of their new housing for lower income families (see Chapter 12). During the early to mid-1980s, the New Jersey economy boomed, migration (of both people and jobs) into the state grew, and political pressures for more effective planning increased. In response, a State Planning Act was passed in 1985, establishing a State Planning Commission and its staff arm, the Office of State Planning.

The Commission was charged with preparing the primary instrument for coordinating planning and growth management in the state – the State Development and Redevelopment Plan. The statute provides that the plan shall protect the natural resources and qualities of the state, while promoting development in locations where infrastructure can be provided. It also establishes statewide objectives in a variety of areas including land use, housing, and economic development. The plan is intended to be used to guide the state's capital expenditure.

The plan was prepared according to procedures spelled out in the Act. First, a preliminary plan was approved by the Commission. This was then used in an interactive planning process called *cross acceptance* which was intended to integrate municipal, county, regional, and state land use plans as well as the capital facility plans needed to assure efficient services. This process was the crucial mechanism for obtaining support for the plan from the local governments whose cooperation is essential for its implementation. New Jersey has a strong tradition of home rule, and the 567 municipalities are very suspicious of state action in the land use field. It was therefore essential that the plan preparation process should involve the active participation of the municipalities (see Box 11.5).

The process was an involved one; there was even a *Cross Acceptance Manual* prepared by the Office of State Planning. However, in essence the idea was simple: the authorities that need to coordinate their activities were given a mechanism by which they could talk until agreement or compromise was reached. As was expected, not all was plain sailing. Nevertheless, a sufficient level of agreement was reached to permit the plan to be finalized, vague though it is in important respects.

The New Jersey State Development and Redevelopment Plan was approved in 1992, after a long period of debate and public hearings. It establishes statewide goals and objectives for a wide range of policies including land use, housing, economic development, transportation, recreation, and historic preservation. The plan embraces the concept of growth areas, though it is coy about identifying these (except in the case of the older cities.) Several hundred other

---

BOX 11.5   CROSS ACCEPTANCE IN NEW JERSEY

The term cross acceptance means a process of comparison of planning policies among governmental levels with the purpose of attaining compatibility between local, county, and state plans. The process is designed to result in a written statement specifying areas of agreement or disagreement and areas requiring modification by parties to the cross acceptance.

In general, cross acceptance involves comparing the provisions and maps of local, county, and regional plans and regulations with the goals, objectives, strategies, policies, standards, and maps of the preliminary state plan.

locations are identified as areas where development, redevelopment, and economic growth are considered to be in the public interest, but these are not actually designated, and no growth targets are established. There is thus a high degree of uncertainty in the plan – a result of the acute political difficulty in obtaining agreement among conflicting interests.

Much of the recent growth in New Jersey has been along transportation corridors, and this pattern is likely to continue in the future. The plan takes this fact as a basis for a major strategy of developing centers in the prosperous corridors. These centers are not envisaged as an elongation of the corridors: on the contrary, they are to be high-density consolidations around existing development. Their attraction is that of good transportation (which an elongation of a corridor would jeopardize). The development of centers provides the opportunity for enhancing the transportation advantages.

Major features of the plan are its emphasis on mixed-use centers and on the expansion of existing urban areas. These are considered to have sufficient capacity to meet the anticipated population growth in the state up to the year 2010. An impact assessment study, undertaken by Robert Burchell of the Rutgers University Center for Urban Policy Research, concludes that the implementation of the plan would save some 130,000 acres of land at no appreciable increase in the cost of development. However, the plan is not self-implementing, and though there are procedures for certification of the consistency of local plans with the state plan, it is unclear how this will work out. The mechanisms for implementation are as uncertain as the provisions of the plan. Indeed, it is not at all clear what the long process of plan preparation has actually achieved. Certainly, there is nothing equivalent to the implementation provisions to be found in Oregon.

CONCLUSION

The status of planning . . . has been substantially altered by the adoption of a state land and growth management system.

DeGrove 1984: 389

State involvement in growth management has increased significantly as the need to overcome the inherent problems of local land use planning has become apparent. Though it is premature to declare the 'revolution' suggested by Callies and Bosselman, this involvement and the ways in which it has been sustained, developed, and imitated do constitute a remarkable change in the attitudes of states to land use planning.

In addition to the six states discussed here, several more have passed or proposed legislation. For example, Rhode Island passed a Comprehensive Planning and Land Use Regulation Act in 1988 (DeGrove and Miness 1992). This requires consistency between every local government's comprehensive plan and the state's comprehensive plan. Local plans are reviewed by the Department of Administration for consistency with the Act. Any disagreements are decided by a Comprehensive Appeals Board, which can if necessary substitute a plan of its own (a unique provision). Maine also passed a Comprehensive Planning and Land Use Regulation Act in the same year which is very similar to that of Rhode Island. Washington State followed in 1990 with its Growth Management Act which requires comprehensive plans for populous and other fast-growing local governments. Other states with growth management programs include Georgia and Maryland. The list continues to increase.

Growth management has become concerned with far more than channeling urban growth in desirable directions. It has necessarily involved a large number of regional policy issues. These range from concerns for the protection of land *against development* (including agricultural land, natural resources, fragile environments, and amenity) to concerns for *the promotion of development*, such as housing, transportation, and economic growth. (Some of these have been discussed in this chapter, and others are discussed in later chapters.) This broadening of interest is not accidental: growth management is inherently a governmental process which involves many interrelated aspects of land use. The process is essentially coordinative in character since it deals with reconciling competing demands on land and attempting to maximize

BOX 11.6 ELEMENTS OF GROWTH MANAGEMENT

1 *Consistency* among governmental units.
2 *Concurrency*: requiring infrastructure to be provided in advance or concurrent with new development.
3 *Containment* of urban growth: the substitution of compact development for urban sprawl.
4 Provision of *affordable housing*.
5 Broadening of growth management to embrace *economic development* (the 'managing to grow' aspect).
6 Protection of *natural systems*, including land, air and water, and a broadened concern for viability of the rural economy

Source: Based on DeGrove and Miness (1992): 161

locational advantages for the public benefit. This can be done adequately only if all the relevant factors are taken into account. To illustrate: a narrow approach could lead to a worsening of the problem of affordable housing. This was a widespread concern in the early days: for example, Oregon's growth boundaries were initially criticized for being likely to increase land prices and thus housing costs. In fact, the opposite has occurred because the densities within the boundaries were increased: Oregon's goals were intentionally comprehensive and included issues such as housing, which are essential elements of the well-being of the state. Any growth management approach which omitted concerns for such vitally important aspects of the socio-economic life of the state would be not only inadequate, but also unacceptable.

Acceptability across the spectrum of interests is the key characteristic of successful growth management policies. Securing this acceptability is difficult, enormously time-consuming, and fraught with political problems. Moreover, it is an ongoing process: the determination of land uses, the timing of development, the coordination of development with the provision of infrastructure all involve continuing debate and planning, the achievement of consensus, and the provision of adequate finance. In short, growth management is a major part of the continuing process of government.

The importance of acceptability stems not simply from the dictates of a democratic system, but also from the necessity for cooperation in implementation on a regional basis. Without the necessary cooperation the system will not work. It also needs strong public support both for the policies and the taxation required to finance them. Many of the difficulties facing growth management policies have stemmed from a lack of sufficient support, particularly with funding. The continuing support for Oregon policies is in no small part due to the emphasis placed on citizen involvement. (It is significant that Florida has followed Oregon with its '1000 Friends' who perform an active role in monitoring both local and state activities in growth management.)

One particular problem of acceptability has frequently arisen in connection with state agencies (Wickersham 1994: 543). It is curious but true that a state often has acute difficulty with its own agencies. Having been established with specific goals to do a specific job, they can be loath to compromise their mission by taking on wider considerations. They are specially designed to carry out their particular functions; they have specialist staff for these purposes; they have their own political supporters; and they often resist 'compromising' their work by taking on extra – and perhaps conflicting – objectives. More apparent is the conflict between state (and regional) goals and the objectives of individual local governments. This is the hub of the growth management machine: in the final analysis, it is the local governments which operate most land control policies. They have to be persuaded not only to accept limitations on their actions, but also a subjugation of these to wider interests – hence the importance of 'acceptance', 'conformity', 'consistency' and similar concepts in the lexicon of growth management. Techniques to make these work are limited: they range from bribery to force; but generally they involve a great deal of debate. All planning requires a lot of talk, but none as much as issues which involve reconciling local interests with wider goals.

It is too early to judge the impact of state involvement in land use planning. It is uncertain whether it is an expanding sphere of government which has established strong roots or a temporary burst of activity which will not last. However, there is now much more planning activity by states and much more intergovernmental cooperation than even a decade ago; and there are indications that interest in effective growth management is increasing. The effectiveness itself, however, is less clear. Moreover, the states discussed in this chapter are, of course, exceptional: otherwise it would not be interesting to write about them. Overall, there seems to be some grounds for cautious optimism in a limited number of states.

However, even this cautious optimism has to be qualified by an issue of overwhelming importance: the increased social fragmentation of the metropolitan areas. As the flight to the suburbs continues, the problems from which so many are fleeing thereby get worse. These wider issues of growth management demand a higher political profile than they usually receive. In this respect, there are few grounds for any optimism.

FURTHER READING

The quotation at the head of the chapter is from Bosselman and Callies (1972) *The Quiet Revolution in Land Use Control*. A review up to the end of the 1970s was written by Callies in 1980: 'The quiet revolution revisited'. At about the same time there appeared the full-length study by Healy and Rosenberg (1979) *Land Use and the States*, followed by DeGrove (1984) *Land, Growth and Politics*. DeGrove's original work has been updated in DeGrove and Miness (1992) *The New Frontier for Land Policy: Planning and Growth Management*. A review of state systems is provided by Wickersham

(1994) 'The Quiet Revolution continues' : this covers Florida, Georgia, Maine, Maryland, New Jersey, Oregon, Rhode Island, Vermont, and Washington.

On the individual states DeGrove (1984) provides the most detailed account up to the beginning of the 1980s, while DeGrove and Miness (1992) update this (selectively) to the beginning of the 1990s.

Legal materials are included in Callies *et al.* (1994) *Land Use: Cases and Materials*.

There is a constant stream of books on state growth management policies. The reader should check the latest. Those used in the preparation of this chapter (in order of date of publication) are:

DeGrove (1984) *Land Growth and Politics*;

Porter (1992) *State and Regional Initiatives for Managing Development*;

Buchsbaum and Smith (1993) *State and Regional Comprehensive Planning*;

Stein (1993) *Growth Management: The Planning Challenge of the 1990s*;

Abbott *et al.* (1994) *Planning the Oregon Way*.

Accounts of the policies of individual states are:

California: DeGrove (1984); Stein (1993); Fischer (1985) 'California's coastal program: larger-than-local interests built into local plans'; Fulton (1991) *Guide to California Planning*.

Colorado: DeGrove (1984)

Florida: DeGrove (1984); DeGrove and Miness (1992); Porter (1992); and Stein (1993). Audirac (1990) 'Ideal urban form and visions of the good life: Florida's growth management dilemma'; Koenig (1990) 'Down to the wire in Florida: concurrency is the byword in the nation's most elaborate statewide growth management scheme'.

Georgia: DeGrove and Miness (1992); Buchsbaum and Smith (1993); and Stein (1993).

Hawaii: DeGrove (1984); Callies (1984) *Regulating Paradise: Land Use Controls in Hawaii*; Callies (1994) *Preserving Paradise: Why Regulation Won't Work*.

Maine: DeGrove and Miness (1992).

New Jersey: DeGrove and Miness (1992); Buchsbaum and Smith (1993); and Stein (1993).

North Carolina: DeGrove (1984).

Oregon: DeGrove (1984); Buchsbaum and Smith (1993); and Stein (1993). Rohse (1987) *Land Use Planning in Oregon: A No-nonsense Handbook in Plain English*; Knaap and Nelson (1992) *The Regulated Landscape: Lessons on State Land Use Planning from Oregon*; Oliver (1992) '1000 Friends are watching: checking out the record of Oregon's pace-setting public interest group'; Abbott *et al.* (1994) *Planning the Oregon Way*.

Rhode Island: DeGrove and Miness (1992).

Vermont: DeGrove (1984); DeGrove and Miness (1992); and Porter (1992).

Washington State: DeGrove and Miness (1992).

QUESTIONS TO DISCUSS

1 What are the objectives of growth management policies?

2 Why are growth management policies so difficult to implement?

3 Outline a theoretically effective growth management policy.

4 Why have growth management policies widened to include such issues as housing and economic development?

5 Discuss the importance of citizen involvement in growth management policy-making.

6 Why have states become involved in growth management?

PART V

DEVELOPMENT ISSUES

A very large number of development issues would need to be covered in a comprehensive text. Here a selection of three is made: transportation, housing, and community and economic development. These are three of the most important and difficult of today's development issues. Transportation is the essential 'connector' of activities. Upon its adequacy depends the efficiency and convenience of settlements. The importance of transportation in the urbanization process was discussed in Chapter 2. In Chapter 12, the focus is on its centrality in modern society, the problems to which it gives rise, and the huge and expensive measures needed to deal with these. There is also a caution: our understanding of the complexities of metropolitan areas (within which four-fifths of the population live) is limited: in the current state of knowledge, we are unable to plan transportation systems with confidence about their adequacy or even their effects.

Housing is another important development issue; but it is much more: it provides a home (which two-thirds of households own or are buying); it is a major item of household expenditure (with a median of about a fifth of monthly income); it has a long life and therefore requires continued maintenance; if this is

inadequate there can be drastic neighborhood effects (and these can themselves lead to lower standards of maintenance). There are many other aspects to housing: its location affects households' accessibility to opportunities, it is a major land use, and an important source of municipal revenue. The list could easily be lengthened. Large parts of this book deal with various aspects of housing: urbanization (Part I), land use regulation (Part II), environmental quality (Part III), and growth management (Part IV). In this part, the focus is on the working of the housing market, the provision of affordable housing, and (continuing the discussion on housing discrimination) on access to affordable housing.

Housing is also an issue of – and in – community and economic development. The promotion of community and economic development has for long been a concern of federal and state government. The Community Development Block Grant has been an important measure of support for many years, and it has been embraced in the latest measures of 'community empowerment' introduced by President Clinton. Chapter 14 provides a brief history of some of these endeavors, and highlights the different political philosophies which they reflect.

12

TRANSPORTATION

The ordinary 'horseless carriage' is at present a luxury for the wealthy, and although its price will probably fall in the future, it will never, of course, come into as common use as the bicycle.

Literary Digest 1899; quoted Jackson 1985: 157

THE CENTRALITY OF TRANSPORTATION

Transportation is the lifeline of the economic system. It is the essential means by which activities are linked and thus made possible. Without access, most economic activity could not take place. Transportation is thus essential, not for its own sake, but because it provides access. Since it thus serves other activities, its character is determined exogenously. Of crucial importance is the pattern of land uses: this is the major determinant of transport needs. The more that activities are dispersed, the greater is the amount of transportation required to access them. Nevertheless, transport does not simply follow activities: its potentialities facilitate and limit the development and spread of activities. In one sense, it can be said that the history of both economic and urban development is a reflection of the history of the development of transportation.

The course of urbanization (and the disastrous attempt to arrest urban decline by improving access *to* the city) has been outlined in Chapter 2: here the concern is with contemporary problems and policies relating to the operation and planning of urban transportation systems. The chapter opens by summarizing some major transportation trends. This is followed by an analysis of a number of ways

in which traffic might be restrained – through land use planning; by direct controls, by demand management, and by congestion charging. Finally, a brief account is given of a remarkable congressional initiative to tackle transportation problems through comprehensive state and regional planning.

Some of the issues discussed in this chapter are developed further, within different contexts, in other chapters. Thus, the role of land use planning policies in the restraint of traffic growth is dealt with in the discussion of growth management in Part IV. The chapter on development charges discusses the use of transportation impact fees and similar charges on developers.

THE ARITHMETIC OF TRANSPORTATION

The United States has always been concerned to have efficient transportation systems and, to a large extent, it has succeeded. There are nearly 200 million motor vehicles in use (a fourfold increase since 1950). Nine out of ten households have a motor vehicle, and a half have two or more. Four-fifths of passenger traffic is by private auto. Commuting by car is the most usual means of getting to work: three-quarters travel alone (now technically known as SOV travel –

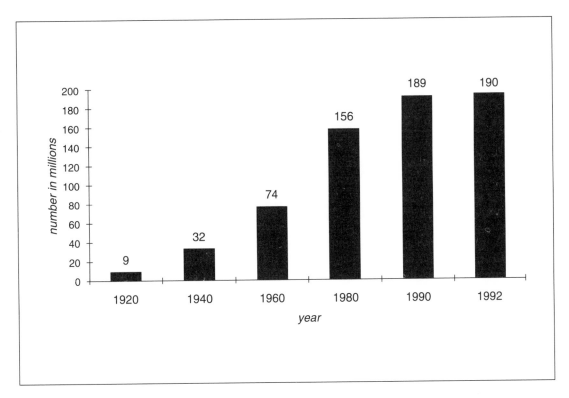

Figure 12.1 Motor Vehicle Registrations, United States, 1920–92
Source: Historical Statistics of the United States, part 2, p. 176; *Statistical Abstract of the United States 1994*, Table 1009

in a *single occupancy vehicle*), and 13 per cent share the ride. Despite the amount of attention given to commuting, only a fifth of auto trips are for this purpose: shopping is almost as important, while personal and pleasure purposes each account for around a fifth of journeys.

Public transit declined with the growth of car ownership, but in more recent years has increased somewhat. Only 5 per cent of work journeys are made by transit. Two-thirds of these are by bus. Railroads are statistically insignificant in the national total of work trips, but they carry over a third of freight (measured in ton-miles). The other major freight carriers are trucks (over a quarter), and oil pipelines (almost a fifth).

Current commuting patterns are much more complex than used to be the case when employment was concentrated in cities. Today there is more commuting from suburbs to suburbs than from suburbs to central cities. This dispersal has been made possible by the flexibility provided by auto travel. It is also highly dependent upon auto travel since (to use transportation terminology) reverse-direction and circumferential commuting poses serious difficulties for public transit.

TELECOMMUTING

It may be that the continued increase in commuting could be reduced by the growth of telecommuting, though this is quite uncertain. In 1992, 30 per cent of the labor force worked at home for at least part of the time. Most of these were self-employed or simply working after regular hours, but a growing number are full-time employees who would otherwise be

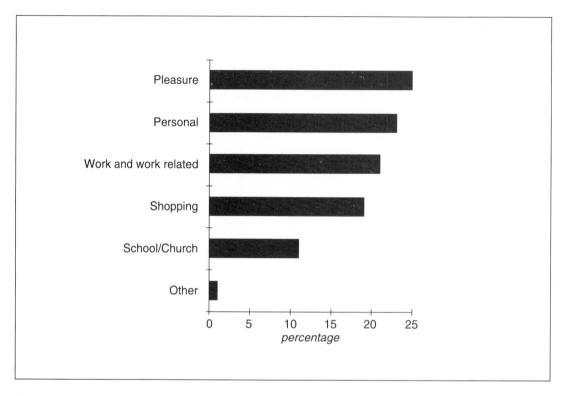

Figure 12.2 Trip Purposes, 1990
Source: DOT Nationwide Personal Transportation Study, NRC 1994: 2. 25

commuting. They are able to work at home because of huge advances in sophisticated telephone and computer systems. In a real sense, telecommunications services can be substituted, partly or completely, for transportation to a conventional workplace.

The extent of telecommuting, the forms it takes, and the implications it has for transport, work, and life-styles is not clear. There are also definitional problems as, for example, with commuting to regional telework centers (which generally appear to reduce travel, though they do not eliminate it). Some use of telecommunications may be *additional* to work undertaken at the office, or simply a more efficient means of evening and weekend office 'homework'. Telecommuting is a diffuse activity, often undertaken on an informal basis. As such, it is not well captured in current statistics; and its transportation impacts are not easily measurable. Not surprisingly, any estimate

of future trends is hazardous, both because of the paucity of information on the current situation and the difficulty of prediction. But there is no doubt that it could bring about great changes in transportation, as is also possible with other dimensions of telecommunications – telebanking, teletaxes, tele-education, teleshopping, and so forth. Telecommuting is officially accepted as a congestion-reducing 'travel demand management' measure eligible for federal funding under various state and federal programs (some of which are discussed later).

A study published in 1993 by the federal Department of Transportation estimated that the number of telecommuters might increase from 2 million in 1992 to between 7.5 and 15 million in 2002. On the definitions and assumptions used, this would involve between 5.2 and 10.4 per cent of the labor force at the later date. The effect could

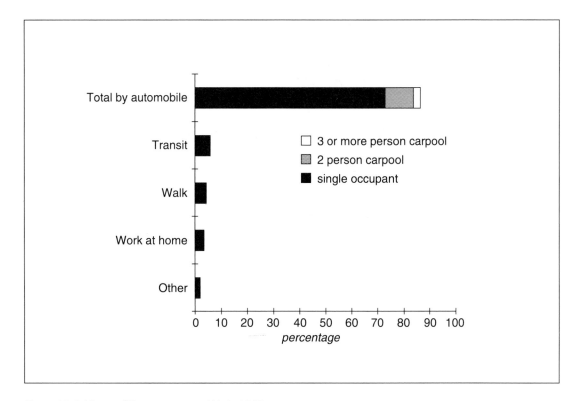

Figure 12.3 Means of Transportation to Work, 1990
Source: Federal Highway Administration. *New Perspectives in Commuting 1992*; American Public Transit
Association, *Transit Fact Book 1993*, p. 76

be to save up to 35 billion vehicle travel miles. These and other figures are reproduced in Box 12.1.

There is no way of knowing whether such a scenario will develop. The observable effect might be largely one of an increase in convenience, efficiency, and opportunities. There are many ways in which the possible benefits might be nullified. The beneficial effects of transportation programs to date (whether in easing congestion or in reducing air pollution) have been overtaken by increases in the number and use of motor vehicles. The most that has been achieved has been a slower rate of traffic growth. This, of course, is better than nothing, but it indicates how difficult it is to bring about significant improvements. Moreover, if telecommuting did have an impact on the reduction of congestion, this might simply attract more traffic to the less congested roads. This is an

illustration of Downs' 'convergence' theory which is outlined below (pp. 163–4).

SUBURBANIZATION AND TRANSPORTATION

Historically, a city has been a node of concentrated functions with very high accessibility. Technological changes in production, energy, and transportation have dramatically reduced the advantages of the city, and a huge amount of activity has moved out to suburban locations. Even more significant has been the new growth which has located in the suburbs. At first sight, it might be expected that this would ease the urban transportation problem since there would be less traveling into city centers, and workers would have a shorter journey to their suburban homes. In

BOX 12.1   PROJECTED TELECOMMUTING AND ITS TRANSPORTATION IMPACTS

	1992	2002
Number of telecommuters (m)	2.0	7.5–15.0
as proportion of labor force (%)	1.6	5.2–10.4
proportion working at home (%)	99.0	49.7
proportion working at telework center (%)	1	50
Average days per week telecommuting	1–2	3–4
Saving in vehicle miles traveled (bn)	3.7	17.6–35.1
as proportion of total passenger vehicle miles (%)	0.23	0.7–1.4
as proportion of commuting vehicle miles (%)	0.7	2.3–4.5
Saving in emissions (%)		
NO_x	0.23	1.1–2.2
HC	0.31	1.4–2.7
CO	0.36	1.7–3.4

Source: DOT 1993: viii–ix

fact, other factors have intervened. First, central city employment has not declined very much (in some areas it has increased slightly). Second, though many suburban residents commute to suburban job locations, their journeys are not necessarily shorter: in many cases they are much longer (whether the commute is in the same suburb or to a different one). Third, very high levels of automobile use have led to increasing congestion: suburban journeys can now be as congested as those to the city. Other factors include the design of shopping malls, office and industrial parks, and the wide range of suburban employment centers which have often been explicitly designed for automobile use: so much so that they can discourage other forms of access. This, together with abundant car parking (typically free) is a major incentive to automobile transport.

The growth of employment in the suburbs has given rise, not only to circumferential commuting, but also to 'reverse' commuting – traveling from the city to the suburb. Travel distance to work has also been increased by a 'jobs–housing mismatch': the lack of affordable housing in areas close to employment

centers which compels households to seek cheaper, far-distant housing locations. Restrictive zoning plays an important role in this. Overall, the pattern of commuting is complex. As a result, dealing with congestion has become extremely difficult.

In principle, there are two major ways in which the problems can be approached: by minimizing traffic generation through land use planning measures, or by directly controlling traffic through regulatory or economic measures.

TRAFFIC RESTRAINT THROUGH LAND USE PLANNING

Since transportation is a function of land use, it seems obvious that one way of effectively reducing transportation problems is by imposing tighter land use controls. However, there are several difficulties here. First, the agencies which deal with land use (mainly local governments) have little or no responsibility for transportation; and the agencies which determine transportation policies (state or regional bodies)

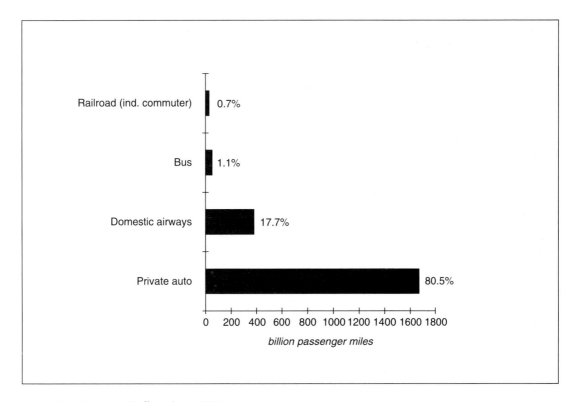

Figure 12.4 Passenger Traffic Volume, 1992
Source: Statistical Abstract of the United States 1994, Table 994

normally have no say in the determination of land uses. Second, the separatism of these agencies is reinforced by the different skills, training, and interests of local land use planners and transportation engineers. Third, as a result of the independence of local governments, action by one to restrict development (of suburban employment centers, for example) would be seized upon by other local governments in the region as an opportunity to secure development for their areas. Only a strong state or regional planning presence could deal with this type of problem (if the necessary political inclination existed). But, fourth, even if controls were imposed stringently, there are serious doubts on how effective they would be in the long run. Downs has argued that growth is impervious to local public policies: even if all local governments banned further development, migrants would still find a way in (see Box 12.2).

Although there may be little likelihood of affecting transportation by growth control measures, there is the alternative of planning land uses to minimize their transportation effects, or to make transit viable. This is not a new idea, of course. The traditional central business districts did precisely this: concentrated employment, service, retail, and other functions were served by highly developed transit systems.

The difficulty arises in implementing such a strategy when the predominant form of development is one of highly dispersed land uses. Owners of the sites selected for concentrated development may be delighted, but those which are to be 'protected' from development may see matters very differently.

Contrariwise, there can be strong objection to development, whether for NIMBY or other reasons. Thus, attempts to secure a better 'jobs–housing' balance (by building affordable housing close to

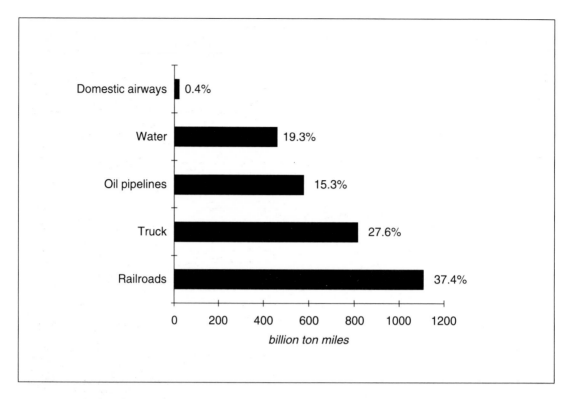

Figure 12.5 Freight Volume, 1992
Source: Statistical Abstract of the United States 1994, Table 994

employment centers) can raise the wrath of those who are already living in the area. It is instructive to note the experience of Bay Area Rapid Transit (BART) in implementing its policy of developing 'transit villages' – housing development around BART stations. Though operating under special legislation that allows the designation of station-area redevelopment districts, BART has sometimes faced considerable opposition from existing nearby residents. More generally, BART has also had difficulty in deciding whether it should give priority to better parking or more development around its stations (Knack 1995).

This illustrates some underlying problems faced by any policy of attempting to influence patterns of behavior by land use changes: how is the choice to be made between competing desirable objectives, and how effective are the plans likely to be? These problems are difficult enough with undeveloped sites; they are greatly compounded in areas where a pattern of uses is already established. There are difficulties in determining what it might be desirable to do, quite apart from the practicalities. An illustrative example is given in Box 12.3.

DESIGNING SUBURBAN CENTERS FOR TRANSIT

In his 1989 study *America's Suburban Centers*, Cervero pointed to the need to design these centers to encourage, or at least facilitate, commuting by transit. One way is to avoid single-use developments in favor of mixed-use developments, which are much more user-friendly than those devoted to only one use (such as offices): in the absence of other uses (such as shops,

Plate 24 Interstate 210, Los Angeles
Courtesy Trevor Warr, Viewfinder Colour Photo Library

restaurants, and banks), suburban workers are forced to have their cars for use during the day for meals, banking, and other personal errands. Surveys have shown that these needs for a car during the day can be a significant factor in determining travel by automobile.

Additionally, since mixed uses peak at different times, they can give rise to economies in parking provision. This can have the further advantage of reducing the scale of a development and making walking more attractive. For such reasons, mixed-use developments can have market advantages.

However, the importance of design (as distinct from use) should not be exaggerated. A study for the US Department of Transportation showed that transit-friendly design features are not in themselves

sufficient to have any significant impact on the transit ridership. They can be a useful complement to other measures, but they are not sufficient to lure commuters out of their cars (Cervero 1994).

Reviewing these various possibilities, it seems clear that, in the short run at least, little relief from traffic congestion is likely through land use planning measures. The alternative is to operate controls directly over roads and traffic, by management, regulation, or economic incentives. Nevertheless, high density residential development centered on a transit facility such as a railroad station makes sense in its own terms, even if its impact on the overall transportation system is slight. At the least, it provides an alternative to the typical suburban/commuter type of development. The market success

---

BOX 12.2   LOCAL POLICIES CANNOT CONTROL GROWTH

Every US metropolitan area has at least some communities encouraging further growth. Even if none did, newcomers would continue to arrive anyway if they believed good economic opportunities were available there, as history has repeatedly proved. Such immigrants would either live on the outskirts of the metropolitan area in unincorporated places with no antigrowth policies, or they would illegally double and triple up in dwelling units within communities that had formally banned further growth. These observations lead one to conclude that growth is impervious to *local* public policy.

Source: Downs 1992: 33

**BOX 12.3   CONTROLLING TRAFFIC BY REDUCING COMMERCIAL
            DEVELOPMENT – GOOD INTENTIONS IN LOS ANGELES**

In 1986 the voters of Los Angeles approved a measure which reduced allowable development on most land zoned for commercial development. This particularly affected the 'strip-commercial' zones along the main streets and boulevards. The intention was to reduce the traffic congestion caused by such development.

But what was to happen to the development pressures involved? Would they simply disappear, or would they emerge elsewhere – and, if so, what would the effect then be?

Martin Wachs pointed out that much of the demand for commercial development would be redirected toward the regional centers which were exempt from the downzoning, and also to the outlying suburban centers which were beyond the jurisdiction of Los Angeles. The result could be a lengthening of journeys to work and shopping in these more distant locations.

Thus, communities which have experienced commercial downzoning in order to reduce the number of trips destined for them may well experience increases in through trips which will be in the future destined to the areas which are allowed to develop.

Regrettably, the downzoning may deprive the city of tax revenues which might be used to relieve traffic congestion through construction programs, while not relieving it of the traffic which downzoning was intended to prevent.

Source: Wachs 1990: 249

---

of a number of these has introduced a new term to the planning lexicon: *transit villages* (Knack 1995).

INCREASING THE SUPPLY OF ROAD SPACE

Before discussing ways of restraining traffic, it is necessary to inquire whether it is not possible simply to increase the supply of road space to accommodate increased numbers of vehicles. This can be done either by new road building or by measures which increase the carrying capacity of the existing roads. The first can have spectacular results, but these are often short-lived. Traffic seems to increase faster than new roads can be built. A major reason for this, of course, is the continued growth in car ownership and use (itself in part stimulated by new roads). This stems from the huge advantages of the automobile for personal mobility and the increase in auto ownership and use. There are many issues here, including the increasing difficulties of managing *without* an auto (because of the wide dispersion of activities, and the inability of public transit to serve these); the large investment that has been made in roads; the availability of parking (provided by employers, shopping centers, etc.); the relatively low cost of auto travel (most roads can be used without direct payment); and so on.

There is, however, a limit to the extent to which the supply of road space can be continually increased. As this has become apparent, and as concern has grown about the cost and impacts of road building, increasing ingenuity has been devoted to making roads able to carry more traffic, and to reducing some of the commuting demand. Some of the techniques are discussed below, but it has to be said at the outset that none have proved particularly effective. Sooner or later any freeing of road space is taken up by increased traffic. So common has this been that it has been suggested that an underlying principle is at work.

DOWNS' PRINCIPLE OF 'TRIPLE CONVERGENCE'

This has been elegantly set out by Anthony Downs (1992) in his theory of 'triple convergence', which is based on the simple fact that since every driver seeks the easiest route, the cumulative result is a convergence on that route. If it then becomes overcrowded, some drivers will switch to an alternative route which has become relatively less crowded. These switches continue until there is an equilibrium situation (which, like any human equilibrium, is not stable – conditions constantly change). On this theory, building a new road or expanding an existing one will have a 'triple convergence': first, motorists will switch from other routes to the new one ('spatial convergence'); second, some motorists who avoided

Plate 25 Monument Valley, Arizona
Courtesy Sarah Boait, Viewfinder Colour Photo Library

the peak hours will travel at the more convenient peak hour ('time convergence'); third, travelers who had used public transit will switch to driving since the new road now makes the journey faster ('modal convergence').

The eventual outcome depends upon the total amount of traffic (actual and potential) in relation to the available roads. If the increase in traffic stimulated by the new road is modest, there will be an observable benefit for all. Though peak-hour traffic may be congested, this is simply because so many drivers are traveling at the time which is most convenient to them. (There may, however, be a loss to transit passengers if the 'modal convergence' leads to a reduction in service.)

TRANSPORT DEMAND MANAGEMENT

Since it is so difficult to change transportation and land use systems, considerable thought has been given to the alternative of 'managing' the transportation system either by physical changes to roads or by influencing traffic behavior.

Some measures are simple, such as phasing traffic lights, changing two-way streets into one-way, controlling street parking, and carefully programming road repairs to ensure minimum disturbance to traffic. Another is to increase the occupancy of autos: most peak-hour commuters travel alone ('lone rangers'). Congestion could be significantly reduced if there were more sharing. To encourage this, some areas have reserved lanes for *high occupancy vehicles* (HOV). The theory here is that the higher speeds achieved on an HOV lane will encourage drivers to change to HOV driving: that is, they will arrange to share their journey to work with others. (Definitions of HOV vary; it can be as low as two: the driver and a passenger.) HOV lanes, of course, are of particular value to buses, and thus provide an incentive to transfer to transit (if the bus goes to a location which is convenient for the commuter). The idea is an attractive one, though the removal of a lane from general use means that the other lanes become more

crowded. This naturally causes annoyance (if not fury) to lone drivers and tempts them to trespass on the HOV lane. This, in any case, is a temptation which is overcome only if there is strong enforcement (at least when they are introduced) and high fines. HOV lanes are more successful if there is an added incentive to use them, as with employer ride-share programs.

PARKING POLICIES

An apparently simple means of reducing traffic congestion is to eliminate the high tax-free subsidies granted by employers to commuters by way of free parking. Some 90 per cent of American auto commuters park free at work. This significantly reduces the real cost of commuting. For instance, it has been estimated that, in Los Angeles, the effect of free parking for the average SOV commuter (i.e. a commuter in a single occupancy vehicle) is to reduce the cost from $6.07 to $1.75 a day – a reduction of 75 per cent (NRC 1994: 2. 518). Several studies have shown that the elimination of employer-paid parking has reduced SOV commuting significantly. Other studies have compared employees who receive employer-paid parking with the groups who pay for their own parking: the share of SOV trips was much smaller for those who had to pay – ranging from 19 per cent to 44 per cent less. There is much evidence of a similar nature.

Free employer-provided parking not only generates SOV commuting: it is also unfair to non-auto commuters who receive no corresponding benefit. One way of rectifying this (in addition to amending the tax code to take account of the benefit) is to require employers to offer a 'parking cash-out program' which would enable employees to obtain cash in lieu of free parking. Such a scheme is in operation in California where employers in any area designated by the Air Resources Board as a 'non-attainment area' are required to offer employees a cash allowance equivalent to the parking subsidy (see Box 12.9 and discussion in Chapter 15).

It is also to be noted that the provision of 'free' parking for employees is expensive: estimates vary around $2,000 a year in the Washington DC area (MacKenzie *et al*. 1992). However, though this seems persuasive in central business districts, the calculus is not so readily acceptable in the suburbs, where there is abundant cheap land.

It is necessary, however, to distinguish between central city and suburban areas. In central areas, there are public transit systems, and land for parking is expensive. The opposite is the case in the suburbs. In any case, why should individual employers try to deny their employees the convenience and low direct cost of SOV travel? Traffic congestion has to be very severe before the car-using public will accept restrictions on the basis of seemingly theoretical arguments.

One of the advantages of parking as a policy instrument is that it is very flexible. For example, charges can be varied by size of vehicle, time of arrival or departure, and duration. Targeted groups can be charged lower (or nil) rates, e.g. the disabled, local residents, and emergency staff. In non-commuter car parks, it is common to charge higher rates for longer periods of parking, but this can encourage 'reparking'. Moreover, there is a case for charging short-term parkers at a high rate since they create more travel.

It should be noted that, however effective parking controls may be in restraining trips to a local area, they do not deter through traffic. Indeed, if parking charges have the effect of reducing local congestion, there may be an increase in through traffic. The same applies with measures to prevent obstruction by autos parked in the street.

Parking measures are the most effective among the many possibilities of affecting transportation demand, but they can be more effective if they form part of a wider approach aimed at securing the benefits of cooperation, or even coordination, among different agencies. Such an approach is the essence of Transport Demand Management (TDM) programs.

TDM PROGRAMS

Transportation demand management can be implemented through voluntary arrangements, or in conjunction with a trip reduction ordinance. However, whatever the legal statutory aspect, it relies essentially on the willingness of both employers and employees to participate. When they do, it is because TDM is cheap, effective, and capable of securing tangible benefits.

At its most sophisticated, TDM operates through an organization of interested parties – employers, developers, members of business associations, land owners, and public bodies including planning and transportation agencies. There are, however, wide variations in organization, scope, and character. Some do little more than act as a source of information for employers and commuters – providing information on alternatives to SOV commuting, for instance. Some are run with no explicit budget or acknowledged cost, while others derive financial support from the private or the public sector (or both). Many have no powers of enforcement; a few have apparently draconian systems for imposing financial penalties. (The Californian South Coast Air Quality Management District can impose a fine of up to $25,000 a day for failure to prepare a plan for reducing vehicle ridership, but this is unusual and is regarded more as an indication of the seriousness of the endeavor.) Some of the main program elements are listed in Box 12.4.

TDM programs have been shown to be worthwhile, particularly with parking controls, where they have had (relatively) the most success – though usually on a site-by-site, rather than an area-wide, basis. Such measures can help if other things remain equal. Unfortunately, they seldom do. Transportation systems are composed of a myriad elements which interact: changes in one element triggers responses in others – often a surge in new traffic to fill a 'space' created by some traffic-reduction measure. In the long run, changes in spatial structure may bring about major improvements, but congestion will remain serious in some areas while drivers do not meet the true costs of their use of roads. If it were to be practicable, the heart of the problem could be reached through the use of congestion charges.

BOX 12.4 TRANSPORT DEMAND MANAGEMENT

An addition to the abbreviations of transportation planning is TDM – transportation demand management. This attempts a cooperative approach to 'the art of modifying travel behavior'. Program elements include:

- reduction of parking provision;
- cash in lieu of free parking;
- car-pool matching services;
- transit information centers;
- alternative work schedules;
- parking management services;
- shuttle services (e.g. to rail station);
- preferential parking for high occupancy vehicles;
- transit incentives;
- guaranteed emergency ride-home program;
- subsidized transit fares;
- subsidized van-pools;
- express bus services;
- alternate work hours;
- home-based telecommuting;
- park-and-ride lots;
- design improvements for local pedestrians;
- showers and lockers for cyclists and walkers;
- secure cycle parking.

Source: Ferguson 1990

CONGESTION CHARGES

The problems of traffic congestion have reached such a point in some areas that more forceful measures are having to be considered. The theoretical basis for congestion charging is essentially simple. Individual drivers are concerned only with the costs they bear. As traffic on a stretch of road increases, each additional car adds to congestion and thus imposes costs on other drivers. Congestion charging translates this cost into individual charges. Auto drivers now have an incentive to take into account the cost of congestion which they are collectively causing. Moreover, since they are forced to bear the cost if they use the congested road, some will find alternative routes or modes of transport, and traffic on the priced road will decrease.

The way in which costs arise at the margin is dramatically illustrated by the estimate of the Bay Area Economic Forum that 'a single driver entering the San Francisco area's congested roads during the peak hours can generate one hour of additional delay for all other drivers there combined'. Though this is a curiously dramatic way of illustrating the point, there can be no doubt that, above a certain level of congestion, the difference between the individual cost and the social cost is enormous. Congestion charges even this out.

Theoretical justifications for charging for the use of roads might be described as overwhelming were it not for the fact that public opposition to the idea is typically even more so. Nevertheless, there is usually far less opposition to *new* toll roads, and even less to bridge and tunnel tolls. This suggests that attitudes are a matter of perception and habit. Roads have traditionally been 'free' to the user: to introduce a charge is to take away a benefit (which, it can be argued, is already paid for in taxes). A new facility, on the other hand, clearly requires new expenditure; and it also brings an equally clear benefit to the motorist.

Major objections are that congestion taxes are politically unacceptable, are difficult to administer, are unfair and penalize low-income motorists. Certainly there is abundant evidence about the political unpopularity of congestion charges. However, as traffic congestion worsens, attitudes to charging may change. There is already some evidence that this is happening (NRC 1994: 1: 64). At the least, some pilot schemes may attract sufficient support for testing the workability and effects of charging.

It has to be acknowledged that there are many uncertainties about congestion charging. The technical problems of administering charges can be surmounted, and it seems evident that even a small switch away from priced roads would have a large beneficial effect on traffic flow. But many effects cannot even be foreseen, let alone measured. The modern metropolitan area is a highly complex urban system (perhaps better described as a multiplicity of interacting systems). Too little is understood about it for confident predictions to be made regarding the effects of introducing new traffic measures. Yet it also is true that there is no solution to the congestion problem without more effective means of restraint – and that congestion pricing currently seems to be one means of achieving this effectively. It is not without reason, however, that policy-makers are cautious. They have to convince a car-owning electorate that the imposition of congestion charges is not just another tax but is likely to be an effective means of improving their transportation situation.

CONGESTION CHARGING AND EQUITY

Convincing a skeptical public that congestion charging will bring about tangible and widespread benefits is difficult, partly because theoretical arguments are insufficient to overcome the doubts occasioned by our degree of ignorance about the working of complex metropolitan systems. A major area of concern is whether a charging system could be designed to be sufficiently equitable. Given the inequities which exist between different socio-economic groups, complete equity is unachievable: any system will benefit the higher income groups most – if only because, in economic terms, their time is more valuable. Of course, there are many ways in which disadvantaged groups could be compensated, but no system could offset this difference. Moreover, though compensation can be devised to benefit *groups*, there is no way in which all the *individuals* in the disadvantaged groups could be recompensed. For example, particularly vulnerable would be low-income working single mothers who can reach their employment only by driving during congested periods. Some may be able to find a substitute in improved (and subsidized) transit, but if they are unable to avoid the particular journey which involves a charge, they could suffer significant hardship.

Nevertheless, it can be argued that the inequitable effects of congestion charging have been greatly exaggerated. One reason for this is that low-income commuter travel patterns are different from those of high-income commuters. More important than income can be origin and destination patterns and the scope for substitutability by time and mode of travel. Improvements to transit and car-pooling alternatives would be of specific value to low-income travelers; and, if necessary, a scheme of rebates might be possible. Kain (1994) has also suggested that those concerned about inequities have failed 'to consider the full range of urban transport technologies and the likely impacts that congestion pricing would have on the level of service provided by these alternatives'.

As with the objections which are made on equity grounds, much of this is theoretical. It is clearly important to ensure that in any scheme of congestion charges particular provision is made for vulnerable groups, and that there is carefully monitoring. An explicit commitment to this is needed at the outset.

BOX 12.5   ROAD PRICING

Popularity of peak-period pricing

Peak-period pricing is well established and acceptable in many areas where demand fluctuates over time. Vacation prices are higher in the holiday season; air fares vary by day and hour as well as by season; telephone calls have peak and off-peak charges; transit and parking charges differ at different times of the day. These charges do not eliminate congestion, but they limit it and divert some demand to less-crowded times. Those who are able and willing to change to the less-crowded times are attracted by the lower cost. Though only a small proportion may change, all benefit by the more even spread of use.

Road congestion pricing

With traffic congestion charges, though it is impossible to make firm predictions, it seems (on the basis of experience with past increases for tolls and parking) that peak charges of 10–15 cents a mile (roughly $2–$3 per daily round trip) would reduce peak travel by some 10 to 15 per cent. The reduction would depend on the design of the charge and the alternative routes and modes that were available. It is expected that most motorists would accept the charge, but some would switch, for example, to other routes, share-riding or transit. (In the longer run there could be changes of job or home.) Though only a small proportion would make a switch, substantial benefits would accrue. The average commuter could save 10–15 minutes per round trip: a time saving of about a fifth. The aggregate value of the time savings would outweigh the charges, and a very substantial income would be available for improvements to the transport system and for special compensatory aid to disadvantaged groups.

Source: National Research Council 1994

There are, of course, long-run implications of changes in the transportation system, though it is extremely difficult to identify these, let alone predict changes. Metropolitan areas have traditionally dealt with congestion by spreading out. It is difficult to predict what might happen if a major change were made in the complex of forces which have produced the familiar suburban pattern.

For the individual commuter, the cost of the journey to work is part of the price to be paid for the favored life-style provided by low-density suburbs – travel costs are lower than housing costs! Anthony Downs has expressed the point neatly in his *Stuck in Traffic* (1992): see Box 12.6.

Whether charges are extended beyond the relatively small mileage of tolls currently levied (which now includes the Dulles Greenway – see Box 12.7), there is a clear necessity for transportation planning to be on a more comprehensive basis. Transportation systems operate over wide areas: both

control measures and investments can be adequately planned only over the wider region where their impacts are felt. The interconnections of transportation networks spread far beyond the locality in which they are made. These simple points present administrative, financial, and political difficulties which governments have been loath to grasp. By the end of the 1980s, however, it was clear that significant federal action was both needed and, more surprisingly, accepted. The result was the passing in 1991 of a radical piece of legislation.

THE INTERMODAL SURFACE TRANSPORTATION EFFICIENCY ACT 1991

The Intermodal Surface Transportation Efficiency Act is usually referred to by its acronym ISTEA – commonly pronounced as 'iced tea' (which is certainly

---

BOX 12.6   DOWNS' ADVICE TO THE WEARY COMMUTER

My advice to American drivers stuck in peak-hour traffic is not merely to get politically involved, but also to learn to enjoy congestion. Get a comfortable, air-conditioned car with a stereo radio, a tape player, a telephone, perhaps a fax machine, and commute with someone who is really attractive. Then regard the moments spent stuck in traffic simply as an addition to leisure time.

Source: Downs 1992: 164

BOX 12.7   THE DULLES GREENWAY

The Dulles Greenway, opened in 1994, is the first private toll road to be built in modern times. The fourteen-mile road cost $326 million. Users pay a toll of $1.25, but save thirty minutes on a one-way journey. The developers have the right to operate the road for 42.3 years, after which it passes to the state. Additionally, the rate of return is capped at 18 per cent. It is hoped that, by the year 2000, the Greenway will attract 68,000 cars a day.

Source: Urban Mobility Corporation 1995

---

more memorable than its title). It is a remarkable piece of legislation which will either bring about a revolution in the way transportation investments are planned and implemented (as Congress intended) or it will go down in history as one of the most ambitious of congressional fantasies. Its objectives (see Box 12.8) embrace an extraordinary degree of coordinated planning. The planning process is now concerned with broad issues of overall transportation and environmental efficiency rather than narrow matters of highway construction. This is facilitated by the flexibility of federal funding. Many funds which were previously restricted to categorical programs can now be switched to provide the mix of projects which will best meet air quality, congestion, mobility, or other problems. The Act requires the preparation of *state transportation plans* and various other plans and *transportation improvement programs*. In metropolitan areas, plans are carried out jointly by the state and the *metropolitan planning organizations* (MPOs) under the terms of a formal agreement. The MPO is 'the forum for cooperative transportation decision-making for an urbanized area'. In the largest metropolitan areas (those with a population of over 200,000) the MPOs are also known as *transportation*

management areas (TMAs), and have additional responsibilities, particularly in connection with clean air (discussed in Chapter 15).

Curiously, but perhaps wisely, the term 'intermodal' is not defined, but it encompasses all transportation modes, including airport system plans, state rail plans, and port system plans. Public participation is a requirement of the planning process: plans are to involve all transport users. This 'proactive public involvement process' involves all 'affected public agencies, representatives of transportation agency employees, private providers of transportation, other interested parties affected', and specifically 'those traditionally underserved by existing transportation systems'.

The issues to be covered in the planning process are set out at length and in detail. They include *congestion management strategies* (including ridesharing, and pedestrian and bicycle facilities), the effects of transportation policy decisions on land use and development, the consistency between transportation plans and programs and land use plans, preservation of future rights-of-way, and 'the overall social, economic, energy, and environmental effects of transportation decisions'.

Plate 26 Tram, San Diego
Courtesy Charles Edwards, Viewfinder Colour Photo Library

**BOX 12.8   INTERMODAL SURFACE TRANSPORTATION EFFICIENCY
ACT**

The National Intermodal Transportation System shall consist of all forms of transportation in a unified, interconnected manner, including transportation systems of the future, to reduce energy consumption and air pollution while promoting economic development and supporting the Nation's pre-eminent position in international commerce.

This remarkable initiative on the part of Congress has been prompted in part by environmental considerations. Both the Clean Air Act Amendments and ISTEA preclude the construction of new highways in areas which fail to meet federal air quality standards. In these areas (which include many major metropolitan regions), alternative (and 'specific') measures have to be proposed for reducing automobile travel (see Box 12.9).

Traffic congestion is both intensely frustrating and very costly – estimated at over $40 billion a year (not counting the cost of environmental damage). It is becoming clear that the metropolitan areas simply cannot build their way out of the

Plate 27 Metro Train, Washington, DC
Courtesy G. Weinbren, Viewfinder Colour Photo Library

BOX 12.9   MANDATORY REDUCTIONS IN TRAFFIC

Instead of providing new capacity, those areas not in compliance with the Clean Air Act Amendments must propose specific measures for reducing automobile travel through measures such as trip reduction ordinances, employer-based transportation management, transit improvements, pricing, traffic flow improvements, and parking management. Employers with more than 100 employees in the ten metropolitan areas rated as severe or extreme 'non-attainment' are required to submit plans by 1994 that will result in reduction in the number of employees driving to work alone.

Source: National Research Council 1994: 1. 20

---

problem and that insufficient relief can be obtained from traffic management measures. The peak period of congestion is lengthening and on some metropolitan highways congestion lasts throughout the day. Innovative ways of managing road systems and

of demand management have brought little relief. The result has been increased interest in exploring the validity of the theoretical advantages of congestion pricing (at least on the part of transportation experts!). Interestingly, there are indications that

Plate 28 Metro Center Station, Washington, DC
Courtesy Rick Buettner, Viewfinder Colour Photo Library

business leaders, who have for long opposed congestion charges, regard it as a lesser evil than the
imposition of demand management schemes such as
trip reduction programmes (NRC 1994: 1. 21). It is
also significant that ISTEA provides federal funding
for a pilot congestion pricing program.

One of the many uncertainties facing the
implementation of ISTEA is that of forging the
necessary links between land use planning and
transportation planning. The Act does not provide
the metropolitan planning organizations with any
new legal authority in this area; instead it lays great
emphasis on a partnership of all relevant agencies to
promote area-wide interests and goals. Nowhere
is the political nature of the planning process more
evident than on this issue. Though the new system
will be greatly concerned with technical issues of

great complexity, the more troublesome problems
will lie in devising methods of communication,
mediation, and decision-making which are acceptable to the multiplicity of agencies, authorities, and
interests in a metropolitan area.

CONCLUSIONS AND UNCERTAINTIES

There is no simple solution to the problems of congested transportation. There are too many barriers:
of ignorance (of ways of reducing pollution); of cost
and tax implications (of major road and transit
developments); of public attitudes (which set limits
to what is politically possible); of governmental
machinery (which, in land matters, is essentially local

and self-centered); and of understanding the sheer complexities of metropolitan areas. In the long run, changes in land use patterns may bring about a significant change in travel behavior, but this will take a very long time. Established land uses are vast compared to the incremental changes which can be brought about by new development; and there must be doubts as to whether it is possible to be sure that these changes could be effectively planned and implemented to affect travel behavior in intended ways.

Nevertheless, there are many ways in which conditions can be improved. The simplest is by action on free parking provided by employers – if there is sufficient support for this. Congestion charging could be very effective, though its promises are latent until public opinion is more favorable – which could follow from increasing congestion and pollution. Other policies which can help are ride-sharing, park-and ride connections to transit systems, a speedier system of dealing with the aftermath of road accidents, further development of TDM, planning high-density residential development at transit stations.

One final caveat is necessary. Not only is there considerable ignorance and uncertainty about many of the issues discussed in this chapter: there is also a wide variation among areas in their pattern of land uses, transportation systems, economic profile, income distribution and many other matters. This variation may also be matched by differences in culture and attitudes. It follows that the points made in this chapter are not necessarily equally relevant in all areas, or even in those metropolitan areas which are the main centers of urban congestion.

FURTHER READING

A non-technical, clear and interesting book on traffic congestion is Downs (1992) *Stuck in Traffic: Coping with Peak-hour Traffic Congestion*. This is a refreshing, contentious analysis of the congestion problem which argues that the problems are not serious enough to precipitate appropriate action – which, in any case (in

Downs' view), is not likely to prove effective. Downs' earlier discussion of 'convergence ' is in 'The law of peak-hour expressway convergence' (1962).

Moore and Thorsnes (1994) *The Transportation/Land Use Connection* provides a succinct outline of this subject and a summary of recent research. It also illustrates how (in the Portland Metro region) transportation and land use planning can be integrated.

Transport Implications of Telecommuting is a review of the field carried out for the US Department of Transport and published in 1993. Two recent studies of telecommuting are Handy and Mokhtarian (1995) 'Planning for telecommuting', and Mitchell (1995) *City of Bits: Space, Place, and the Infobahn*.

Robert Cervero's *American Suburban Centers: The Land Use–Transportation Link* (1989) convincingly supports the author's contention that 'the low-density, single-use character of many suburban work centers was a root cause of the congestion problems being faced in suburbia'. Peter Calthorpe (1993) presents a design for *The Next American Metropolis: Ecology, Community, and the American Dream*. This includes concepts of 'sustainable communities', 'pedestrian pockets', and 'transit-oriented development'.

Issues relating to congestion charges are fully examined in a study commissioned by the federal government following the passage of the Intermodal Surface Transportation Efficiency Act of 1991: National Research Council (1994) *Curbing Gridlock: Peak-period Fees to Relieve Traffic Congestion*. Volume 1 consists of the report and recommendations; volume 2 contains a set of commissioned papers. A short review of 'what it really costs to drive' is given in MacKenzie *et al.* (1992) *The Going Rate*.

At the time of writing, it was too early for any assessment of the working of ISTEA, but there has been a great deal of discussion. Particularly useful is National Research Council (1993) *Moving Urban America*.

A review of TDM is given by Ferguson (1990) in his

article 'Transportation demand management: planning, development and implementation'.

Wachs (1990) details the experience in one state where some efforts were made to regulate traffic by land use controls:'Regulating traffic by controlling land use: the South California experience'.

QUESTIONS TO DISCUSS

1 In what ways has transportation changed in recent decades?

2 Do you think that telecommuting will have major impacts on transportation?

3 How far can land use planning controls be used to restrain traffic?

4 How effective are transit villages in a comprehensive transportation plan?

5 Critically discuss Downs' 'Principle of Triple Convergence'.

6 Discuss the role of transport demand management as a solution to traffic congestion.

7 Discuss the argument that parking controls are the single most effective means of traffic restraint.

8 'If drivers paid the proper prices for their use of roads, there would be no traffic congestion problem.' Discuss.

9 How viable is comprehensive transportation planning?

13

HOUSING

a decent home and a suitable living environment for every American family . . .

Housing Act 1949

THE COMPLEX OF HOUSING

Housing is of central importance in both the national economy and the individual's standard of living. It is a major land use, and its location is a crucial factor in the economy of cities, in transportation, in local economic development, and in the access to opportunities available to individuals. At the same time, its very high cost, compared with other items of household expenditure, presents some particularly difficult problems of finance. Its long life necessitates continual maintenance to prevent deterioration. The condition of individual houses can have neighborhood effects: poor maintenance can blight nearby houses. Deterioration can also result from neighborhood changes – social, economic, or physical. In addition to providing physical shelter, an individual's position in the housing market can affect social status, capital gains (or losses), and credit availability. The ramifications of this combination of attributes makes housing an extraordinarily complex matter. Its multiple dimensions include locational, architectural, physical, economic, social, medical, psychological, and financial. As a result, 'housing policy' involves very much more than the building of houses. Moreover, the reader who has come this far will not need reminding that discrimination is a major issue in the determination of land uses. This discrimination affects the operation of the housing market and adds greatly to the problems of ensuring the provision of affordable housing.

Housing policy can take a number of forms. The most important has been the devising of mechanisms to facilitate home ownership. The key issue here is that the capital cost of housing is so high that few households are able to purchase a home outright: they typically require the assistance of a financial mechanism to enable them to spread payments over a long period of time. The development of various types of mortgage has enabled a large proportion of households (around two-thirds) to become home owners. (The important role of the federal government in this policy area is outlined in Chapter 2.) Tax benefits have also played a part in making home ownership cheaper and financially attractive (for example, by way of deductions for mortgage interest and capital gains benefits).

Home ownership and rental housing differ in many ways (social, economic, physical, locational), but a crucial difference is that with home ownership the householder obtains the mortgage directly, whereas with rented housing there is an intermediary investor. This has an obvious but important implication: if investors do not foresee a profit, they will not invest. Therefore, if alternative investments are more profitable, or if incomes are too low to enable a rental

Plate 29 Blighted Blocks, Philadelphia
Courtesy Alex MacLean/Landslides

housing investor to make the expected profit, there will be a shortage of rental housing. (Rent controls can have the same effect.)

HOUSING MARKET THEORIES

In the automobile market, those who can afford to do so buy new cars; others buy used ones. As a result cars filter down to lower income groups. Does something similar happen with houses? Not exactly, since older houses may be better than newer ones. But if instead of age one considers quality, there are some similarities. As houses decline in quality they become cheaper and affordable by those with lower incomes. Further, as household incomes increase, better-quality housing can be afforded. Thus housing of declining quality

filters downwards, and households with rising incomes filter upwards.

These simple ideas form the basis for much theorizing about the operation of the housing market. For present purposes, the relevant issue is whether the filtering process works sufficiently well to meet the needs of lower-income households; or does something impede this neat process? Since there is a general shortage of housing for lower-income households, it is self-evident that it does not work sufficiently to deal with all needs. The reasons for this are important. First, many houses do not filter at all, since they remain occupied by higher-income households or are converted to other uses (such as commercial) or are demolished to make way for land use changes (higher-quality uses, highways, etc.). Second, houses must fall very greatly in price (as do motor vehicles) to be affordable by the poor; and this fall in price may well imply a marked fall in quality (the condition of the accommodation). In short, by the time a house filters down to the lowest income level, it may be grossly inadequate, badly maintained, and have a backlog of overdue maintenance costs which cannot be afforded. Indeed, by this point the house may well have become a public health hazard, subject to action by the local authority. Moreover, housing quality typically depends on the level of maintenance of the building and the neighborhood in which it is located. Given a stable neighborhood, a high degree of maintenance, and a continuing program for replacing outworn services and fittings, the life of a house can be infinite, and it can steadily increase in value. More generally, however, changes in neighborhood quality (itself often the cause of reduced maintenance), obsolescence of

internal fittings, changes in fashion, rising standards of heating, cooling, insulation and suchlike, and a host of other factors cause values to fall. These and many other complications do not arise in the motor vehicle market.

This is a highly simplified view: theories of filtering abound; and they conflict disconcertingly. This is partly because, while some analysts view it as a *process*, others view it as an *outcome*. There is no necessary connection between the two. Changes in the process may have differing outcomes; and similar outcomes may result from differing processes, depending on a multitude of variables.

However, the theory that good housing filters down to poorer households 'works' to a limited extent, though it is restricted by a host of factors. One theory rests on the differences in the size of income groups. Lower-income groups are more

Plate 30 Housing Blight, Detroit
Courtesy Alex MacLean/Landslides

numerous than those of higher income, and thus will constitute a large demand for houses vacated by the latter. As a result, the fall in prices will tend to be small. But to the extent that the filtering process is successful, the result may tend to be 'self-corrective': lower prices would reduce the willingness of existing owners to trade, and the supply would diminish. Whatever the validity of such theories, it is clear that filtering cannot meet the housing needs of the poor. Their very poverty makes it impossible for them to pay the costs of decent housing. Filtering stops before housing of adequate quality gets to them. Yet much argument about housing policy is centered on the efficacy of filtering.

Among the many reasons why filtering fails to meet low-income needs is that some housing is abandoned before it becomes cheap enough. At first sight, abandonment seems nonsensical: surely some income is better than none? This is not so if the costs are higher than the income. It can cost more to demolish a building than the resultant vacant site is worth. The unfortunate result is a cancerous growth of decay. The tragedy of the worst inner-city areas is that they are blighted by abandonment and a lack of demand.

LOW-INCOME HOUSING

Low-income households have particular difficulty in affording market rents, and housing policy has for long struggled with the problems to which this gives rise. These problems are exacerbated by several special factors. First, among the many matters which affect the cost of housing are the standards imposed by government on the quality of housing: a host of regulations impose minimum standards of building for health and safety. Some of these (such as adequate sanitation) are totally accepted; others (such as high minimum house sizes) are debatable. But whether acceptable or not, they increase the cost of housing above market levels for the poor. Some analysts go further and argue, as did the report to President Bush of the Advisory Commission on Regulatory Barriers to Affordable Housing (1991) that 'millions

of Americans are being priced out of buying or renting the kind of housing they otherwise could afford were it not for a web of government regulations'.

Second, as this quotation suggests, the specific problem of providing housing for the poor merges into the general problem of affordability, which affects a wider range of income groups. There is no simple cut-off point between the poor and the not-poor. Moreover, in providing housing for the poor at a subsidized cost, there arises a basic unfairness for those of the poor (the majority) who receive neither good housing nor the financial benefits that go with it. This has been one of the reasons why programs of housing vouchers have attracted less hostility than the provision of public housing: they provide rent assistance to renters of private as well as public housing. But there is a more important reason for the opposition to public housing: quite apart from ideological issues, there is typically very strong opposition to the location of public housing. This is the classic case of NIMBY: 'public housing may be all right somewhere else, but not here'. This opposition may be racial, social, or simply a result of the fear that the construction of public housing would lead to a fall in local property values. ('We personally do not object to public housing, but others do – and this will affect house values.') The poor management and the severe problems which have arisen in some public housing projects has created an indelible image of crime-ridden, drug-infested, dangerous, and decayed urban eyesores. Their unpopularity is now widespread, and more effort is being expended on transferring public housing to tenants and other owners than on expanding the supply.

There are two additional points to make here. First, despite the extent of the heated arguments on public housing, the total amount nationally is extremely small: even at its largest, the number was only about 1.5 million units. Today, it amounts to only about 1 per cent of the national total number of houses. Second, though the worst projects have attracted a great deal of attention and supported the stereotype, many public housing projects are of good quality, well maintained, and popular with their tenants.

PUBLIC HOUSING AND URBAN RENEWAL

Public housing policies have been characterized by extraordinarily strong opposition. This has taken many forms: arguments about the sanctity of property rights and the limits to which government should interfere in market forces; fear of undermining individual self-reliance; concern that the private market will be jeopardized by 'unfair competition'; mistrust of the competence of government in such an area; the huge cost which a significant program would involve; and the belief that the needs which could not be met by private enterprise are best left to charity and voluntary effort. Despite the appalling housing condition of the poor, these arguments held sway until tentative initiatives were made in the late 1930s. The Depression years saw a worsening of the housing problem of an increased proportion of the population, and eventually (in 1937) federal legislation was passed which signaled a recognition that both slum clearance and the provision of public housing were legitimate areas of public policy. The substance, however, was thin, mainly because of the bitter opposition of the National Association of Real Estate Boards and kindred spirits.

The opposition continued throughout the 1940s (indeed, it has never ceased). The industry argued that the private market could meet all the nation's housing needs without the intervention of government (though 'aids to private enterprise' such as those provided by the Federal Housing Administration – discussed in Chapter 2 – were championed). The outlook for post-war housing policy was therefore bleak. A housing bill was introduced in 1945, but was killed by vociferous opposition, first in 1946 and again in 1948. It eventually passed as the 1949 Housing Act – the single Fair Deal piece of legislation which Truman managed to get through Congress. The Act embraced the national goal of 'a decent home and a suitable living environment for every American family', but the means to achieve this were effectively denied by Congress.

The legislation authorized the building of 810,000 units of public housing over a period of six years, though the program was slow in starting, and it took two decades before this target was reached. Part of the reason for this was the opposition to public sector activities which, following the passage of the Act, moved from Congress to local areas. But the task was inherently complex: areas had to be selected and designated for acquisition; sites had to be cleared; complicated negotiations were required for federal funding; and arrangements had to be concluded with private investors and developers for redevelopment. The important role given to private enterprise was part of the political price which had to be paid to secure the passage of the legislation. One aspect of this was provision for urban redevelopment by private enterprise with local government supervision and federal–local subsidies that bridged the gap between the market value of land and the (much higher) actual cost of acquisition and clearance.

The concept of linking redevelopment to the politically unpopular provision of public housing gave rise to widespread problems. This was of particular importance since redevelopment was to be 'predominantly residential'. But private developers were not interested in low-income housing (whether subsidized or not). Profits lay in other directions, particularly downtown shopping and commercial centers. These were also popular with local political and business elites. As a result, pressure built up for the rules to be altered. A major change came with the 1954 Housing Act, when the term 'urban renewal' was introduced, indicating that, in addition to redevelopment, the policy now embraced revitalization, redevelopment, conservation, and 'the renewal of cities'. The Act provided that 10 per cent of project grants could be used for non-residential development. The rationale behind this 'was that there were nonresidential areas around central business districts, universities, hospitals, and other institutional settings that certain city interests wished to clear and redevelop for nonresidential purposes' (Weiss 1980: 267). In Mollenkopf's words (1983: 117), 'the 1954 Housing Act shifted urban renewal from a nationally directed program focusing on housing to a locally directed program which allowed downtown businesses, developers, and their political allies, who had little interest in housing, to use federal power to advance their own ends'.

The proportion allowable for non-residential purposes was increased to 20 per cent in 1959 (together with the needs of dominant institutions such as hospitals and universities), and later to 35 per cent. With the administrative latitude allowed, the eventual result of this was to increase the commercial part of urban renewal to one-half of the total. Indeed, by manipulating definitions and procedures, it was possible to force the proportion up to two-thirds.

Despite amendments to the legislation and some notable achievements, for example in improving the physical appearance of hundreds of American cities, urban renewal became subject to increasing criticism. Above all, it failed to help the poor: indeed, it made their position worse.

While urban renewal bolstered central business districts and may even have contributed indirectly to a city's economic vitality, it also dislocated neighborhoods and often created more urban blight than it removed. Urban renewal benefited some center-city businesses and upper-middle-class households who obtained desirable inner-city housing at bargain prices, but frequently the real cost was borne by the urban poor. Downs (1970: 223) estimated that households displaced by urban renewal suffered an average uncompensated loss amounting to 20–30 per cent of one year's income. Moreover, while originally conceived as a means of increasing the supply of low-cost housing, urban renewal actually exacerbated the urban housing problem: more houses were destroyed than were replaced.

Disillusionment with urban renewal led to its decline at the end of the 1960s. It thereby joined public housing as a cause which even its sponsors no longer supported though, of course, its impact has lived on. But the fundamental weakness of urban renewal was that it was conceived in terms of a land use instrument which was to 'save' the declining cities. Though there were pockets of success in commercial (sometimes monumental) centers and middle-class residential areas, these typically had little or no wider effects, except of an undesirable nature such as the displacement of low-income households. Their impact on restraining the exodus to the suburbs was minimal. It is likely that a greater force for effective 'renewal' lay with the increasing number of new immigrants who, like so many before them, sought out opportunities in the cities – opportunities which urban policy had defined as problems.

ALTERNATIVES TO PUBLIC HOUSING

The removal of the federal government from the production of public housing created a void (small though it was) which has in part been filled by state and local efforts. These can take many forms, including public–private partnerships or neighborhood non-profit bodies (Stegman and Holden 1987; Suchman et al. 1990). Nationally, there are bodies such as James Rouse's Enterprise Foundation and the Inner City Ventures Fund. The former assists non-profit neighborhood groups on matters such as access to capital and technical expertise. The Inner City Ventures Fund was established by the National Trust for Historic Preservation to provide financial and technical assistance to non-profit neighborhood organizations for the rehabilitation of historic buildings to be used as affordable housing and commercial properties that benefit low-income residents.

Some community schemes are of a partnership nature, such as two project-based corporate–community partnerships in New York City which were backed by Chemical Bank and Citibank. Such non-permanent partnerships call on financial, technical, and organizational resources to achieve a specific development objective – often the acquisition and rehabilitation of deteriorated property. In Chicago, a housing partnership has involved several types of organization: community, private, and local government. Financial contributions from local employers attract tax benefits as well as positive community relations. (Another housing partnership is noted in Box 13.1.)

A different type of organization is the San Francisco BRIDGE: the Bay Area Residential

---

BOX 13.1   BOSTON HOUSING PARTNERSHIP

Considered a model for the nation, the Boston Housing Partnership builds on strong state and local commitments to and programs that support the provision of low-income housing. The partnership was formed in 1983 under the leadership of William Edgerly, chairman of the State Street Bank and Trust Company, in response to recommendations from a group of public officials, private business interests, and neighborhood organizations. Known as *Goals for Boston*, the group wanted the city to devote more attention to the housing needs of its disadvantaged neighborhoods, and to address both housing abandonment in distressed areas and displacement of low and moderate income households in strong market areas.

BHP members include the city of Boston, major banks; insurance and utility companies; local universities; local businesses; community and housing development organizations; and the Massachusetts Housing Finance Agency.

Source: Suchman *et al.* 1990: 23

Investment and Development Group. This is unlike many local organizations in that it operates over a wide area and, instead of supporting local groups, is a direct provider of housing. Over the decade since its foundation in 1983, it has built some 5,000 dwellings and has an ambitious continuing program. It attracts funding from corporate investors, but also has obtained benefits from tax credits and a variety of innovative techniques, tax-deductible donations, for-profit housing, density bonuses, and other support from its strong connections with the local business community.

There is a large number and variety of these local non-profits, and it is generally agreed that they provide an acceptable and efficient way of providing additional low-income housing. Some of them have concerns and activities which extend well beyond housing: these are discussed in the next chapter.

HOUSING SUBSIDIES

The problem of housing affordability results from the level of market prices being higher than the ability to pay. The gap can be bridged either by increasing incomes or by reducing rents. Either way, a subsidy is required. However, simply increasing incomes does not, in practice, help very much since households with low incomes may (and often do) prefer to spend increased income on other necessities than housing (as the Experimental Housing Allowance Program demonstrated – a finding which raises some interesting policy issues which cannot be pursued here). To target the specific problem of housing costs, financial aid has to be tied to the payment of these costs. If the subsidy is tied into the construction of affordable housing, it obviously benefits only those who obtain such housing (and do not move out of it).

Much current policy assumes that it is better to pay the subsidy direct to needy households, who then have a degree of housing choice. These matters are in practice not as simple as this since problems remain of ensuring housing supply, of combating discrimination, of relating the subsidy to needs while at the same time avoiding disincentives to increase income, and so forth: housing policy issues are never easy. They are also costly (and therefore are vulnerable to public expenditure cuts). There has been considerable experience of such a system with the so-called 'Section 8' certificate and voucher programs. At the time of writing, the current policy is to develop these (see Box 13.3).

The complexities of the housing problem cannot be met by any single policy. Even if the housing certificate program were fully funded, it is still necessary to ensure that an adequate number of housing units are built, and that there are no barriers to access. In addition to the community-based

---

BOX 13.2   INADEQUATE HOUSING ASSISTANCE

The basic problem with housing assistance is that there's not enough of it. The programs reach only about a fourth of the poor; most other renters end up spending more than half their incomes just for shelter. In the 1980s, when growth of the programs slowed, Congress moved to ration the assistance by ordering local authorities to give available units to their neediest applicants first. That sounded like a better idea than it was. It greatly increased the number of very poor people in public housing projects particularly. The projects were transformed, and not for the better.

Source: *Washington Post*, 16 October 1995

BOX 13.3   HOUSING CERTIFICATES

HUD's proposed Housing Certificate Fund, which builds upon the existing Section 8 certificate and voucher programs, will empower assisted families to choose moderately priced housing in the locations that offer them opportunities for upward social and economic mobility. Tenant-based assistance of this kind is less likely than project-based programs to concentrate needy households in high-poverty neighborhoods.

Source: US Department of Housing and Urban Development 1995: 35

---

provision already noted, there are other techniques which can help. These include schemes to entice builders to produce low-cost housing and measures to remove the local barriers to such housing which are erected by municipalities.

The former includes the Low Income Housing Tax Credit Program, introduced in the 1986 tax reforms and made permanent in 1993. These credits are used to raise equity for approved housing construction or rehabilitation. The developments are subject to restrictive covenants which run with the land for thirty years or more and cover such matters as the number of housing units, guidelines relating to rents and eligible household incomes, and criteria for targeting specified needs. About three-quarters of a million rental housing units were allocated through the tax credit program between 1987 and 1994.

The removal of regulatory barriers was the subject of a presidential commission (referred to earlier, p. 178). Its NIMBY report received much publicity, and it was followed up by guidance to the states from HUD. Some states increased their pressure on local governments to ease restrictive zoning policies, and some passed legislation prohibiting discrimination against manufactured housing; but generally the report simply merged with concerns already being expressed about the need to increase the provision of affordable housing. These were given a further push by congressional action: Title I of the National Affordable Housing Act of 1990 requires states to address regulatory barriers in preparing *Comprehensive Housing Affordability Strategies* (now forming part of the *Consolidated Plan* for all HUD community planning and development programs) which are a prerequisite for obtaining Community Development Block Grant funds (discussed in the following chapter). The importance attached to this by Congress is indicated by the introduction, in 1992, of a grant program for the preparation of these strategies. There can be little doubt that the issue of affordable housing has had considerable exposure, but the results of this are less easy to establish.

AFFORDABLE HOUSING AND GROWTH MANAGEMENT

Municipalities represent the interests of their electorates; and since voters object to the construction of low-cost housing, municipalities will often prevent it by one means or another. The reasons have been listed in previous chapters: they include the desire to safeguard the character of the area and its property values, to prevent increased development and the traffic it creates, and likewise to minimize tax burdens by restraining additional infrastructure needs.

Growth management programs may be designed, or used, for such purposes. Indeed, since such programs can have the effect of raising house prices, thus increasing the affordability problem, they may attract considerable opposition from supporters of affordable housing. Debates on state growth management policies can center on this, and a number of states have broadened their land policies to encompass the provision of affordable housing (as discussed in Chapter 11).

As part of its overall statewide planning endeavors, Oregon devised a housing goal (see Box 13.4). A striking feature of this goal is that it is expressed in terms which are clearly translatable into action: this contrasts with the symbolic, empty statements which sometimes pass for 'housing policy'.

The first draft of the Oregon state goals did not address the issue of housing but, as a result of concerns about the negative impacts of growth management policies on the housing market, the specific housing goal was added. Several local governments attempted to circumvent it by changes to their charters, or by introducing conditions which prevented the building of low-cost housing (Abbott *et al.* 1994: 103). These attempts were quashed by the courts. It was held that the goal imposed an affirmative duty on local governments to facilitate the provision of housing at a price or rent which was affordable by current and *prospective* residents. There was thus an obligation to provide for households from outside the area – or, to use the pertinent phrase, for 'regional needs'. There is a long legal history to this of which a high point is the famous *Mount Laurel* case in New Jersey. It is worth looking at this in some detail.

REGIONAL HOUSING NEEDS: THE CASE OF MOUNT LAUREL

From the time of the *Euclid* case (1924–6) to the 1960s, the exclusionary practices of municipalities received little critical attention from the courts or from federal or state governments. The political independence of municipalities was largely sacrosanct. At the same time, policies in relation to both land use and housing were predominantly local in character. There was little concern for regional needs: municipalities were, in the graphic phrase of one critic, 'tight little islands'. Those needs of the exploding cities which could be met by low-density housing were welcomed by suburban municipalities, but low-income housing was nobody's responsibility – in practice even if not in theory. There were, of course, numerous critics, but they had little political impact.

Similarly, efforts by the federal government to promote 'fair housing' and 'equal opportunity'

BOX 13.4   OREGON HOUSING GOALS

To provide for the housing needs of citizens of the state:
buildable lands for residential use shall be inventoried and plans shall encourage the availability of adequate numbers of needed housing units at price ranges and rent levels which are commensurate with the financial capabilities of Oregon households and allow for flexibility of housing location, type and density.

policies, to persuade local governments to include a 'housing element' in their local plans, to introduce *area housing opportunity plans* and suchlike, all proved to be generally ineffective. They never had more than a marginal effect on strong market forces and the powerful underlying social attitudes. It was against this background that the New Jersey court made a frontal attack on the discriminatory practices of the local governments in that state in a set of cases collectively known as *Mount Laurel*.

Mount Laurel is a most pleasant area which has attracted large numbers of people from nearby Philadelphia and Camden. Its population doubled between 1950 and 1960 (from 2,800 to 5,200), doubled again between 1960 and 1970, and grew by another half by 1985 (to 17,600). In common with many of the 567 municipalities in New Jersey, Mount Laurel imposed minimum lot size and other restrictions. There was no doubt as to the intention, as well as the effect, of these: it was to keep out low-income families and other unwanted groups. Activist groups of academics, lawyers, and others (including Paul Davidoff's Suburban Action Institute, the National Committee Against Discrimination in Housing, and the stalwart lawyer, Norman Williams) had already taken legal against other municipalities, sometimes with success – in legal, if not in practical terms). Mount Laurel, with a blatantly exclusionary zoning policy, was an obviously important target, and in 1975 a number of public interest groups brought a case against the township. The case eventually found its way to the New Jersey Supreme Court.

The *Mount Laurel* saga was long drawn out, and the case came before the court three times. On the first occasion, the court proclaimed the doctrine that a municipality's land use regulations had to provide a 'realistic opportunity for the construction of its fair share of the present and prospective regional need for low and moderate income housing'. Much litigation followed this opinion, and there were varying inter-pretations of what it really meant. Indeed, far more time and effort was spent on this than on actually doing anything substantively, and it became clear that the court had not provided any effective remedy. Developers (who were keen to build for a lower-

income market) were largely powerless against municipal stalling tactics. Mount Laurel's response to the court was, to put it mildly, niggardly: it rezoned twenty acres on three widely scattered plots owned by three separate individuals, two of whom were not even residential developers. Its zoning ordinance entirely prohibited the construction of apartments, townhouses, and mobile homes.

The case returned to the court, which took a most militant approach. It was determined to introduce effective means to compel municipalities to provide 'a realistic possibility' for a 'fair share' of housing opportunities for lower-income households. This it did by requiring that, where a municipality refused to fulfill 'its constitutional obligation' (of permitting low-income housing) the court itself would take over the responsibility. That this involved overriding the local zoning ordinance and various other matters concerning the actual provision of housing did not deter the court: it would if necessary employ a court-appointed master to deal with zoning approvals and 'the use of effective affirmative planning and zoning devices'.

Mount Laurel was a truly remarkable case. It was tantamount to a usurpation of municipal powers (it was certainly attacked as such by both local and state government). However, the court was at pains to stress that, though it would have preferred not to take extreme measures, it had no alternative. It was now up to the legislature to tackle the difficult political problem from which it had so far shied.

Not surprisingly, the reaction against the court's opinion was vociferous, and there was a great deal of activity aimed at both the implementation and the obstruction of the court's wishes. But the growing effectiveness of the court-imposed regime eventually forced the New Jersey legislature to take the political action which was so clearly required and, after long and acrimonious debate, the New Jersey Fair Housing Act was passed.

THE NEW JERSEY FAIR HOUSING ACT

New Jersey's Fair Housing Act was designed to retrieve the role which the court had usurped in relation to the provision of lower-income housing. It was intended, above all, to 'disarm' the judiciary. This achievement was made possible only by crafting the provisions of the Act in such a way that it was more acceptable (or, to be more precise, less unacceptable) than the system being operated by the court. Above all, the 'judicial monster' created by the court was replaced by a system of voluntary compliance. This, of course, implied a considerable weakening in the system of control over exclusionary zoning. (Some idea of the depth of feelings aroused is given by the public statement of one senator who proclaimed: 'I would rather go to jail than allow my community to be overrun with Mount Laurel housing.')

The Act established a Council on Affordable Housing charged with carrying out 'the Mount Laurel obligation'. It also transferred most existing (and future) suits to the Council. Transfer was dependent upon a municipality including in its zoning ordinance a 'housing element' which contained a fair-share plan. If accepted by the Council it is approved (certified), and the municipality is then shielded from builders' suits – unless a builder can provide 'clear and convincing evidence' that the zoning ordinance is exclusionary.

As a part of its fair-share plan, a municipality can transfer up to half of its fair-share obligations to another municipality by payment through a *regional contribution agreement*. Many suburban communities have negotiated such transfers to the older urban areas of the state, such as Newark. The transferred funds can be used by the receiving municipality for new building or rehabilitation. The funds are raised in various ways, mostly through agreements with developers, who make the payments in lieu of building affordable units in their new developments.

The Fair Housing Act tips the balance of advantage greatly in favor of a municipality. All it has to do is to satisfy the Council that it is making an appropriate fair share allocation (under provisions much less demanding than under the court's system).

HOUSING MEASURES IN OTHER STATES

Less dramatic, though equally interesting, measures have been taken in other states. Massachusetts was one of the first when, in 1969, it passed its 'anti-snob' law which provided a process of appeal for developers against obstructive municipalities. This raised huge opposition from the local governments who succeeded for many years in rendering it ineffective. However, a less confrontational approach by the state (involving mediation in cases of dispute) led to changes which made the scheme less unacceptable.

BOX 13.5   NEW JERSEY REGIONAL CONTRIBUTION AGREEMENTS

The transfer of millions of dollars from suburban and rural municipalities to the more urbanized areas of the state is perhaps one of the most significant and unanticipated effects of the Fair Housing Act. This substantial affordable housing program has provided an average annual subsidy of approximately $13.5 million over a six-year period to municipalities with the greatest proportion and number of low and moderate income households in the state.

By the end of 1994, 29 agreements had been signed, involving 19 receiving and 33 sending municipalities. At an average of more than $19,000 per unit, 4172 units have been or will be built or restored to a standard condition at a total cost of approximately $81 million.

Source: New Jersey Council on Affordable Housing *Annual Report* 1995

A similar system introduced by Connecticut in 1989 provides for appeal to the courts rather than to a state administrative body. Supporting this are 'negotiated investment strategies' and 'regional fair housing compacts' aimed at increased affordable housing provision.

Another approach is to require municipalities to adopt a 'housing element' in a master plan as a prerequisite to the use of zoning powers. The New Jersey Fair Housing Act amended the Municipal Land Use Law to this effect.

Such initiatives by the states have had a limited effect. Complex unpopular laws seldom attain their objectives: there are too many ways in which they can be circumvented.

THE FEDERAL FAIR HOUSING ACT

At the federal level, the Civil Rights Act of 1968 was quickly used to outlaw racial discrimination in the sale of houses, but discrimination through zoning proved to be a much more difficult issue. A famous case is that of Arlington Heights, which, concerned a proposal by a religious order. It owned eighty acres of land near the center of the village (a suburb of Chicago) and proposed to develop this for a federally subsidized multi-family racially integrated housing project. A rezoning was necessary which, following fierce public opposition (only twenty-seven of the village's 64,000 residents were black), was denied by the local plan commission.

The case came before the courts and eventually to the Supreme Court of the United States. There were a number of complexities (as usual) which are ignored here: the essential point is that the court held that there simply was not sufficient proof that discrimination was a motivating factor in the village's decision. In the court's words, 'official action will not be held unconstitutional simply because it results in a racially disproportionate impact . . . proof of racially discriminatory intent or purpose is required'.

As an interpretation of the equal protection clause of the Constitution, this decision was technically correct: there must be *intent* to discriminate. But on any broader approach, it gave support to blatant discrimination. That this is so is demonstrated by a later development of the same case when it came to be dealt with under the Fair Housing Act. This legislation differs significantly from the general constitutional requirements in that it requires proof only of a racially discriminatory *effect*. It was clear to the parties concerned which way the wind was blowing; a settlement was reached out of court, and the project went ahead.

Clearly, the specific provisions of the Fair Housing Act are a stronger tool to combat housing discrimination than the Constitution. Regrettably, they have not so far proved to be adequate.

INCLUSIONARY ZONING

Where a local government supports the provision of affordable housing, there still remains the problems of ensuring that it is provided. This can be difficult in high-cost areas. However, as already outlined in Chapter 7, there are a number of incentives which can be offered to developers (sometimes with acceptance being a condition of approval). One of these is 'inclusionary zoning'.

The essential feature of inclusionary zoning is that it seeks the provision of lower-cost housing either by offering a developer a higher density in return ('incentive zoning' or 'bonusing'), or by a mandatory requirement that a certain proportion of units are affordable. Its purpose is generally to increase the provision of lower-cost housing anywhere in the municipality (to 'open up the suburbs' to lower-income households). Alternatively, it may seek to ensure the provision of housing in a central area which is predominantly commercial, thus 'bringing life back to the city after office hours' or increasing the supply of lower-cost housing in an area which is being gentrified.

Typically, however, the objective is simply to provide 'affordable' housing. This has become more prominent because the cost of housing has increased more rapidly than incomes, thus presenting increasing numbers of middle-class households with an affordability problem which had previously been confined essentially to the poor. According to the President's 1995 National Urban Policy Report, the proportion of householders owning their homes *declined* between 1980 and 1995, from 65.6 per cent to 64.2 per cent. Though this might appear a small drop, it is noteworthy for three reasons: first, it represents 1.4 million renters 'who would otherwise have become home owners'; second, it was the downturn of a trend which had shown a very long-term upward movement; third, and probably most important, there are the political implications: 'the affordability problem' was now expanding. This political perception is perhaps the most significant – more than outweighing considerations of alternative rationales for the decline (such as the possibility that a 'saturation' level for home ownership has been reached, or that changing attitudes are emerging, possibly as a result of demographic changes).

Thus, the lower-income middle class have come into competition with the poor for the limited benefits of programs to deal with issues of housing affordability. There are no prizes for correctly guessing who usually wins.

Inclusionary housing, like any housing policy, is liable at some point to stumble over the law. Objectors will complain that the policy is unconstitutional, or that the administration is unfair. There seems to be no limit to the barriers which opponents can erect against attempts to meet the needs of minorities: there is always a chance of winning.

CONCLUSION

The provision of housing for lower-income groups has never been a popular policy among the electorate. As a result, the role of the federal government shifted from directly supplying housing to indirectly promoting production by subsidizing other agencies, offering tax incentives, and providing rent subsidies to tenants. Additionally, and independently, it has promoted 'fairness' in the private market.

Mount Laurel is unusual in that it assumed epic proportions, but its elements are repeated time and time again across the nation, typically with the same result – the exclusion of minorities. *Mount Laurel* does demonstrate, however, that, no matter to what extremes the courts are prepared to go, they are no match for the wiles of municipalities whose power base extends from their tiny townships to the state capital. As in other fields, the courts are 'of limited relevance', to use the term from Rosenberg's 1991 book *The Hollow Hope: Can Courts Bring About Social Change?*. They are constrained by the limited nature of constitutional rights, the lack of judicial independence, and the judiciary's lack of powers of implementation.

The Fair Housing Act offers little more encouragement. Discrimination in housing has proven a far more intractable problem than the sponsors of the Fair Housing Act of 1968 anticipated. Amending legislation, passed in 1988, provides enforcement 'teeth' which the earlier Act was lacking, but it remains to be seen how strong its bite is.

Inclusionary zoning offers a different approach, but it is (at least so far) slight in its impact, even when it is implemented with Californian enthusiasm. However, it does have the advantage that it adopts the coinage of American zoning: money. Thus those concerned in dealing with it, either to promote it or to avoid as much as they possibly can of it, are talking the same language. This chapter, like others in the volume, provides testimony to the weakness of government in the face of strong socio-economic forces. Exclusionary zoning is prevalent because it is widely desired by those who have acquired, or are in process of acquiring, their share of the American Dream. Like all dreams, it is easily shattered: hence the vociferous opposition to any development which carries such a threat.

Many of the problems discussed in this chapter go far beyond any concept of 'housing'; and they are far too deeply embedded in the socio-economic structure to be significantly touched by current urban policies. Whether well-funded broader programs could achieve more is an open question. Unfortunately, too many are unaffected by these deeper problems to be moved to do anything about them; and even violent riots attract only temporary attention. There is, however, a glimmer of hope to be seen in an approach which is based on community involvement and development. This is discussed in the following chapter.

FURTHER READING

A useful short survey of theories of filtering is given in Bourne (1981) *The Geography of Housing*. The classic work is Grigsby (1963) *Housing Markets and Public Policy*. See also Grigsby *et al*. (1987) 'The dynamics of neighborhood change and decline'.

Inclusionary zoning is examined at length in Mallach (1984) *Inclusionary Housing Programs: Policies and Practices*. See also Merriam *et al*. (1985) *Inclusionary Zoning Moves Downtown*. See also Burchell *et al*. (1994) *Regional Housing Opportunities for Lower-income Households: An Analysis of Affordable Housing and Regional Mobility Strategies*.

A review of the Mount Laurel saga is given in Berger (1991) 'Inclusionary zoning as takings: the legacy of the *Mount Laurel* cases'. Generally on discrimination in housing, a good account is to be found in Danielson (1976) *The Politics of Exclusion*. Up-to-date discussions are to be found in Keating (1994) *The Suburban Racial Dilemma*; Yinger (1995) *Closed Doors, Opportunities Lost: The Continuing Costs of Housing Discrimination*; and Galster (1966) 'Racial discrimination and segregation', in his edited collection of essays on *Reality and Research: Social Science and US Urban Policy since 1960*. State programs are discussed in Stegman and Holden (1987) *Nonfederal Housing Programs: How States and Localities Are Responding to Federal Cutbacks in Low-income Housing*.

President Clinton's National Urban Policy Report, *Empowerment: A New Covenant with America's Communities* (US Department of Housing and Urban Development 1995) is a political statement outlining the urban policy agenda (and extolling its superiority over previous efforts).

The effect of regulation on the provision of affordable housing is dealt with at length in Advisory Commission on Regulatory Barriers to Affordable Housing (1991) *Not in My Back Yard*; see also Council of State Community Development Agencies (1994) *Making Housing Affordable: Breaking Down Regulatory Barriers. A Self-assessment Guide for States*; and Lowry and Ferguson (1992) *Development Regulation and Housing Affordability*.

The role played by *Public–Private Housing Partnerships* is detailed in a report of that title by Suchman *et al*. (1990). Private housing provision is discussed in a number of reports from the Urban Land Institute, including Porter (1995) *Housing for Seniors*, and Urban Land Institute (1991) *The Case for Multifamily Housing*. On manufactured housing, see Allen *et al*. (1994) *Development, Marketing, and Operation of Manufactured Home Communities*, and Suchman (1995) *Manufactured Housing: An Affordable Alternative*.

A good overall review of affordable housing policies is White (1992) *Affordable Housing: Proactive and Reactive Planning Strategies*. This also has a useful bibliography.

Housing programs change with bewildering frequency, and most published material is rapidly outdated. The student should take care to use the most recent publications.

QUESTIONS TO DISCUSS

1 What are the important features of housing? How do these affect land use and land use policies?

2 'The problem of housing affordability affects not only the poor but also many middle-income households.' Discuss.

3 What are the problems which faced the public housing program? Are these problems soluble?

4 Discuss the various ways in which the gap between market rents and affordable rents can be bridged.

5   Compare the ways in which motor
    vehicles and houses 'filter down' to
    lower-income groups.

6   Can discrimination be effectively
    controlled by legislation?

7   What does the *Mount Laurel* case tells
    us about the limits of court action?

8   How can states compel or persuade
    local governments to permit affordable
    housing?

14

COMMUNITY AND ECONOMIC DEVELOPMENT

The days of made in Washington solutions, dictated by a distant government, are gone. Instead, solutions must be locally crafted, and implemented by entrepreneurial public entities, private actors, and a growing network of community-based firms and organizations.

President Clinton 1995

CHANGING PERSPECTIVES

Public policy, like clothing, has its fashions. Services in kind and in cash alternate in popularity. Top-down and bottom-up policies change in the world of accepted program design. Discretion constantly vies with flexibility for the dominant factor in public policy. The level of government which is to be pre-ferred as the leader in policy moves with bewildering frequency among the federal, state, and local levels. Yesterday's orthodoxy becomes today's anathema; today's accepted wisdom turns into tomorrow's target of inadequacy. Nowhere are these pendulum swings more clearly seen than in urban policy. In particular, the role of the federal government is always under scrutiny, if not attack: it is always too weak, or too strong; or too weak *and* too strong but in the wrong places.

President Johnson's policies for 'the great society' and the 'war on poverty' embraced a positive role for the federal government. This 'creative federalism' was superseded by President Nixon's 'new federalism' which, he claimed, was to 'start power and resources flowing back from Washington to the states and communities and, more important, to the people all across America'. President Carter viewed this as a federal government's 'retreat from its responsibilities,

leaving state and local government with insufficient resources, interest, or leadership to accomplish all that needed to be done'. He sought the best of all worlds in his 'new partnership' with every level of government as well as the private and community sectors. President Clinton's national urban policy is characterized by an emphasis on the empowerment of communities.

Such rhetoric always exaggerates, but it does reflect shifting views on the relative importance of differing features of policy, and alternative adminis-trative responsibilities. New administrations attempt to correct the shortcomings of their predecessors while, at the same time, strengthening the areas of proven success. In order to stimulate new endeavors to tackle old problems, and to marshall general support, the newness may be exaggerated and can be more striking in its packaging than in its content. This in part reflects the character of democratic politics, but it also masks the true uncertainties which policy-making involves. A government which stressed these uncertainties would be regarded as weak; governments have to display an appearance of conviction that their policies really will work.

The Clinton policy of community empowerment bears many features of previous approaches, but is essentially another attempt to redefine the roles of the different governmental, private, and voluntary bodies that are involved in community and economic development. It is useful to examine some of the attempts made by previous governments to tackle the difficulties that this involves.

THE WAR ON POVERTY

Prior to the Kennedy and Johnson administrations, explicit federal urban policies had been largely restricted to urban renewal and public housing. Disenchantment with these spread at the same time that urban problems grew – problems which ranged from race, civil rights, poverty, and violence to state and local government finance. A response to these problems emerged as public concern developed, and as the political scene changed – with Kennedy's rediscovery of poverty in the early 1960s, and Johnson's overwhelming election victory in 1964. The era of the 'Great Society' was at hand. To an unprecedented extent, federal policies were developed to reach 'deeply into the urban social and political fabric' (Mollenkopf 1983: 95).

As with so much in this field, the sequence of events, the policy initiatives and their impact comprise a complex and confused story. Federal programs proliferated on a bewildering scale: more than tripling in the 1960s. These covered the whole spectrum of public policy, from food stamps to regional development, from the 'War on Poverty' to health services, from education to model cities and the Community Action Program (designed to provide power to inner-city residents to improve their neighborhoods). These new programs, each with its own budget and bureaucracy, 'generated a massive federal administrative structure and a significant transformation of federal–state–city relationships' (Frieden and Kaplan 1977: 3). Such a change demanded a greater degree of coordination among federal programs (and agencies) than had ever been required before.

Two responses emerged. First, the arguments which had been deployed for several years in favor of the establishment of a new federal department for urban policies gained the support which they had previously lacked, and the Department of Housing and Urban Development (HUD) was created in 1965. Second, a task force appointed by President Johnson to advise on the organization and responsibilities of the new department proposed that the coordinative role of HUD should be directed through a 'model cities' program toward the poverty areas of central cities. This started a new chapter in urban policy.

THE MODEL CITIES PROGRAM

The birth, brief life, and death of the model cities program constitutes a fascinating case study in the making of public policy. Here, a few features are highlighted: a fuller story is given in Frieden and Kaplan (1977).

From its inception, there was debate on how many model cities there should be. One school of thought opted for a very small number: hence its original designation as 'demonstration cities' – a term which was quickly abandoned when it became associated with urban riots. Others, as indicated above, envisaged model cities being the major channel for aid to poverty-stricken urban areas. In the event, the need to obtain political support for the legislation led to an increased number of cities – eventually 150. The idea of 'demonstration' cities was thus killed. But, more than this, Congress was not willing to see funds diverted from other programs into model cities; and so it became committed largely to a new categorical program rather than to a mechanism for reforming existing grants-in-aid.

There were also apprehensions about the concentration of power in a single agency which this implied: it was felt that this could lead to an uncontrollable, autocratic and overpowering bureaucracy. (See also Box 1.1, p. 9.) Moreover, the redistributive features of the model cities program implied that other federal agencies would be expected to divert some resources from their traditional clients. This 'went against the grain of normal agency behavior, congressional grant-in-aid policies, and ultimate reliance on established

> ## BOX 14.1   MODEL CITIES BILL 1966
>
> The purposes of this title are to provide additional financial and technical assistance to enable cities of all sizes [to implement] new and imaginative proposals and rebuild and revitalize large slums and blighted areas; to expand housing, job, and income opportunities; to reduce dependence on welfare payments; to improve educational facilities and programs; to combat disease and ill health; to reduce the incidence of crime and delinquency; to enhance recreational and cultural opportunities; to establish better access between homes and jobs; and generally to improve living in such areas.

interest groups that benefited from existing programs'. Such were the considerations which doomed the original ideas underlying the model cities program. Frieden and Kaplan concluded that 'if the designers of future urban policies take away any single lesson from model cities, it should be to avoid grand schemes for massive, concerted federal action'. The program, together with urban renewal and other community development programs administered by HUD, was folded into the Community Development Block Grant at the end of 1974.

THE NEW FEDERALISM

While Johnson's policies embraced a positive role for federal government ('federal activism' or 'creative federalism'), Nixon (1969–74) promoted a 'new federalism' which, he claimed, would bring about a return to the original conception of federalism as envisaged by the Founding Fathers. Its main feature[5] was intended to be the replacement of large numbers of categorical programs (and all the controls which accompanied them) by block grants.

This was, in fact, a reaction against the federalist policies of the previous Democratic administration. One highlight of this was Senator Muskie's extensive congressional hearings launched in 1966. The Senator observed that what had been created was almost 'a fourth branch of government, but one which has no direct electorate, operates from no set perspective, is under no specific control, and moves in no particular direction' (Haider 1974: 60). Virtually all the new programs were 'functionally oriented, with power, money, and decisions being vertically dispersed from

program administrators in Washington to program specialists in regional offices to functional heads in state and local governments'. From the perspective of the Advisory Commission on Intergovernmental Relations (*Annual Report* 1970), this left 'cabinet ministers, governors, county commissioners, and mayors less and less informed as to what was actually taking place, and [made] effective horizontal coordination increasingly difficult'.

Given this background, there was a great deal of support for Nixon's proposals. In particular, city mayors saw them as a means of obtaining additional assistance with their fiscal problems and of enabling them to recover some of the power they had lost in the Johnson years.

Nixon's urban aid strategy had two major elements. First, and most innovative, was 'general revenue sharing' which provided federal funds on the basis of a formula encompassing population, incomes, urbanization, and tax effort. The essential policy objective was to allow localities to take spending decisions on the basis of their knowledge and understanding of local needs. Second, 'block grants' were extended by the merging of groups of categorical grants. The best known of these is the Community Development Block Grant (CDBG) program.

Nixon's ideas were never implemented to the extent which he had envisaged, mainly because of congressional opposition and the political impact of Watergate. General revenue sharing was abolished in 1986. However, the CDBG proved so popular with local political constituencies that it survived constant financial cutbacks, though in an attenuated form. It is appropriate to examine this program more fully.

COMMUNITY DEVELOPMENT BLOCK GRANTS

The three-year $8.6 billion Community Development-ment Block Grant (CDBG) program was signed into law by President Ford shortly after his inauguration in August 1974. The Act folded seven categorical programs administered by HUD (including urban renewal and model cities) into this single grant program which was directly targeted on cities, particularly those showing signs of social and economic distress. It was intended to achieve a balance between providing maximum flexibility for local decisions and securing the national purpose of developing 'viable urban communities by providing decent housing and a suitable living environment, and expanding economic opportunities, principally for persons of low and moderate income'.

This has never been an easy balance to attain. At first, there was minimal federal control: eligible local governments simply requested the allotment that was due on a predetermined formula. HUD officials checked entitlement and issued approvals. Any assessment of the value was undertaken later. Local governments took full advantage of their freedom to decide on the allocation of funds and, not surprisingly, there was a number of highly publicized cases of expenditure which *prima facie* seemed inappropriate. Tennis courts took pride of place in these indictments. For instance, Little Rock, Arkansas, used $150,000 from its CDBG to construct a tennis court in a wealthy section of the town. Chicago used $32 million for snow clearance. Other criticized schemes included golf courses, polo fields, and wave-making machines.

There is nothing surprising here: if local governments are given freedom to allocate funds as they wish, they will do precisely this. A requirement that 'maximum feasible priority' be given to projects benefiting low- and moderate-income families allowed a good deal of leeway. Nevertheless, grants were distributed according to a formula based on population (25 per cent), housing overcrowding (25 per cent), and poverty (50 per cent). The formula was changed in 1977 to direct resources from high-income suburbs and urban counties to needy central cities – though not with complete success. However, HUD studies showed that 62 per cent of benefits went to lower-income groups. The pattern of expenditure remained fairly constant through 1987: about a third went to housing, a fifth to public facilities and improvements, and a similar proportion to economic development. The remainder went on acquisition and clearance, administration and planning, and other activities. The pattern has changed over time, and in the 1990s housing was taking a larger share (around two-fifths). Most of the housing expenditure was on rehabilitation (Urban Institute 1995).

As is not uncommon with public policies, different sources provide different conclusions on the effectiveness of the CDBG; but it does seem that, despite an attempt at targeting needy areas, the CDBG benefits were spread widely, and became even more so after the 1974 legislation gave more discretion to cities in the allocation of funds. The increase in benefits going to wealthier areas was a result of local politicians using their discretion in favor of pleasing influential sectors of their electoral constituencies. Local discretion increased still further under the Reagan administration.

CARTER'S NEW PARTNERSHIP

In March 1978, Carter submitted to Congress his *National Urban Policy Report*, containing proposals for 'a comprehensive national urban policy'. Reviewing previous policies, he noted that during the 1960s the federal government had taken 'a strong leadership role' in identifying and dealing with the problems of cities. This proved to be inadequate because the federal government alone had neither the resources nor the knowledge required 'to solve all urban problems'. During the 1970s, federal government 'retreated from its responsibilities', leaving state and local government with insufficient resources, interest, or leadership to accomplish all that needed to be done. The lessons had been learned: 'These experiences taught us that a successful urban policy must build a partnership that involves the leadership of the federal government and the participation of all levels of government, the private sector, neighborhood, and voluntary organizations and individual citizens.' The 'new partnership' thus involved a positive role for the federal government, together with incentives to state and local governments, and to the private sector.

Carter's policy consisted of a large package of existing legislation and new proposals, with an emphasis on the stimulation of private investment. However, there was no suggestion that the migration from the northern cities should be stemmed, even if this were thought to be desirable. Local economic development was not conceived of as a way to stem powerful forces of change, but to assist declining cities to 'a new stage of urban development' of which the main features were decentralization and the dispersal of population and economic activity.

Among Carter's specific policy initiatives was the Urban Development Action Grants (UDAG) program, which passed Congress with relative ease. The bounties of this program were distributed extensively, and it therefore proved widely popular. UDAG was aimed at the stimulation of private investment to create jobs in distressed communities by schemes agreed between the private and governmental sectors. The grants were intended to create leverage on private money, particularly in distressed cities. Unfortunately, this was easier said than done since the targeting of distressed cities was not the same thing as the alleviation of distress: 'Cities that provided the best investment opportunities – where private funds were more available – were not likely to be severely distressed' (Barnekov *et al.* 1989: 79). There has been much controversy over the success (measured in differing ways) of the UDAG program. There was, however, no doubt about its popularity with developers, builders, urban chambers of commerce, and pro-development mayors. To them, the UDAG program seemed to offer benefits as profitable as those of urban renewal.

An unintended result of the policy of promoting private development (through UDAG and other programs) was increased competition among cities; this led to escalating subsidies and an increase in federal controls (thus reducing local discretion). Indeed, 'regulation gradually emerged as the key strategy for implementing the new generation of urban aid programs. The creeping growth of the new rules . . . gradually shifted power back to the federal government' (Kettl 1981: 123).

The major legacy of the Carter administration was its reorientation of policy toward the stimulation of private investment. By its last year, the Carter administration's policy had shifted away from urban issues to much broader concerns for economic growth.

This emphasis on economic development as the foundation of federal policy was embraced and increased by Reagan. His administration brought about the dramatic change of raising unfettered economic forces to be the mainspring of 'policy' – a policy of 'do nothing'. This had the powerful (but highly controversial) support of the President's Commission on A National Agenda for the Eighties (which Carter had found too extreme). This report is of importance not only because of its place in the history of urban policy but also because it continues to represent some widely held views about the objectives and limitations of policies directed at urban conditions. It is therefore worth examining in some detail.

NATIONAL AGENDA FOR THE 1980s: URBAN AMERICA

The essential message of *Urban America* was that unfettered market forces would benignly bring about an efficient and equitable urban settlement pattern, with the economy operating at such a high level that many 'social' problems would disappear (or at least be reduced to a level which the enhanced resources of a liberated economy could meet). Such problems as persisted should be approached directly by 'people policies' (as distinct from 'places policies'). Above all, policies which tied people to declining areas should be avoided: 'urban programs aimed solely at ameliorating poverty where it occurs may not help either the locality or the individual if the net result is to shackle distressed people to distressed places'. Such policies were inherently wasteful. By contrast, 'a federal policy presence that allows places to transform, and assists them in adjusting to difficult circumstances, can justify shifting greater explicit emphasis to helping directly those people who are suffering from the transformation process'.

The report strongly criticized the concept of a national urban policy: 'efforts to revitalize urban areas through a national urban policy concerned primarily with the health of specific places will inevitably conflict with efforts to revitalize the larger economy'. They will therefore do more harm than good. The forces underlying urban change are 'relatively persistent and immutable', and thus are highly resistant to public policies which try to stem them or harness them to policy goals which are not consistent with wider economic development purposes (see Box 14.2).

Though the report did not enter into much detail about the translation of principles into practice, it did list 'prime candidates' that should be 'scrutinized for eventual reduction or elimination', such as economic development, community development, housing, transportation, and development planning. Also suggested was a scrutiny 'for major restructuring or elimination' of such programs as 'in-kind benefits for the poor (such as legal aid and Medicaid), the growing inventory of subsidies that indiscriminately aid the non-poor as well as the poor (such as veterans'

benefits), protectionist measures for industry (trade barriers for manufacturers and price supports for farmers), and minimum wage legislation'.

Though the philosophy of *Urban America* was very much to the liking of the new President, the report was never explicitly accepted by him. Given the number of constituencies which would have been affected by the 'scrutiny' list, this is hardly surprising. But, as we shall see, President Reagan moved forcefully to develop policies which bore a strong resemblance to it.

THE REAGAN YEARS

Reagan's pursuit of privatization was in the tradition of previous administrations, but he gave it a particular twist: so much so, in fact, that the difference became one of kind rather than of degree. The policy was of the utmost simplicity (some would say simplemindedness): free rein to private forces was the key to economic growth and thus to urban regeneration and the solution of many social problems. Government intervention was not only inadequate: it was counterproductive. Government action was no solution: it was part of the problem.

Reagan's first major policy statement was made in an address to Congress in February 1981. This *Program for Economic Recovery* was, as its title suggests, focused on economic matters. Most of the address

---

BOX 14.2   THE 'URBAN AMERICA' PHILOSOPHY

The federal government can best assure the well-being of the nation's people and the vitality of the communities in which they live by striving to create and maintain a vibrant national economy characterized by an attractive investment climate that is conducive to high rates of economic productivity and growth, and defined by low rates of inflation, unemployment, and dependency.

People-oriented national social policies that aim to aid people directly wherever they may live should be accorded priority over place-oriented national urban policies that attempt to aid people indirectly by aiding places directly . . . A national social policy should be based on key cornerstones, including a guaranteed job program for those who can work and a guaranteed cash assistance plan for both the 'working poor' and those who cannot work.

Source: President's Commission on a National Agenda for the Eighties 1980b: 101

dealt with general economic policy issues: proposed limitations in the growth of federal expenditure, reductions in tax rates, 'an ambitious program of reform' to reduce federal regulatory burdens, and the establishment of a monetary policy 'to provide the financial environment consistent with a steady return to sustained growth and price stability'. Urban affairs arose only incidentally – which was precisely what was intended. There was no 'urban policy', other than cuts in programs, and an emphasis on the stimulation of national economic growth (neatly expressed by one of Reagan's senior officials in an article tendentiously entitled 'A positive urban policy for the future' – see Box 14.3). Programs regarded as counterproductive were reduced or completely eliminated, such as the Economic Development Administration, the Urban Development Action Grant, and subsidized housing. Much of this was, of course, along the lines proposed in the *Urban America* report.

ECONOMIC DEVELOPMENT POLICIES

Economic development policies have played a major, though variable, role in public policy. Their rationale has been a matter of wide and generally inconclusive debate, and the crucial question remains: how far, and in what ways, can public policies promote economic development?

Economic incentives have figured significantly throughout the country's history: Alexander Hamilton received a tax exemption from New Jersey in 1791 for a manufacturing company he owned. All states now use development incentives of one kind or another, though there is increasing concern about

their effectiveness, and state politicians have begun to voice doubts which echo the longstanding skepticism of economists.

It seems self-evident that firms seeking a new location will be influenced by the level of local taxes and any economic incentives offered by government. There is, however, little definitive evidence on the matter. On the contrary, there is abundant if not entirely conclusive evidence that neither local taxes nor financial incentives play a significant role in attracting economic growth. The traditional location factors (which vary according to product) are the significant ones. These include proximity to markets and materials, energy and transportation costs, labor availability and costs, the economies and diseconomies of agglomeration, and a host of more elusive qualitative factors. In fact, the number of factors that can be relevant is so large that it would take a heroic feat of economic analysis to isolate their relative importance; and many will be specific to particular places and times.

Moreover, it is generally held that competition by incentives can be a zero-sum game, with jobs merely being shunted among different parts of the country. This argument is persuasive at the national level, but at the state and local levels it can be argued that there is a political imperative to offer incentives, even if they simply counteract the efforts of other states (Wolman 1988). An interesting twist to the debate has been given by Bartik (1991), who has argued that the areas with high unemployment are likely to be more aggressive in the use of incentives and, thus, that more jobs will be created in the most needy areas.

Whatever judgment is made on the arguments, there can be little doubt as to the need for incentives

---

BOX 14.3   NATIONAL ECONOMIC GROWTH AS URBAN POLICY

Improving the national economy is the single most important program the federal government can take to help urban America; because our economy is predominantly an urban one, what's good for the nation's economy is good for the economies of our cities, although not all cities will benefit equally, and some may not benefit at all.

Source: Savas 1983

COMMUNITY AND ECONOMIC DEVELOPMENT

to be carefully evaluated. Quite apart from their efficacy, there is an important question as to whether the money spent on incentives would not be more effectively devoted to investment in education, housing, infrastructure, or any of many other services and amenities which make a location attractive. More accurately, local economic development policies are likely to be more effective if they encompass both the direct attraction (and retention) of business and all the other things which make a place an attractive working and living environment.

There is increasing concern about the shortcomings of economic incentives. The National Governors' Association has recommended that cost-benefit analysis be used to determine whether an incentive provides a positive return, and, if so, whether a better return could be gained from alternative investments. This may be too academic an approach, but several states do now require that incentives be examined to ensure that there is a net benefit. Others have introduced sanctions against firms that fail to produce the benefits promised. In such ways are state policies beginning to change.

Other changes can be seen in a more sophisticated approach to economic development (see Box 14.4), and in a concern to ensure that other state policies do not badly affect the economy (as with redevelopment schemes which bring about a physical improvement at the cost of a loss of jobs). More attention is now being given to the potential of local enterprise and to 'capacity building' in areas such as education and training. Such thinking is particularly relevant in inner city areas where the locational advantages can be capitalized (Porter 1995).

ENTERPRISE ZONES

Despite this new and growing concern to fashion economic incentives more carefully, a long-standing debate on enterprise zones continues. The introduction of legislation establishing seventy-five of these was the one and only urban policy initiative attempted by the Reagan administration. Based on an idea imported from Britain, this had three

distinctive features: first, the primary aim of enterprise zones was to be the economic improvement of poor neighborhoods; second, community institutions were seen as crucial to economic development; and, third, small businesses were to be favored over large ones (Green 1991: 32). Enterprise zone benefits have mainly taken the form of reduced taxes and, since (until recently) these did not explicitly appear in the federal budget, they had an obvious political attraction. However, this difference is one of appearance only: forgone revenues have the same effect as a straight subsidy. (The subsidy is given by way of non-collection of tax, rather than as a payment after the tax has been collected.) Such 'tax expenditures' are now included in the budget, and thus enterprise zones might have lost some of their attraction. Curiously, this does not seem to have dampened support for them: perhaps tax expenditures are simply less politically sensitive.

Though the Reagan enterprise zone concept was an attractive one, Congress failed to pass the necessary legislation. The reasons were partly procedural, partly technical, and partly political. Above all, one question proved difficult to answer: would enterprise zones create new jobs, or would they merely attract jobs from somewhere else? More surprisingly, the enterprise zone policy failed to be passed by the Bush administration, though this was because of a presidential veto of the legislation which contained it. It was finally enacted by the Clinton administration in 1993.

The abbreviation is the same but the words it stands for are changed: EZs are now *empowerment zones*. Thus, a concept espoused for a decade and broadly supported is retained. However, the concept has also been widened, and it is accompanied by *enterprise communities*. The new EZ/ECs thereby have the combined advantages of retaining established support and constituting a new initiative. The initial funding (tax waivers and block grants) amounted to $2.5 billion, with six urban EZs receiving $100 million each, three rural EZs $40 million each, and ninety-five ECs a modest $3 million each. Pressures to increase the number of areas was largely resisted (there were over 500 applications), though an

---

BOX 14.4   IOWA NEW JOBS AND INCOME ACT 1994

The general assembly finds that the public and private sectors should undertake cooperative efforts that result in improvements to the general economic climate rather than focus on subsidies for individual projects or businesses. These efforts will require a behavioral change by both the state and business, balancing short-term self interest with the long-term common good.

Source: Quoted in Gilbert 1995: 440

BOX 14.5   THE CLINTON EZ/EC INITIATIVE

The Clinton Administration's EZ/EC initiative differs fundamentally from previous proposals for 'enterprise zones', which relied almost exclusively on geographically targeted tax incentives to create jobs and business opportunities in distressed communities. The EZ/EC program combines federal tax incentives with direct funding for physical improvements and social services, and requires unprecedented levels of private sector investment as well as participation by community organizations and residents. This collaborative strategic planning and co-investment exemplifies the federal government's emerging role as a catalyst for local change, and exemplifies the larger principles of President Clinton's Community Empowerment Agenda:

> We need to do more to help disadvantaged people and distressed communities . . . There are places in our country where the free enterprise system simply doesn't reach. It simply isn't working to provide jobs and opportunity . . . I believe the government must become a better partner for people in places . . . that are caught in a cycle of poverty. And I believe we have to find ways to get the private sector to assume their rightful role as a driver of economic growth.

Source: US Department of Housing and Urban Development 1995: 45

---

additional six areas were added in new categories of *supplemental empowerment zones* and *urban enhanced enterprise communities* (see Box 14.6). The 400 applicants who were not successful at this stage are continuing to press their claims.

The important new features incorporated in this program are: grant funding (in addition to the tax benefits); a strategic plan for the coordinated economic, human, community, and physical development of the area; together with pledges of support from state, local, community, and private sources such as foundations, academic institutions, and local businesses. Unlike the earlier proposals, under which areas would have been selected by the federal government on the basis of statistics of distress, potential EZ/ECs were selected locally. The federal government then determined which of these had made the most persuasive bids. Those that were

successful were judged to have demonstrated a commitment to a thoroughly considered strategy of local initiative. In the words of the American Association of Enterprise Zones, 'the new policy's strategic planning requirement sends the signal that recovery depends on local initiative, and holds out federal assistance as reinforcement, not as the agent of change' (Cowden 1995: 10).

There are some striking similarities between this program and Johnson's model cities initiative, with localities making plans and Washington responding. There are also some lessons to be learned from the experiences of the earlier program, including the need to ensure that benefits are widely spread geographically (covering a good majority of the congressional districts), that adequate time is allowed for the program to get under way (resisting the natural desire for 'instant gratification' which

BOX 14.6   AREAS SELECTED UNDER THE EZ/EC PROGRAMS 1995

Urban Empowerment Zones ($100m)
Atlanta, Baltimore, Chicago, Detroit, New York, and Philadelphia/Camden

Supplemental Empowerment Zones ($125m and $90m)
Los Angeles and Cleveland

Rural Empowerment Zones ($40m)
Kentucky Highlands, Mid-Delta (Mississippi), and Rio Grande Valley (South Texas)

Urban Enhanced Enterprise Communities ($25m)
Boston, Houston, Kansas City, and Oakland

Enterprise Communities ($3m)
95 areas across the country

demanded premature judgments of model cities), and that careful monitoring is undertaken to establish what works and under what conditions (Hetzel 1994).

COMMUNITY DEVELOPMENT

'Community empowerment' is at the center of the Clinton urban policy. The EZ/EC program embraces this as an essential element: in the words of a White House statement, the program 'is designed to empower people and communities all across this nation by inspiring Americans to work together to create jobs and opportunity'. Other programs are similarly based, as with the new program for *community development banks*. Resisting the claims of the existing banking system that it can serve local communities well, the program involves the establishment of new financial institutions to provide much-needed credit to poor areas which have been neglected by the traditional banks (despite the 1977 Community Reinvestment Act which requires financial institutions to meet the credit needs of their entire communities, including low- and moderate-income neighborhoods). An effective way for these obligations to be met is through public–private partnerships. Additional support for these is provided by increased funding ($690 million over a five-year period) for the Community Development Block Grant. The Clinton administration estimates that this will create an additional 60,000 jobs. The salient feature of the CDBG is its flexibility and adaptability to local needs and initiatives. (Its adoption by six presidents is a testimony to its popularity.) It is an excellent tool for community development. The new version of the CDBG forms part of a *consolidated plan* for all HUD community planning and development programs (see Box 14. 7)

Housing is an important element in the development of community programs. It constitutes one of the most serious urban problems, but it is also a problem which communities have demonstrated an ability to tackle (though whether they can operate on a scale which will make a significant impact on housing conditions must be doubtful). Community development corporations (CDCs) have also operated successfully in other areas, such as starting small businesses, promoting training schemes, and providing child care. Since they are essentially local organizations, they range widely in character, initiative, and success; and local power structures and planning offices differ in their willingness to cooperate with them. While some have little interest in them, others provide a great deal of support. Some cities are noted for their positive encouragement of community development, as with the 'equity planners' of Cleveland, Dayton, Portland, and other cities discussed in a

Plate 31 Kenyon-Barr Project, Cincinnati: Densely Populated Black Residential Area Prior to Clearance
Courtesy Cincinnati Historical Society, B-86-045

book by Norman Krumholz and Pierre Clavel (1994). They coin the term to describe 'professional urban planners who, in their day-to-day practice, have tried to move resources, political power, and political participation away from the business elites that frequently benefit from public policy and toward the needs of low-income or working-class people of their cities'. These activist professionals are called equity planners 'because they seek greater equity among different groups as a result of their work'. Their work reflects the same philosophy as that promoted by Davidoff (whose 'advocacy planning' is discussed in Chapter 1).

Community development corporations are often supported by local or national foundations such as the Ford Foundation (see Box 14.8). The National Congress for Community Economic Development, in a 1995 report, *Tying It All Together*, estimate that there are over 2,000 CDCs in operation, of which about two-thirds are in urban areas. They have built some 400,000 units of affordable housing and 23 million square feet of commercial and office space, and have created more than 67,000 full-time jobs. Though they have received much support from foundations and local organizations, their principal source of income is the CDBG.

Plate 32 Kenyon-Barr Project, Cincinnati: Area after Redevelopment
Courtesy Cincinnati Historical Society, B-89-109

SUPPORT FOR SOCIAL CHANGE

There is abundant evidence that community organizations can make a significant contribution to the welfare of communities. Moreover, being local they can aim at objectives which a locality wants, rather than have the community saddled with government programs which may not represent its priorities or its wishes. They work for the direct provision of locally wanted facilities and services, instead of relying on the trickle-down effect of 'economic developments' such as high-rise office towers, sports stadiums, convention centers, cultural megapalaces,

and other manifestations of the 'edifice complex' (Squires 1989: 289). They focus on the basic needs of the poor.

Nevertheless, the inadequacy of resources available for this type of community activity, its inherent limitations in relation to stronger economic and political forces, and the deep-seated nature of problems of race, class, and poverty all point to the need for public policies as major forces for change. Some of the difficulties of devising such policies have been discussed in this chapter, and the power of the forces of discrimination have been dealt with at various points in this book. The question

---

BOX 14.7   CONSOLIDATED PLAN FOR COMMUNITY DEVELOPMENT

The Consolidated Plan is a creative approach to community development that encourages communities to work in collaboration to develop a comprehensive vision for action to achieve community objectives.

The Plan consolidates the planning, application and reporting requirements of HUD's programs: Community Development Block Grant, Emergency Shelter Grant, HOME Investment Partnerships, Housing Opportunities for Persons with AIDS program, as well as Comprehensive Housing Affordability Strategies.

The Plan seeks to promote a comprehensive approach to address urban problems, reduce paperwork, improve accountability to achieve results, and includes strong elements of citizen participation. A basic premise of the consolidated planning process is that local jurisdictions and citizens, not Washington, know what is best for their own communities.

*Source:*HUD Fact Sheet 1995

BOX 14.8   FORD FOUNDATION COMMUNITY DEVELOPMENT PARTNERSHIPS

By funding local community foundations that act as intermediaries in attracting and allocating funds to community development organizations, Ford's Partnership model seeks to enhance the capacity of CDCs and stimulate local support systems. Partnerships help to increase the visibility and credibility of CDCs so that they can expand their base of local support. The Partnerships augment CDC funding with vital technical assistance and training. Ford sees the Local Partnerships aiding CDC capacity by:

- brokering technical resources;
- creating local project financing mechanisms;
- sensitizing commercial financial institutions to CDC projects;
- accelerating project approval through local government;
- experimenting with means to address broader financial and social issues that impede the scale and impact of physical development activities;
- generating broader CDC support;
- disseminating the CDC model to a wider audience; and
- changing public perception of the role of local CDC initiatives.

*Source:*Rutgers Center for Urban Policy Research, *CUPR Report*, 1995

---

remains as to whether there is sufficient public understanding and support for attempts to overcome them.

FURTHER READING

There are many books on the development of urban policy. An overview is given in Robertson and Judd (1989) *The Development of American Public Policy*; and Judd and Swanstrom (1994) *City Politics*. More-detailed studies include Frieden and Kaplan (1977) *The Politics of Neglect: Urban Aid from Model Cities to Revenue Sharing*; and Kaplan *et al.* (1970) *The Model Cities Program: The Planning Process in Atlanta, Seattle, and Dayton*. Gelfand (1975) *A Nation of Cities: The Federal Government and Urban America, 1933–1965* is a particularly good account of the period covered.

Barnekov *et al.* (1989) *Privatism and Urban Policy in Britain and the United States* is more than a comparative study; it explores the implications and outcomes of the dominant cultural tradition affecting urban policy: a tradition that relies on private initiative and competition as the main agent of urban change.

Norman Krumholz, a former planning director of Cleveland, has written extensively and eloquently on 'equity planning'. See particularly Krumholz and Forester (1990) *Making Equity Planning Work*; and Krumholz and Clavel (1994) *Reinventing Cities: Equity Planners Tell Their Stories*. These works are in the tradition of Paul Davidoff's 'advocacy planning'; Krumholz contributes to a series of article on Davidoff in the spring 1994 issue of the *Journal of the American Planning Association* (60: 129–61). A collection of essays is edited by Squires (1989): *Unequal Partnerships: The Political Economy of Urban Development in Postwar America*.

A major text on local economic development is Blakely (1994) *Planning Local Economic Development: Theory and Practice*. See also Blair (1995) *Local Economic Development: Analysis and Practice*. A detailed evaluation of the CDBG program has been undertaken by the Urban Institute and published by HUD: Urban Institute (1995) *Federal Funds, Local Choices: An Evaluation of the Community Development Block Grant Program*.

QUESTIONS TO DISCUSS

1 **What are the objectives of community and economic development?**

2 **Compare the varying approaches to economic development taken by different administrations, and outline their strengths and weaknesses.**

3 **Discuss how far a national urban policy is feasible.**

4 **Are incentives for economic development justifiable?**

5 **What are the ingredients for a successful local economic development policy?**

6 **Compare the merits of local community development and local public policy.**

7 **Describe the enterprise zones initiatives. Do you consider them to be effective?**

8 **Why is urban policy so difficult?**

PART VI

ENVIRONMENTAL PLANNING

Environmental policy encompasses a huge field, ranging from the disposal of household refuse and toxic waste to the protection of endangered species, from clean air and clean water to the control of vehicle emissions, from soil erosion and desertification to wetlands – to name but a few issues. Legislation abounds at both federal and state levels: the federal environmental legislation alone encompasses over a hundred statutes which have been passed during the last sixty years. More than a dozen federal agencies have major environmental responsibilities, and every state has an administrative organization for environmental protection. Any comprehensible account clearly has to be highly selective. The academic writer has the luxury, denied to the policy-maker, of being able to omit important relevant matters, and to choose those which are thought to illustrate adequately the nature and problems of environmental policy. The state level can also be largely ignored, ostensibly on the ground that state laws mirror or supplement federal provision. The states are, however, vital to the implementation of federal policies.

Even with major omissions, the discussion here is very long. To make it less daunting, the main discussion of environmental policies (Chapter 15) is divided into five main sections. It starts with an outline of the growth of diverse environmental concerns and their culmination, during the 1970s, in the burgeoning of 'environmental policy'. The first of this new generation of policies was, ironically, an act to force government agencies to take environmental issues into account in all fields of public policy. Following a review of this National Environmental Protection Act, three substantive areas of environmental policy are summarized: air, water, and waste. This is followed in a separate chapter by a discussion of the problems and limits of environmental policy. More obviously than in many areas of public policy, the constraints imposed on the environmental policy-maker are very apparent. At the same time, the fundamental importance of underlying values is clear: paradoxically, though this is a field involving much scientific expertise, many issues are too uncertain to be left to experts.

15

ENVIRONMENTAL POLICIES

ENVIRONMENTAL CONCERNS

The first great fact about conservation is that it stands for development.

Gifford Pinchot, 1910

ENVIRONMENTAL AWARENESS

Words change their meaning over time. Nowhere is this clearer than in the environmental field. Gifford Pinchot is sometimes referred to as 'the father of conservation', yet, as the quotation shows, he used the term in a way which is quite different from that of today. Pinchot was responsible for the establishment of the US Forest Service in 1905, and had strong views on the need for managing the forests in the interests of long-term commercial development. Forests had to be managed like any other crop. Wanton exploitation was inefficient: good management involved sensible conservation.

Pinchot had a strong influence on forestry policy, but his was not the only view being expressed about natural resources and the environment. Then as now, attitudes toward the environment varied widely. The traditional view had been that nature had to be conquered. Nature had to be defeated, or at least tamed. Land and natural resources seemed limitless: why conserve them? – there was always more over the next ridge. The cornucopia of the New World presented a huge market place for exploitation, development, and profit.

Pinchot's was only one of many voices speaking out against this innocent profligacy. His concern was utilitarian. Others had more romantic, artistic, religious, and transcendental visions. John James Audubon presented the beauty of birds in his paintings (even if he shot them first); Ralph Waldo Emerson warned that 'nature cannot be cheated' (though he believed that nature had a capacity for self-healing); Henry David Thoreau embraced nature's role for the spiritual nourishment of humans (and, though he presented no ideas for implementing his philosophy, it later became the intellectual foundation of the movement for wilderness preservation). Later writers provided the beginning of a basis for action, though it was many years before this could be seen. Remarkable among these was George Perkins Marsh, who in 1864 published *Man and Nature; or, Physical Geography as Modified by Human Action*. Far from the seducing idea of nature being self-regenerating, Marsh stressed the irreparable damage which human activity could inflict upon the land. His immediate influence was small (though Pinchot used some of his ideas), but he set out the fundamental ideas of what we now know as ecology. He prompted an increasing realization of the interconnectedness of things.

More successful in getting action were those who focused on specifics – like John Muir who, in addition to espousing the intrinsic value of the wilderness, campaigned for the Yosemite national park. Though his success had more to do with his close friendship with President Theodore Roosevelt than the force of his arguments, he greatly strengthened the campaign for national parks as well as the promotion of tourism as an economic incentive for preserving such areas. He also helped in the formation of the Sierra Club (1892), which grew into a major force in the campaign for preserving wilderness.

However, 'the economics of superabundance' largely prevailed until the New Deal of the 1930s saw some lurches in a new direction: the Tennessee Valley project, the establishment of the US Soil Conservation Service, the expansion of the public domain with new forests, and the abortive attempt to bring all federal land responsibilities together in a Department of Conservation. Though World War II intervened, the conservation ethic was now on firmer ground, and events of the post-war years gave it a strong forward impetus. These events ranged from worrying disasters to a flowering of books and articles on the environment, and from congressional action (tentative at first, but growing in strength) to bureaucratic activism.

Attitudes evolve over time, and it is seldom possible to point to a date when change can be said to have taken place or emerged. By common consent, Earth Day 1970 is the convenient date marking the culmination of a series of changes. This environmental celebration was preceded by a flood of writings critical of the way in which the environment was being maltreated. Among these were the works of Lewis Mumford (on urbanization), René Dubos (on drugs and their effect on micro-organisms and 'the chain of life'), Aldo Leopold (on a land ethic), Paul Ehrlich (on overpopulation), and Rachel Carson's *Silent Spring* (on pesticides). There were many more.

Environmental awareness was increasing in other ways. Pressure groups campaigning for change mushroomed at local and national levels. David Brower achieved fame by his aggressive leadership of three bodies. First, he gave the Sierra Club a new lease of life. His style, however, proved too much even for the rejuvenated organization and he was forced to resign.

He then established a new body with a name taken from a quotation of John Muir: 'The earth can do all right without friends, but men, if they are to survive, must learn to be friends of the earth.' Friends of the Earth (which bred parallel organizations in several countries) soon achieved a high profile, but Brower ran into further difficulties and he moved again – this time setting up the Earth Island Initiative. Whatever Brower's shortcomings, there is no doubt that 'he helped rekindle the transcendental flame lit by Thoreau and Muir, and played a major role in pulling the old preservationist movement out of the comfortable leather armchairs of its clubrooms and into the down-and-dirty arena of local and national policy-making' (Shabecoff 1993: 101).

Other groups were established at this time, including the Conservation Foundation, the Natural Resources Defense Council, and the Environmental Defense Fund. By the end of the 1960s, the number of members in organizations such as the National Audubon Society, the National Wildlife Federation, the Sierra Club, and the Wilderness Society were increasing dramatically. Environmental concerns were moving to center stage.

Against such a turmoil of activity, it is unlikely that any one factor can be identified as the most important; but it is generally accepted that the emblem of the time is the unlikely one of a book on pesticides. Rachel Carson's *Silent Spring* (first published in 1962) is an extraordinarily eloquent, moving, and lucid presentation of the environmental dangers of manufactured poisons (such as DDT, which was banned in 1970 under the Clean Air Act). Its dramatic message, expressed in almost poetic terms, gives it a place among the great books of the century.

THE FIRST EARTH DAY

However compelling Carson's arguments may now seem, they did not precipitate rapid action: the forces ranged against environmental policy were too powerful. But the increasing public awareness of environmental problems gradually put the environment on the political agenda. A number of environmental disasters added to the growing concern – and, in turn, made the public sensitive to disasters that previously might have had little publicity beyond their immediate locality: a huge spill off the California coast which sent vast quantities of crude oil on to the beaches of Santa Barbara and neighboring towns; the bursting into flames of Cleveland's Cuyahoga River; fish killed by toxic waste in the Hudson River; beaches fouled by garbage: reports multiplied as environmental concern grew.

Then 22 April 1970 saw a remarkable series of activities throughout the nation, ranging from teach-ins to litter collection, and also including the pouring of oil into a reflecting pool belonging to the Standard Oil Company of California as a protest against oil slicks. Earth Day 1970 brought together a wide range of supporters of environmental protection. It signaled an important change in environmental politics. Concern for the environment was no longer restricted to a few: its broad base demanded a political response which Congress was quick to recognize. So many politicians joined in the Day's activities that Congress was forced to close down.

Congress had already passed the National Environmental Protection Act (which required all federal agencies to take account of environmental factors); many more environmental laws followed. This period of frenetic law-making was quite exceptional: it was contrary to the normal incremental approach which distinguishes the political process. Nor was the passing of legislation the only evidence of the new environmental activism. The designated area of wilderness increased from 10 million acres to 23 million acres. Seventy-five parcels of land were added to the National Park Service over the same period; and the National Wildlife Refuge System grew similarly. In such ways were environmental policies pursued.

THE REAGAN YEARS

The Reagan years saw a halt to environmental policy initiatives – except those which involved a reversal of previous policies. The agenda had changed: it was now 'regulatory reform'; the shackles on the American economy imposed by previous administrations were to be removed, thus releasing the inherent powers of private enterprise. Environmental deregulation was the overriding objective of those appointed to the senior positions in the Reagan administration, and they set about their tasks with vigor. Their successes were significant, though less far-reaching than they had anticipated. They found that the public was not as enamored of the implications of deregulation as had been thought, and the very achievements of deregulation prompted a resurgence of environmental concern. Moreover, the separation of powers among the branches of American government ensured that moderating influences would be significant. Indeed, some areas of environmental policy, such as clean air and water, were actually strengthened during these years. Before the Reagan administration came to its appointed end, the reaction was abundantly clear, and positive environmental action was at the forefront of domestic policy. Public opinion was seen to triumph over even a popular president's agenda.

Though much effort was expended in the battle between Reagan's onslaught and the defenders of environmental policy, these years also witnessed a re-evaluation of the adequacy and viability of the extraordinary range of policies which had been introduced over the preceding two decades. Though largely ignored by the Reagan administration, this period of heartsearching was a useful investment of time (Vig and Kraft 1990: 18). In the first place, it was apparent that the legislation had made remarkably optimistic assumptions about the speed with which the technical problems posed by compliance could be solved. Second, the administrative and compliance costs were also underestimated. Added to these difficulties were the legal challenges used by the affected industries to avoid the costs of compliance, as well as the time and effort involved for the regulatory agencies. In short, the legislation posed problems of implementation that were unanticipated. There was thus a ready-made agenda for a new administration.

The environment was among the salient issues of the 1988 election, with both Bush and Dukakis vying for the honor of being the true leader in the field. Following his election, Bush quickly moved to establish himself as a real friend of environmentalists and, though his record was patchy, he returned environmental policy to center stage. Despite his support for Reagan's underlying beliefs in the efficacy of the market place, he espoused a number of environmental causes, particularly the strengthening of the clean air policy.

The second Earth Day was celebrated twenty years after the first, in 1990. It seems that environmental protection is firmly established as a central feature of domestic (and, less certainly, foreign) policy. How far this is deeply entrenched, however, is another matter. As Walter Rosenbaum has noted, 'the political ascendance of American environmentalism has occurred during two decades of almost uninterrupted domestic economic growth'. There has been no pitting of job losses against environmental losses, except in a number of localities – and where this has occurred, jobs have typically won (Rosenbaum 1995: 342).

THE NATIONAL ENVIRONMENTAL PROTECTION ACT

The National Environmental Policy Act has a potentially important role to play in an integrated effort to achieve sustainable development
National Commission on the Environment 1993

ENVIRONMENTAL CONTROL OF FEDERAL PROGRAMS

Much environmental policy takes the form of controls operated by government over the actions of private bodies. In modern societies, however, a great deal of activity is undertaken by government itself, operating through a profusion of agencies. The range and diversity of this activity is enormous – from the development of natural resources to the building of roads; from the dredging of harbors to the administration of national parks; from the promotion of funding of social programmes to the conduct of military operations. Thus the governmental machine which is responsible for protecting the environment from unacceptable private actions is itself responsible for a huge number of activities which can equally affect the environment – and, in some cases (as with nuclear power or military investments), can be particularly hazardous to the environment. In the real world, there is no guarantee that a governmental agency will act in a way which safeguards the environment. On the contrary, there is abundant evidence that, given the choice, a governmental agency will seek to achieve its specific, narrow objectives without regard for wider public considerations. Concern for environmental matters will normally be ranked as subordinate to the objectives for which the agency has been established. Any doubt about the validity of this contention would be settled by examining the Department of Energy's 'gross mismanagement', deliberate deception, and suppression of information on its fourteen military nuclear facilities – including the concealment of major accidents (Rosenbaum 1995: 122).

Thus there is a real problem: how is the environment to be protected from unacceptable actions on the part of government? Or to put the matter more vividly, even if more loosely: who controls the controllers? There is, unfortunately, no ready solution to this conundrum. The art of government is not akin to driving a machine: it is a highly diffuse process which at best is extremely difficult to manage, and at worst is beyond control. It necessarily operates by dividing its functions into manageable parts with specific responsibilities. An agency established to carry out a particular function cannot be required to give a higher priority to the protection of the environment: this would compromise its very *raison d'être*. All that can be done is to devise a mechanism which obligates governmental bodies to give full and serious attention to environmental factors in the course of carrying out its functions.

This is what the National Environmental Policy Act (NEPA) does (and many states have similar provision in relation to state government). It requires all federal agencies 'to the fullest extent possible' to carry out their functions in accordance with environmental policies which are set out in broad terms in the Act. For this purpose, there are a number of 'action-forcing' procedures, of which the most important is the requirement for an environmental impact statement (EIS) in connection with any federal action 'significantly affecting the quality of the human environment'.

This is a *procedural* requirement: it is for the agencies themselves to determine the implications of the EIS for the proposed action. At first sight, this may seem to give agencies a remarkable degree of freedom to interpret NEPA as they wish, since it is each agency itself, and not any superior controlling body, that has the responsibility for deciding what action, if any, is required following an EIS. Though this is true, there are several qualifications. Federal agencies have to abide by regulations made by the Council on Environmental Quality: a body established by NEPA to oversee the implementation of the Act. These regulations have to be followed rigorously. The procedural rules relating to an EIS (which are outlined below) dictate a process of thorough examination and reporting which the agencies must follow *to the fullest extent possible*. Public involvement plays an important role in this process, and appeals can be made to the courts. NEPA involves a complex process in which power does not rest in any single place: the agencies, the public, the courts, the Council on Environmental Quality, and other government departments including the Environmental Protection Agency all play a role. There is thus a typical system of dispersed power.

FEDERAL ORGANIZATION FOR NEPA

There are two federal agencies which have the responsibility for the working of NEPA: the Environmental Protection Agency (EPA) and the Council on Environmental Quality (CEQ). Though they have different functions, it is simpler to regard the two bodies as sharing responsibility for the development and oversight of national environmental policy.

The Environmental Protection Agency is the larger federal regulatory agency, and it has very wide-ranging environmental management responsibilities, including the responsibility for EIS review. It was established by executive order of President Nixon in 1970, and it carries a huge administrative burden. Among its many functions, EPA receives all environmental impact statements and checks them for completeness. More significantly, it reviews statements for their adequacy 'from the standpoint of public health or welfare or environmental quality'. Any EIS that is judged to be unsatisfactory is referred to CEQ for resolution.

CEQ is responsible for environmental policy coordination, monitoring of and reporting on environmental quality, and the working of NEPA. As part of the Executive Office of the President, it has a position of power and can act as 'the environmental conscience of the executive branch', though how real this is will depend on the stance of the President. (A current-awareness service is required to keep up to date on such matters.) The Council has several functions, including advice to the President on environmental matters and the production of an annual report; the monitoring of environmental trends; and overseeing federal agencies and their compliance with environmental policies. Its regulations define the ways in which NEPA is implemented. These regulations set out the details of the environmental review process.

The procedures have to be taken seriously, and environmental values must be pursued 'to the fullest extent possible'. Thus, to quote from one of the multitude of court cases on the Act (*Calvert Cliffs*), there is no 'escape hatch for footdragging agencies': NEPA imposes a duty on federal agencies which cannot 'be shunted aside in the bureaucratic shuffle'.

THE ENVIRONMENTAL REVIEW PROCESS

NEPA requires agencies to integrate environmental considerations into their operations. Any federal action is subject to NEPA if it can have a significant impact on the environment. This includes the application of policies, new legislative proposals, adoption of plans and programs, and the approval of specific projects. This broad approach covers not only direct action by a federal agency, but also any action taken by other public or private bodies which are funded by, or require the approval of, the agency.

There is no exemption from the NEPA mandate unless Congress has explicitly made an exception (as it did with the closure of certain defense bases) or unless there is a 'clear and unavoidable conflict in statutory authority', e.g. where an agency is statutorily required to take action so rapidly that it is impossible to prepare an EIS within the time frame. The courts have made it clear, however, that they will not allow this to be a loophole; and they have been kept busy dealing with unacceptable claims by agencies.

A special procedure applies to cases in which it is unclear whether or not an EIS is required. These arise where it is not evident that an action will have a significant effect on the environment, or where identified significant effects can be 'mitigated' satisfactorily. The procedure involves preparing an *environmental assessment* (EA). This must provide sufficient evidence to demonstrate that the action will not (or will) have a significant environmental impact. If the EA shows that there will be no significant impact, the agency is not required to prepare an EIS. If, on the other hand, the conclusion is that there will be a significant impact which cannot be mitigated, a full EIS follows. Where mitigation is decided upon, the agency must conform to additional requirements for public review. A *finding of no significant impact* is known in the trade as a FONSI – or, as the case may be, a 'mitigated FONSI'. Such is the way in which the rhetoric of environmental policy is translated in the language of bureaucracy.

According to CEQ regulations, 'the primary purpose of an EIS is to serve as an action-forcing device to insure that the policies and goals defined in the Act are infused into the ongoing programs and actions of the federal government' (Bass and Herson 1993: 122). It is thus not the production of a passive documentary analysis of environmental impacts: it is

intended as an important part of the decision-making process. It must not be used to rationalize or justify a decision already taken. It therefore has to be carried out in advance of a decision – and early enough to influence that decision. Indeed, its coverage extends beyond the proposed action to embrace reasonable alternatives. Moreover, a process known as *scoping* is required: this is a public process which should start very soon after a decision has been taken to carry out an EIS. It seeks the views of the public and other agencies on what is to be covered in the EIS and the alternatives to be examined. Generally, a draft EIS is required, and public hearings may also be necessary. Whatever the detailed requirements, the overriding purpose is to ensure that there is the widest possible input from other agencies and the public.

It is important to appreciate that at no point do the procedures provide for an adjudication on an EIS. The system is essentially one of publicity, inquiry, discussion and negotiation. When all this is completed, the agency takes its decision (which could be to reject environmentally preferable alternatives because of overriding non-environmental matters). The decision has to be presented in a manner laid down in the regulations: a written *record of decision*, which is available to the public, has to explain the decision, the alternatives that were considered, and any mitigating measures which have been adopted.

Agencies are not forced to accept a view which they think conflicts with their own interpretation of the findings of the EIS; nor is an agency required to adopt an environmentally preferred alternative or measures of mitigation (though explanations are required in the EIS). NEPA deals only with procedural matters: it imposes no duties concerning the protection of the environment. Moreover, there are no statutory powers of enforcement. Such enforcement as exists lies with the courts, and therefore in the last resort with the alertness and resources of the public in general and of interest groups in particular.

Congress has thus clearly relied upon lawsuits as an important way of obtaining agency compliance with NEPA. Against this background, it is not surprising that the courts have had a heavy load of cases dealing with NEPA issues. Thousands have

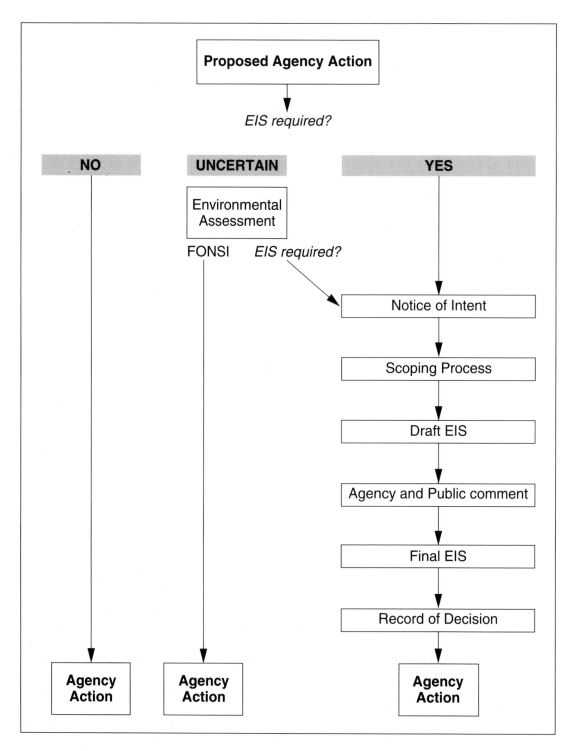

Figure 15.1 The NEPA Environmental Review Process
Source: Based on Bass and Henderson 1993: 15

been filed, and almost every federal agency has been involved in litigation. Some have been more involved than others: transportation has been a particularly lively area.

With the legal process forming such a significant part of the NEPA system, the issue of who may bring a case before the courts is an important one. This question of *standing* has a long legal history, and the court's views have changed over time (Findlay and Farber 1992: 2). Briefly, to have standing under NEPA, plaintiffs must be able to assert that there has been an injury to a part of the environment which they use. The injury has to be shown to be causally related to the allegedly illegal action on the part of a federal agency. Thus, when Walt Disney Enterprises wanted to build a $35 million resort in an area of great natural beauty in the Sierra Nevada Mountains, the Sierra Club was able to claim that its members would be injured by the development since they would no longer be able to roam through an unspoilt wilderness.

There are several points to note here. First, there was an environmental interest at stake and, since NEPA protects environmental interests, the plaintiffs had standing. Second, the plaintiffs were asserting their own interests – not those of a third party; they could thus claim direct 'injury'. (Related to this is the rule that it is not acceptable to claim a generalized interest, such as 'humankind' or 'the poor' or 'recreationalists'.) Third, the 'injury' was remediable by a favorable judgment. These are three requirements for a case to be successfully prosecuted.

Generally, the Supreme Court has consistently ruled that an agency decision cannot be set aside because of its effect on the environment. This follows logically from the position that NEPA is purely procedural. It can be argued that the court has been unduly narrow in its approach to this issue: a triumph of form over content. Nevertheless, the EIS process is important not only for facilitating legal challenges, but also for serving broader purposes. It alerts environmentalists to the disclosed implications of administrative issues which otherwise might not have been apparent; it serves as an early warning system for newly arising issues; and it compels agencies to

carry out discussions and maintain contact with environmental groups. In these ways, EIS has helped to bring environmentalists into the policy arena: they now have a role which has been legitimated and facilitated. This success of environmental groups was noted by business interests who followed a similar pattern of organization, research, and lobbying. Particularly noteworthy has been the effectiveness of the not-for-profit legal foundations (such as the Pacific States Legal Foundation) which have participated in both negotiation and litigation. The EIS thus has brought about a greater degree of participation in the process of designing environmental regulations. It has thereby subscribed to the democratic principle that all who are affected by public policy should play a part in determining what this should be.

CLEAN AIR

'The purity of the air of Los Angeles', an enthusiast wrote in 1874, 'is remarkable. The air when inhaled, gives to the individual a stimulus and vital force which only an atmosphere so pure can even communicate.'

Boorstin 1973

TECHNICS AND POLITICS

The cleanliness of air might seem, at first sight, to be essentially a technical subject. Nothing could be further from the truth: it is a highly political matter. This is partly because the technicalities are highly problematic, but also because measures to cleanse air (whether by removing contaminants or by preventing them from entering the air in the first place) involve costs and benefits which arise in different areas and therefore affect people differently. These distributional effects arise in all areas of public policy, but they are particularly troublesome with air pollution since the underlying scientific base is weak: there is a degree of ignorance about air pollution that must surprise the layman. The huge area of ignorance means that inadequate facts are capable of widely differing interpretations – the perfect base for lengthy and frequently inconclusive political argument. As will be shown later, these difficulties are by no means confined to clean air, but extend over wide areas of environmental policy.

EARLY CLEAN AIR POLICIES

Policies for clean air have a long history. It was as early as 1881 that Chicago and Cincinnati passed laws to control smoke and soot from furnaces and locomotives. A hundred other cities followed in the next thirty years; by 1950 the number had risen to over 250. State action came much later, with Oregon being among the first in 1952; but by 1970 all the states had passed air pollution control legislation. The federal government came on to the scene in 1955, and several acts were passed in the following years. At first, the federal role was restricted to providing financial and technical aid to the states. In the mid-1960s, however, a more positive federal role emerged in relation to cross-boundary pollution and the setting of emission standards for motor vehicles. The 1967 Air Quality Act went further, and promoted the planning of air pollution strategies. Air quality regions were to be designated to cover areas (within state borders) of interconnected air pollution problems. The states were to establish air quality standards for these areas and develop plans for their achievement. Standards were to be based on advice from the Department of Health and Welfare on the health effects of common pollutants.

It soon became apparent that this system would not work effectively. The Department of Health and Welfare was tardy in designating air quality regions and in preparing the advice the states needed; and even where the necessary guidelines were produced, few states developed plans. By 1970, not a single state had devised a complete program for dealing with any pollutant. Moreover, the automobile industry had proved to be adept in circumventing emission controls. Against a background of mounting public concern about the environment (dramatically

evidenced by Earth Day), it was clear that stronger federal action was needed.

POLLUTION AND ECONOMIC DEVELOPMENT

Much of the problem facing the states stemmed from their concern to safeguard their economic development. Any individual state that took positive action to control air pollution could be at an economic disadvantage, since new industries would naturally select the cheaper locations of the states which had no controls. Whether or not this was actually a significant factor in location decisions, it was certainly seen as such. There was also some concern, particularly in the case of industries operating in several states, that there would be widely varying standards. Nowhere was this more important than with vehicle emissions, and manufacturers pressed for national standards.

The problems went much wider than this, however; and they still do. First, of course, is the perennial confrontation between environmental and industrial interests. Environmentalists can evoke powerful images of ecological devastation (such as the *Exxon Valdez* oil disaster) and risk to health. Equally, those opposing environmental controls are able to rouse the spectre of declining economies and the loss of jobs.

Second, there are major regional conflicts, which also remain as part of the permanent political landscape. Particularly striking is the conflict between the Midwest and the Northeast, the root of which is the competitive production of coal for power plants. The Midwest produces low-cost coal which has a high sulphur content while the Northeast produces coal which has a low sulphur content but a high cost. Sulphur is a major cause of acid rain, and it is a target of much clean air policy. The sulphur content of Midwestern coal can be reduced, but this involves a cost which reduces its economic advantage. The two regions have diametrically opposed interests in the control of sulphur. Any measure which imposes costs on power plants reduces the economic advantage of the Midwest mines and increases that of the Northeast. Thus the political conundrum arises as to what controls are to be operated and who is to bear the cost.

This is the classic problem with air pollution controls; and it has other dimensions. High chimney stacks for coal-fired power stations using Midwestern coal are not very expensive (even when they are higher than the towers of New York City's World Trade Center). They considerably reduce the pollution in the Midwestern states; but they do so by transporting the pollution to other areas, predominantly in the East. As a result, a part of the real cost of the use of Midwestern coal is borne in the Northeast. But it is not only the Northeast which receives the pollution: it is spread over a very wide area. It is also mixed with other pollutants produced both locally and further afield. The resulting cocktail is made up of a variety of pollutants, in differing amounts, from different areas: it is impossible to determine the origin of the ingredients. The areas which suffer the impact of the pollutants, whether they be Northeastern cities or Northern lakes and forests, naturally view the problem as one of wide geography which should be equally widely shared. But 'clean' states do not see it this way. The states in the Sunbelt, for example, have a good proportion of modern industrial and utility facilities. Their investment in clean air has already been made: why should they contribute towards the cost of cleaning up in the dirty states?

The issues are further complicated by attitudes to the control of new pollution sources. It would seem common sense to impose stricter standards on new industries than on those already existing, since they can meet them more easily. Introducing pollution control measures when a plant is being built is considerably cheaper than adding them later. There is thus a national gain at a relatively low cost. Such arguments are attractive to older areas because they see the extra costs as reducing the competitive advantage of clean states. The clean states, on the other hand, argue that the burdens of cleanup should be borne by the areas where the emissions are the greatest. These and similar arguments raged in the debates on the 1970 Clean Air Act which finally imposed different requirements for new and existing plants.

THE STRUCTURE OF CLEAN AIR CONTROLS

Though the legislation has been amended considerably since 1970, this basic division continues. All new plants are required to conform to *new source performance standards* which are devised separately by EPA for each industry to take account of costs and the 'best available technology'. Existing plants in areas which violate *national ambient air quality standards* are subject to locally determined controls. These standards constitute acceptable levels of pollution in the ambient (outside) air. In effect, they are national objectives which are to be met at some future date (which is subject to postponement when they prove unattainable). They relate to six major pollutants: carbon monoxide, ozone, particulate matter, sulphur dioxide, nitrogen dioxide, and lead.

There are many other substances that pollute the air, but little is known about the effects these have on health. The degree of ignorance is quite alarming: even the extent of pollution is very uncertain. Though EPA is required to identify and designate air pollutants and to establish emission standards for them, progress has been slow. This has been largely because of a lack of research. The needed research is laborious, expensive, and slow to produce results; and the 'results' tend to be of sufficient uncertainty as to create lengthy debate, particularly on the part of the firms that are responsible for the pollution and for cleanup costs.

ACID RAIN

Rain is naturally acid, due to the presence of natural elements, but the degree of acidity can be increased by pollution to such an extent that it results in harmful environmental effects. These effects are caused by acid deposition which may be borne by air, dew, fog, and wind, as well as by rain. The term 'acid rain' is therefore not quite accurate, but it is typically used in preference to acid deposition.

The effect of acid rain varies according to local conditions such as the wetness of the ground, the rate

---

BOX 15.1   ACIDITY

The major chemicals that produce acid rain are sulphur dioxide, nitric oxide and nitrogen dioxide (known collectively as nitrogen oxides) and hydrocarbons (and also, to a lesser extent, ammonia and carbon dioxide). Acidity is measured on a logarithmic pH scale which goes from 0 to 14. Pure water has a pH of 7; any reading above this is alkaline and any reading below it is acidic. Rainwater has a normal acidity of 5.6; damage to the natural environment is associated with a pH of 5.0 or lower; at the extreme level of 3.0 extensive damage to buildings will occur over time. A very extreme reading of 1.69 was recorded in 1984 south of Los Angeles, in the Californian town of Corona del Mar. To appreciate how severe this was, it can be compared with battery acid, which has a pH rating of 1. (Since the pH scale is logarithmic, a pH of 3.0 is ten times stronger than one of 4.0.)

---

of run-off, and the vegetation cover. In urban areas, the erosion effects of acid rain differ according to the types of building materials: limestone is affected much more than granite or even sandstone. High acidity, of course, is a killer: large areas of forest and lakes in the Northeast of the USA and in Canada have suffered greatly. High smoke stacks have been responsible for carrying pollutants away from the source to far-distant areas.

The burning of fossil fuels (mainly coal and oil) is a major factor in the creation of acid rain (and thus, as frequently happens, energy policy and environmental issues and policies intertwine). Conventional power stations are the worst offenders in producing sulphur dioxide and nitric oxide. The USA is second only to the former USSR in sulphur dioxide emissions, the principal offending states (in decreasing order) being Ohio, Pennsylvania, Indiana, Illinois, Missouri, Wisconsin, Kentucky, Florida, West Virginia, and Tennessee (Pickering and Owen 1994).

OZONE

Ozone is a gas which is both enormously beneficial and dangerous. In the stratosphere, it protects the earth from harmful rays; at ground level, it is noxious to humans. Its chemical composition – three atoms of oxygen (O_3), compared with oxygen's two (O_2) – makes it highly reactive. It readily combines with other substances, often thereby forming a most obnoxious mixture. Smog is its public image. Unlike

air pollutants such as carbon, nitrogen and sulphur oxides, ozone is a 'secondary' pollutant formed by reaction between primary pollutants and natural constituents of the air. These 'precursor' pollutants consist mainly of nitrogen oxides and *volatile organic compounds* (VOCs) including hydrocarbons such as benzene. The largest causes of ozone are motor vehicles, 'small stationary sources' such as paint shops, and large refineries and chemical plants. Indeed, much of modern economic activity creates ozone, but it is also produced as a result of natural processes. In short, the production of ozone is extremely varied and complex, and there is much about it which is not understood. As a result, ozone is particularly difficult to regulate. Virtually every major urban area in the USA fails to meet the ozone quality standard (which many experts consider is too low for the protection of human health). Ozone can cause bronchitis, asthma, and other pulmonary diseases; indeed, it can impair many bodily functions and even give rise to heart failure. It is also harmful to trees, crops, and aquatic systems.

Improved vehicle emission controls have greatly reduced the impact of individual vehicles, but the improvement has been more than offset by increased traffic. It seems that there is little prospect of significant quick (or even slow) solutions to the problem, though the regulatory mechanisms affecting ozone have been strengthened and widened in the 1990 Act.

Plate 33 Los Angeles Smog
Courtesy Trevor Warr, Viewfinder Colour Photo Library

STATE IMPLEMENTATION PLANS

Responsibility for enforcing air quality standards lies with the states, operating through their *state implementation plans*. These plans allow states to adjust their controls according to the cleanup costs in their areas (and they operate in conjunction with state provisions). They impose pollution limits for geographical divisions called *air quality control regions*, detail the arrangements for control, and set out measures for cleanup (and for emergencies). The plans are subject to the approval of EPA and have the force of law. States are induced to formulate and operate their implementation plans by the power of the federal purse: grants for highway and sewer works, for example, can be withheld from recalcitrant states.

Air pollution varies not only among industrial plants but also, of course, among (and within) individual states. Thus there arises the question of how to treat areas which already meet national standards. Are they to be required to make their contribution to the national cleanup effort by aiming for higher standards? This may make sense nationally, but it is likely to be opposed locally: why should costs be incurred in an area which is already in conformity with national standards? The additional costs of being required to operate more stringent controls than elsewhere would (so it could be argued) unfairly create barriers to the economic development of the area. There is an alternative: the area could be allowed to *increase* its pollution as long as this does not bring it below the national standard. The political difficulty here (much stressed by the environmental lobby) is that such a policy for clean air would deliberately and explicitly be promoting dirty air in some areas.

The 1970 Act did provide for some differentiation in regional standards which became a natural target for litigation. A case brought by the Sierra Club was decided in their favor, and EPA was barred from approving any state plan which permitted increased levels of pollution in the clean areas. Though the decision was applauded by environmentalists, it aroused violent objection from groups concerned with economic development in the West. The latter claimed, for example, that it 'precluded further development of vast energy resources' (in New Mexico); and 'continued poverty in many rural areas' (of Utah) (Ackerman and Hassler 1981). A Bill to amend the 1970 Act was killed in 1976 by a filibuster mounted by Western senators who argued that it would seriously affect economic development and energy exploitation.

The problem of the clash between clean air policies and local economic development is one which seems likely to remain insoluble. An argument that clean air policies themselves create employment (which they certainly do) carries little weight: even if the new employment is created in the areas which suffer because of pollution controls, it would be a very happy coincidence if it provided jobs for the those displaced. The intractability of the problem is evidenced by a provision in the 1990 Act for a programme of compensation for workers who lose their jobs as a result of clean air policy.

AREAS OF SEVERE POLLUTION

In addition to the withdrawal of federal funds from states which failed to produce satisfactory implementation plans, the 1970 Act made provision for the imposition of federal plans. Though EPA was (for strong political reasons) very reluctant to act in default, it was forced to do so in a number of cases because of lawsuits by environmental organizations. (This was done in Phoenix, Chicago, and several places in southern California.)

The 1990 Act took a different approach. It was recognized that some parts of the country would have great difficulty in meeting clean air standards, and that the imposition of national standards would be unworkable (as the failure of the earlier state implementation plans testified). The solution adopted was essentially one of differential standards devised on the basis of the severity of the pollution problem. Areas with acute difficulties are designated *non-attainment areas*. This status is determined on the basis of pollution criteria for ozone, particulate matter, and carbon monoxide. Each non-attainment area has a deadline for meeting the standards, and the programmes for this (state implementation plans) are geared to the severity of the problems: the more acute it is, the more aggressive must be the program. Of particular importance here is the permit system which was introduced by the 1990 Act. This requires *all* major stationary sources to have an operating permit from the state specifying all the conditions that are imposed on that source. These include not only the amount of allowable emissions, but also the requirements for monitoring and maintenance.

VEHICLE EMISSIONS

Vehicle emissions are a major source of air pollution: they account for over a half of all air pollution. Though a great deal of effort has been made to reduce it, real achievements in making vehicles cleaner have been offset by increases in vehicle use. Thus, though cars produced in the mid-1990s will emit between 70 and 90 per cent *less* pollution than their 1970 counterparts, vehicle travel has more than doubled during this period, partly as a result of urban development patterns. At the same time, concern has grown about other, previously unrecognized, environmental threats such as acid rain and global warming. The political response has been a major strengthening of vehicle emissions control. Its political salience is epitomized by the fact that, instead of standards being set in the usual way by EPA, Congress has actually written them into the legislation.

Controls have been operated by placing most of the responsibility for reducing emissions on manufacturers and, consequently, standards have been nationally operative, irrespective of any differences in local air quality. This has now changed, and increasing responsibility has been placed on vehicle owners through state vehicle inspection programmes. State Implementation Plans can include these and other measures such as discouraging traffic by the control of parking and encouraging car pooling (discussed in Chapter 12). The 1990 Act provides for the phasing out of lead and also for tighter tailpipe (exhaust) standards. EPA is required to study the desirability and feasibility of further changes in standards. Cars produced since 1994 are required to be equipped with 'onboard diagnostic systems' which feature dashboard warning lights to show when emission control equipment is malfunctioning. Stiff penalties face 'backyard mechanics' who tamper with emission controls. In areas with severe pollution problems special regulations apply.

California, which has acute air quality problems, has gone further than the federal government in its efforts to reduce emissions. The so-called 'California Pilot Program' requires manufacturers to produce specified numbers of 'clean' (i.e. electric) cars. These have to meet severe standards; how they are met will depend on the results of a research and development effort (that so far have been disappointing). A number of states are adopting policies based on those of California, which are sometimes described as 'technology-forcing': the standards involved are stricter than can be met with existing technology. Evidence that such an approach can be effective (even if not as quickly as its protagonists would wish) is provided by the successful development of catalytic convertors; but it seems dangerous to rely on a technological quick-fix to environmental problems – a point which is developed later in this discussion.

THE 1990 ACT AND ITS FUTURE

It remains to be seen how the wide-ranging and complex provisions of the 1990 Act work out in practice. Congress, having labored mightily in producing the Act, has passed responsibility to the EPA and the states. Their previous record has been uneven but, given the changes in public attitudes and the strengthened legislative framework, it may be hoped that implementation will improve. In addition to the provisions outlined above, mention should be made of the increased penalties for emission offences, most of which have been made criminal felonies. Moreover, the courts have increased powers to compel EPA and the states to comply with legal requirements; and there are extended citizen suit provisions. It has also been suggested that the workers' compensation program may bring labor and environmental interests together in a way which has hitherto been impossible because of the specter of job losses (Bryner 1993). On the other hand, Congress would be running true to form if it has set clean goals at an unrealistically high level.

CLEAN WATER

Ever since chemists began to manufacture substances that nature never invented, the problems of water purification have become complex and the danger to users of water has increased.

Carson 1962

THE SUCCESSION OF PROFESSIONAL APPROACHES

Water policies illustrate rather clearly the role of fashion in the environmental policy field. Unlike clothing, however, these fashions acquire a life of their own, and persist even when it has become clear that a replacement is timely. That is because they have the backing of experts (Tschinkel 1989). The first expertise to hold sway in the development of water policy was that of physicians for whom the problem of polluted water was solved by washing it away. By the middle of the nineteenth century, the common method for disposing of household wastes was by way of the storm drain. This proved a most effective way of reducing the spread of cholera. Unfortunately, since the raw sewage was dumped into areas from which drinking water was obtained, there were the side effects of some danger to public health. The solution to this was chlorination, which again was effective: typhoid was virtually eliminated. This public health achievement, however, was not without its own problems. Large quantities of diluted sewage cause eutrophication and contamination of bodies of natural surface water, thus wasting enormous quantities of both water and nutrients, and also raising the question as to whether the chlorine itself has harmful effects.

The problems of the next stage – reducing the quantity of nutrients and pathogens entering water systems – was dealt with by the engineering profession with a huge programme of sewage treatment plants. Though this has partly (but not completely) solved one water pollution problem – that of *point pollution* – it has been quite inadequate to deal with wider water pollution problems. Despite the enormous cost involved, EPA's Water Quality Inventory reveals that a third of all surface waters do not meet water quality standards. The reason for this is that much water pollution arises, not from a particular 'point' such as a sewer or discharge pipe or ditch, but from *non-point sources*. These include agricultural runoff, leaking gasoline storage tanks, landfills, abandoned mines, runoff from irrigation and from salted roads, seepage from septic tanks, and hazardous waste sites. This frightening mixture of pollution sources, which was dramatically brought to public attention by Rachel Carson's *Silent Spring* (1962), presents serious difficulties for scientists. Previous generations of public health and engineering specialists may have had supreme confidence in their remedies: their contemporary counterparts have no grounds for any such confidence. Though biologists may be 'waiting in the wings to solve these problems', it is apparent that there is a high degree of ignorance on both the causes of and the cures for these problems (Tschinkel 1989: 161). (To complete the record of professional succession, it may be noted that, given the degree of scientific ignorance, lawyers

are now the dominant professional actors on the environmental stage.)

From this brief account of the fate of successive generations of experts, it is clear that water pollution presents a wide range of difficult analytical, technical, and political problems. As is typical of environmental issues, there is a paradox here: the complex of issues requires breaking down into its constituent elements but also demands a large degree of policy coordination. Without the former, important factors may be missed; without the latter, programmes may have serious shortcomings. Regrettably, the current situation is far short of the ideal. To give one example: involved in some way with water policy are twenty-seven federal agencies, over 59,000 water supply utilities, fifty state governments, and thousands of local governments and special districts (Smith 1995: 110).

FEDERAL WATER POLICY

Given the general abundance of water in the USA, it is not surprising that there has been no long history of water policy. The first congressional Act dates from 1948, when federal research and funding was introduced. Later legislation extended federal responsibilities and, in 1965, states were required to establish water quality standards. It was, however, the 1972 federal Water Pollution Control Act (strengthened by later amendments) which established the current regulatory framework. There is a separate Safe Water Drinking Act which regulates public water supplies.

Pollution from point sources is controlled by a system of EPA permits: the National Pollution Discharge Elimination System. These are source-specific: each has to be individually determined. Enforcement is undertaken by the states whose pollution control programmes (which are grant-aided) have to be approved by EPA. Municipal sewage treatment plants are required to meet EPA standards. Pollution from non-point sources is mainly regulated by the states, as is ground water.

WATER QUALITY STANDARDS

The 1972 Act was a prime example of 'technology-forcing' legislation. It embodied the policy (mainly for point source pollution): 'that the discharge of toxic pollutants in toxic amounts be prohibited'. The aim was for 'fishable and swimmable' waters by 1983, and complete elimination of discharges into navigable waters by 1985. The Act set standards relating to water quality and effluent limits. The latter were based on the principle that all pollution is undesirable and should be reduced to the maximum extent that is technologically possible. Imperious deadlines were set to meet the standards based upon concepts of the 'best practicable technology', the 'best available technology economically achievable', and suchlike.

With hindsight it is obvious that the standards were far from 'technical'. They had to be interpreted,

and this involved negotiation with representatives of polluting industries. Since the character of the pollutants and the polluting process varied among industries, individual standards were required for different types of industry. Contrary to the intention, the realization was essentially political and, not surprisingly, the goals were not met. Nevertheless, the legislation appears to have had some positive effect in the control of water pollution, though it is difficult to assess how much. Evidence on trends in water quality are elusive, adequate data are sparse, and it is impossible to isolate the effects of particular programmes. A National Water Quality Inventory Report to Congress is prepared every two years: this summarizes the information collected by the states. The information is incomplete, but it shows that about two-thirds of the waters assessed are of sufficient quality to support uses such as fishing and swimming, and therefore meet

Plate 34 Run-off Enters Merrimack River, Nashua, New Hampshire
Courtesy Alex MacLean/Landslides

the goals of the Clean Water Act. It is clear, however, that there are some serious problems, for example of groundwater pollution (particularly from agricultural run-off) and of discharges from inadequate municipal sewage treatment facilities.

From inadequate information, all that can be said with any degree of certainty is that the nation's waters do not appear to have deteriorated in quality. Given the extent of population and economic growth, this is some degree of achievement even if it is far from the ambitious original hopes. Matters could have been far worse. Unfortunately, there are indications that some of the problems could increase. This will become clear from an examination of several areas of water pollution control policy.

MUNICIPAL TREATMENT PLANTS

The improvement of municipal sewage treatment plants has for long been a major target of federal policy, and some impressive progress has been made. Unfortunately, serious problems remain (see Box 15.2). With high rates of pollution from agricultural run-off, urban streets, large-scale destruction of wetlands, floodplains, coastlines, and suchlike, 'we are actually going backwards in our efforts to restore the health of our aquatic ecosystems' (Adler 1994: 19).

Finance for the provision and maintenance of municipal plants has not been adequate, despite huge expenditure. Part of the problem has been that, though grants were available for capital costs, the states received no support for running costs; the result was the building of 'Cadillac projects without the funds or technically qualified operators to maintain them, and some plants operate substantially below design capacity' (Ingram and Mann 1984). Voting capital funds is always politically more popular than supporting running costs.

Grants were initially at a high level (75 per cent of construction cost); but this was reduced to 55 per cent in 1981, and replaced by a loan program in 1987. The need for further upgrading of treatment plants remains. The inability of plants to cope with waste water is a primary cause of pollution. In 1989, EPA reported that over two-thirds of the nation's 15,600 wastewater plants had 'documented water quality or public health problems', and it estimated that $83 billion was needed to bring plants up to the required standard (Smith 1995: 112).

NON-POINT POLLUTION

Almost by definition, non-point pollution is problematic: there is no easily recognizable point at which

---

BOX 15.2   PROGRESS WITH CLEAN WATER – SOME INDICATORS

The federal government invested $56 billion in municipal sewage treatment from 1972 to 1989, with total federal, state, and local expenditure of more than $128 billion. By 1988, 58 per cent of the US population was served. This improved treatment resulted in an estimated reduction in annual releases of organic waste by 46 per cent, despite a large increase in the amount of waste treated. The same measure viewed from the opposite direction, however, shows a glass only half full. In 1988, public sewer systems serving 26.5 million people provided only minimum treatment, and 1.5 million people had no treatment at all, with raw sewage discharge into public waters.

In 1990, the Clean Water Act's 'swimmable' goal was met in about three-quarters of our rivers and estuaries, more than 82 per cent of our lakes, and almost 90 per cent of our ocean waters . . . But this leaves a large number of water bodies which are unsafe for swimming – one out of ten ocean miles, and one in five lake acres. Closer analysis indicates that many more waters are not really safe for swimming.

Source: Based on Adler 1994: 10–11

it can be controlled. Its origins are diffuse and varied. It results from the way in which an industry is structured, managed, and operated: thus improvements may require major changes. Added to this is the fact that the industries concerned are politically powerful: agriculture, construction, forestry, meat packing, shipping, and many others have strong support in Washington and in the state capitals. Bringing influence to bear in these quarters (influence which may spell jobs) is a far cry from diverting a sewer pipe. Coupled with technical difficulties, these problems have retarded progress with the abatement of non-point pollution.

Agriculture is the worst non-point offender: agricultural run-off contains many toxic constituents from pesticides, fertilizers, animal waste and similar materials. To abate these requires changes in standard agricultural practices. States have been unwilling to take an aggressive approach, and have tended to adhere to policies of gentle persuasion. There is, however, an Agricultural Water Quality Incentives Programme which provides technical assistance and subsidies for measures which reduce source contaminants.

GROUNDWATER POLLUTION

Groundwater – despite its name – is water which flows under the ground. Most (98 per cent) of the global amount of available fresh water is groundwater stored in aquifers – the pores and cavities of rock strata. In the United States, about a quarter of fresh water comes from this source – and much more in the arid West. Pollution of groundwater is a mounting problem of alarming proportions. It comes from numerous sources: agricultural activities (particularly fertilizers and other chemicals used in modern agriculture), discharges from sewage works, urban run-off, oil discharges, waste sites of many types, acid rain; indeed, all the pollutants which are discussed in this chapter (and many more) find their way into groundwater. Thus all measures which go toward reducing pollutants also help to protect groundwater, including the Superfund legislation

examined in the next section. (This is an illustration of the interrelatedness of environmental issues.)

One of the many problems with pollution generally is that chemicals can be in use for many years before their hazardous nature is appreciated, and thus the required measures of environmental protection are retarded. As a result, pollution builds up and becomes more difficult to tackle. Even when a pollutant is recognized as such, its presence may remain undetected for many years. This is partly because of the unknown dangers in the thousands of abandoned hazardous waste sites, and also because the slow rate of movement of groundwater increases the difficulty of detecting pollution. There are added complications caused by the variability of ground conditions.

All this adds up to an emerging problem of frightening dimensions. It is well recognized by scientists ('groundwater pollution could become one of the scourges of the age': Hiscock 1995: 246), but it does not have the political salience of other types of pollution.

SAFE DRINKING WATER

The fragmented nature of water pollution control is illustrated by the existence of a separate Safe Drinking Water Act. Originally passed in 1974, following public concern about harmful chemicals in drinking water supplies and the inadequacy of state programmes, the Act required EPA to monitor and regulate twenty-two water contaminants. There were additional discretionary powers to extend controls over further contaminants. Between 1974 and 1986, EPA introduced regulations relating to the twenty-two specified contaminants, but progress beyond this was very slow. Frustrated by the tardy rate of progress, Congress eventually adopted a more directive approach. An amending Act of 1986 required EPA to regulate eighty-three specified contaminants and also an additional twenty-five every three years. At the same time, EPA's responsibilities were increased and strengthened.

In 1993, there were about 200,000 public water

systems regulated under the Safe Drinking Water Act. These served 243 million people (the remainder obtained their water from private wells). A very large number of systems are small and have difficulty in shouldering the financial burdens of compliance. This difficulty is significantly increased by the 'twenty-five every three years' mandate, and is criticized by EPA as adding to the regulatory burden and detracting from the implementation of priority contaminants. In some cases, 'contaminants have been forced onto regulatory schedules that out-pace EPA's ability to develop needed technical information, some regulations have unquantified benefits, yet impose significant costs'. It thus seems that the forced pace of regulation imposed by Congress has not worked well. The EPA (US Environmental Protection Agency 1993) has concluded that 'a fundamental reform' of the legislation is needed which would focus on priority public health threats. There is fragmentary evidence that these threats are real. Since the sources of this evidence include the General Accounting Office (GAO) and EPA (as well as a Ralph Nader study), this gives rise to some concern (Rosenbaum 1995: 227).

THE LIMITS OF THE REGULATORY APPROACH

There has for long been criticism of the favored regulatory approach. Though some failures in implementation are undoubtedly due to intrinsic difficulties, many argue that 'a major share of the responsibility for the slow rate of progress must be assigned to the inappropriate incentive structures created by the regulatory approach to pollution control' (Freeman 1990: 145). The advantages of alternative approaches are discussed in the last section of this chapter.

It would be foolhardy to include in this discussion policy changes that are under consideration at the time of writing. The political uncertainties are far too great. Suffice it to say that issues being addressed include the control of polluted run-off, watershed management, further restriction on the discharge of toxics, and a strengthening of enforcement procedures. The focus is on the protection of water, rather than on treatment after it has been polluted.

WASTE

The sedge is wither'd from the lake,
And no birds sing.
John Keats, *La Belle Dame Sans Merci*

THE NATURE OF WASTE

Humans are 'wasteful': large quantities of the by-products of economic processes are not used and, in fact, are perceived as having no use, either because no one has thought of a use, or because any use that has been considered is judged to be uneconomic. But this is determined at least in part by the way in which the costs of production and their unwanted byproducts are calculated and shared. If a firm is free to dump its unwanted byproducts without regard to the costs imposed, it has no economic incentive to find uses for them. The cost is passed on to those who suffer from environmental pollution and degradation, or to those who have to shoulder the burden of cleanup. Moreover, these costs can be much higher than those which would have been involved in introducing more efficient (less polluting) methods of manufacture or systems of recycling. One reason for this is that biological and chemical processes acting upon waste can render it far more harmful than it was in its initial state – and therefore costly to treat. Another is that regulatory systems can be enormously expensive and, as we shall see, not always effective.

Waste mattered little in primitive societies: little of it was produced, and it caused no harm, natural processes being sufficient to perform a self-purifying function. As wealth and populations grew, waste increased and natural processes became insufficient, especially in urbanizing areas; innovative but more harmful manufacturing systems were introduced; and waste of greater toxicity was produced. The

Plate 35 Trash Trucks Dumping and Tractors Compacting, Tullytown, Pennsylvania
Courtesy Alex MacLean/Landslides

development of new goods involved a widening range of manufacturing processes in which the constituent materials were selected solely for their ability to contribute to producing the wanted good: whether they also produced unwanted byproducts was incidental. Byproducts constituted 'waste' which was to be got rid of in the cheapest way possible. Waste was even to be seen as a sign of wealth: a coal tip growing on the edge on a mining community or a chimney pouring out black smoke from an urban factory were indications of prosperity. The environmental impacts were ignored, or regarded as incidental, or simply (as also with industrial diseases) not understood.

What is waste in one system of economic production is not so in another. The squatter settlements on the edges of third world cities eloquently demonstrate the potential value of urban 'waste':

materials jettisoned as being of no use in the city are put to good use in providing shelter and primitive amenities. In the affluent cities themselves, simple incentives for recycling can transform something which is 'waste' into a marketable commodity. Well-known examples are 'deposits' on returnable bottles and cans, and policies for the preferential use of recycled paper in government departments which, given their prodigious use of paper, is not to be underestimated. The 'value' of waste paper is nicely illustrated by a report from California which notes that

> a ton of loose office paper can be sold for $30. Bale the paper and the market price rises to $150. Pulp the paper and the market price reaches $570. Convert the pulp to writing paper and the price can climb to $920 a ton.

(Schwab 1994: 47)

The opposite also holds: a marketing system which puts a premium on attractive, well-packaged consumer goods creates enormous quantities of waste packing materials. Even a rule introduced for the benefit of the public health – requiring food to be wrapped, for example – can increase waste packaging. Convenient new packaging (from the ubiquitous plastic bag to the polystyrene supports in boxes of consumer durables) creates new and problematic forms of waste. Equally convenient new throw-aways create increasing waste-disposal problems: 1.6 billion pens, 2.6 billion razors, and 16 billion diapers, for instance, are added each year to the mountains of municipal waste in the United States. The diaper is a particularly large problem which makes up between 3 and 4 per cent of the solid waste collected by municipalities; one study estimates the cost of 'disposal' at $4 billion a year – almost the same as the value of the market (Cairncross 1992: 215). On the other hand, all these modern inventions make life easier, sometimes very obviously so as with disposable diapers, sometimes less obviously as with the greater convenience and shelf-life of packaged foods. Indeed, better packaging might lead to less waste of food.

Sometimes, waste can be turned into a wanted good. In addition to the systems of recycling newspapers, bottles, and cans which households have taken to their hearts, there is a large and increasing number of recycling technologies. Plastics are a case in point: they can now be turned into substitutes for timber, concrete, and other building materials. Perhaps the most striking example of recycling is that of automobiles. The once-common sight of a wretched junkyard of abandoned cars has now largely gone, not as a result of effective environmental programs but because of technological changes in the auto industry. These enable steel to be profitably made entirely from scrap. At the same time, the automobile shredder provides the means of separating the various materials. Unfortunately, this bit of magic is in jeopardy since increasing pressure to increase fuel economy has led to a substitution of plastics for metal. These are not only more difficult to deal with: they also reduce the value of car hulks

and thus the incentive to recycle. It is apparent from this short recital that the idea of 'waste' is not a straightforward one. It differs over time, among countries, between industrial processes, and according to the controls operated by governments. The last point is of particular importance: the quantity and character of waste (and methods of waste disposal) can be affected by public policy. Given appropriate mechanisms (whether regulatory or economic) most waste can be disposed of with ease. There is one exception: hazardous waste, of which a ton per head of population is produced each year in the United States.

HAZARDOUS WASTE

Hazardous waste is, in one sense, easy to define: it is simply waste that is hazardous. But so is smoke, agricultural run-off, and leaking chemicals. Some of these types of pollutant affect air or water more than land, but pollutants can affect any or all of these media. The point does not need to be labored: there are difficulties of defining categories of waste, and these arise in part as a reflection of the approach taken to their mitigation. Some legislation focuses on the source of the waste (as with nuclear waste), some on the medium it affects (air and water), some on its character (toxic). There are also differences in the ways in which different wastes are dealt with: policies for clean air and water, for instance, are focused on making these media clean rather than on disposing of the pollution. Hazardous waste is typically thought of in terms of land pollution, but it can affect all environmental media, and the way in which it is dealt with legislatively and operationally is in part a result of the way in which the problems were initially interpreted and defined. Had the accidents of history been different, hazardous waste might have been viewed differently. It should also be noted that, as interpreted by regulations, the US definition of hazardous waste is by no means all-inclusive: it excludes the wastes produced by households and by agriculture, mining, and drilling operations.

LOVE CANAL

Love Canal was one of a number of disasters that have precipitated major legislative responses (others include Bhopal and the *Exxon Valdez* oil spill). Popular environmental history tells that the discovery of the 'ticking time bomb' of 21,000 tons of chemical waste at the Hooker Chemical site (Niagara Falls, New York) revealed a 'public health emergency' of 'great and imminent peril'; quick action by local residents and rapid response by state and federal agencies led to the evacuation of the residents and to the passing of the Superfund legislation aimed at dealing with similar catastrophes throughout the country. The reality is different and much more complicated. The terms used to describe the nature of the problem (in quotation marks above) have special meanings in their original context which were misinterpreted by the public. This is hardly surprising: who was to know – or believe – that the phrase 'public health emergency' was a jargon term used to ensure that Love Canal would legally qualify for federal emergency relief funds, or that the phrase 'great and imminent peril' was an administrative trigger for the allocation of funds for public health studies? The nuances of the legal and administrative meanings of such terms were not appreciated by the public, or by the residents of Love Canal, or by the press, or by the politicians involved. The result was, indeed, quick action (relatively) and, since it was feared that similar catastrophes might arise elsewhere, national legislation dealing with toxic sites. (One further unfortunate effect of the 'crisis' was that public health studies were carried out hurriedly and inadequately – and with results which were later discredited.)

The legislation, however, was already in the pipeline: Love Canal was not its cause, but it did greatly increase political support for it: members of Congress were quick to see the public reaction to Love Canal and the specter of thousands of similarly abandoned lethal sites throughout the country. More important in the long run, Love Canal had a major impact on the character of the legislation. In particular, the widespread public concern provided EPA with an opportunity to widen its mission and to take on new responsibilities for the public health: it made good use of the opportunity. In doing so, it built its case on the basis of the Love Canal problem (Landy *et al.* 1994: 142).

SUPERFUND LEGISLATION

The difficulty of assessing the scale of the hazardous waste problem complicated the task of devising a sensible regulatory system. There was no way of knowing how many hazardous waste sites there were, what was in them, how dangerous they were, or how much the cost of cleanup would be. Indeed, the only certain thing was that the answers to such questions were unknown. It was accepted that there was a great deal of ignorance, though no one knew just how much. Yet Congress had to give a lead: it did this by requiring EPA to develop a *national priority list* (NPL) of sites in greatest need of cleanup. No indication was given as to how these sites were to be selected, but they were to be eligible for Superfund finance. Beginning with 400 sites, the list was to be added to each year on the basis of information obtained by EPA and the states. Sites not included in the list would fall to the responsibility of the states.

The regulatory system for dealing with hazardous waste was introduced by the Resources Conservation and Recovery Act in 1976. This is an extraordinarily complicated piece of legislation, and its complications were increased by later acts, of which the best known is the 1980 Comprehensive Environmental Response, Compensation, and Liability Act. (Mercifully, even its acronym CERCLA has given way to the popular 'Superfund'.) With this legislation, Congress intended to establish a comprehensive 'cradle-to-grave' system for regulating wastes.

The control of hazardous waste operates over the three main participants in the waste production and disposal process: generators, transporters, and operators of treatment, storage, and disposal facilities (known by their abbreviation as TSDs). A manifest system tracks hazardous waste from its generation to its final disposal: records are kept of each stage of the journey made by the waste from its production to its final disposal. There are heavy fines for violators. This impressive-looking system is less effective than might be expected. In the first place, there is little monitoring. Second, the system applies only to waste which is moved from the site where it is generated. Thus all the waste that is dealt with by the producers is not covered: this is the majority – estimated to be up to 90 per cent. Though some onsite disposal requires a permit, the self-management thus allowed is subject to only very limited monitoring. Inspection by EPA is rare, and state agencies do not have the resources for regular monitoring.

It should be noted that implementation of this system (as with much else in the environmental protection field) depends essentially on the capability (and willingness) of the states. It was assumed by Congress that the widespread public concern about

hazardous waste would prompt the states to set about implementing the scheme with enthusiasm. In fact, implementation has been very varied. Some states have done little, while a few have shown much initiative and (coupled with additional powers from the state legislature) have reached high standards of effectiveness. New Jersey's Environmental Compensation Responsibility Act requires a hazardous waste site assessment on the sale or transfer of any industrial or commercial property. This 'has provided a tremendous impetus for careful site assessment, completely transforming the local real estate market' (Mazmanian and Morell 1992: 88).

More generally, however, the record of the states is disappointing – partly, in their view, because of inadequate federal funding. Progress was particularly disappointing in the early years; but a change began to take place in the early 1980s and, though still

Plate 36 Fresh Kill Landfill, Staten Island, New York City
Courtesy Alex MacLean/Landslides

patchy, state implementation improved. Paradoxically, one result was a *reduction* in the number of landfills. This came about because of the unwillingness of many operators to incur the expenses of the new EPA standards. Rather than upgrade, many landfills simply closed. These closures, together with the expenses incurred by operators who did upgrade and a continuous growth in the amount of waste being produced, resulted in a marked rise in land disposal costs. Increasingly, it began to be realized that there was a far better alternative to disposal, namely waste *treatment*. This was embraced in new legislation – the 1984 Hazardous and Solid Waste Amendments Act. This gave pride of place to waste reduction, followed by recycling, treatment, and – as a last resort – land disposal.

The Act was written in a way which forced implementation: so-called 'hammer provisions' required EPA to introduce new controls by fixed dates, with automatic arrangements. These were designed, not only to overcome any tardiness on the part of EPA, but also to circumvent antagonistic action by President Reagan (who correctly saw that stronger regulation of waste disposal would significantly affect a large number of firms).

LIABILITY AND COMPENSATION

What is particularly frightening about hazardous waste is its 'timebomb' character (to coin the term which quickly stuck to Love Canal). It can be a very long while before the toxic effect begins to appear – by which time it is too late to take preventive measures, and also late for ameliorative action. The sense of apprehension is backed up by other emotions – of confusion, distrust, betrayal, and even treachery. What had happened to the skills and competence of American industry? How could such a successful machinery of production have wreaked such a disaster on an unsuspecting people?

This sense of outrage emboldened Congress to pass some severe penalties on those responsible for producing hazardous waste. Instead of the traditional legal doctrine of negligence, a far more severe

doctrine was invoked: that of strict and several liability. This means that excuses and mitigating circumstances are irrelevant, and that all who have been involved in the generation of the waste are liable. Thus, the common legal immunities are absent. At the same time, the law was made retroactive. The term 'Superfund' by which the waste regulation system is generally known is a misnomer: it gives the impression that there is a huge federal fund available for clearing up hazardous sites. In fact, the legislation is designed to pass on the costs to the maximum extent possible. There are complex provisions intended to identify the responsible waste-producing parties and make them pay for the cost of disposal. Only when no *potentially responsible party* (inevitably known as a PRP) can be identified is federal funding available. This Superfund, financed mainly by a tax on chemical manufacture, was established in 1980 with an initial spending limit of $1.6 billion (increased to $8.5 billion in 1986 following the alarm caused by the Bhopal explosion).

The incredible complexities to which this has given rise have proved hugely profitable to lawyers. Though the courts have taken the very sensible approach that efforts to achieve cleanup have priority over the allocation of costs, the subsequent wrangling over costs can take a very long time. It is not unusual for there to be seemingly endless arguments, not only from those initially identified as being responsible for the hazardous waste (present and past owners and operators, generators, and transporters) but also from those making counterclaims, cross-claims, and third-party claims. The Act expressly authorizes PRPs to make claims against other PRPs; the court allocates costs 'using such equitable factors as the court determines are appropriate'. In spite of the length and cost of litigation, it can be very much cheaper than the cleanup costs. A study of one landfill Superfund site produced estimates of the total cost of cleanup ranging from $50 million to $4.5 billion. As involved attorneys comment, 'with so much at stake, it was very difficult for the various PRPs (site owner, site operator, industrial generators, and transporters) to agree how to proceed with cleanup or to fund cleanup activities' (Muse *et al.* 1995: 135). (Just to illustrate one of the difficulties that can arise, generators can be liable for cleanup costs even if they had no knowledge of the site to which their waste was transported.)

The liability provisions are so strict that firms who might have been willing to pay their 'fair share' voluntarily have been deterred from doing so since they could find themselves forced to pay far more than a fair share – at worst, the whole cleanup cost. For the large firms, Superfund has become 'the legal equivalent of a survivor-pays-all game of roulette'. One state official has wryly observed that a Superfund listing 'can actually be counterproductive in achieving cleanup' of a site (Mazmanian and Morell 1992: 37).

The great emphasis on liability is in striking contrast to the way in which compensation is dealt with: there is, in fact, no provision for compensation of those affected by contaminated sites (local residents, former workers). The absence of victim compensation was deliberate: it was intended to ensure that resources went to cleanup and were not depleted by compensation. But this did not stop victims suing PRPs. The precedent was set by the residents of the area around Three Mile Island (the 1979 near-catastrophic meltdown of a nuclear reactor); levels of compensation reached huge proportions with Love Canal, where the residents finally reached a settlement of $20 million.

In looking at these large issues of industrial hazardous waste, we should not lose sight of the importance of risks which are much closer to home. Many private garages and garden sheds contain an abundance of highly toxic aids to gardening, car upkeep, house cleaning and maintenance. Data on these hazards are largely anecdotal: ignorance of the extent of the problems of storage, use, and disposal is profound. These wastes may present a significant potential risk but, like non-point water pollution and radon, they have less salience than the more dramatic forms of pollution. They arouse little public attention and therefore little action; and, as already noted, household waste is excluded from the definition of hazardous waste. Tragically, they may be a disaster waiting to happen.

These domestic hazards do not fit neatly into the structure of pollution controls. But, given the nature of this structure, this is not surprising. Particularly baffling to a newcomer is the distinction between hazardous waste and toxic substances. Surely hazardous waste *is* composed of toxic substances? Had the history of environmental policy taken a different turn, they might have been dealt with in a comprehensive manner. As it is, they not only have different control systems: toxic waste actually involves two systems of its own – one for pesticides and one for other 'toxic substances'. (In fact, this oversimplifies the situation since there are twenty-four Acts which deal with toxic substances.)

TOXIC SUBSTANCES AND PESTICIDES

There is a sense in which the period following World War II can be described as the chemical age. The huge expansion in the production of chemicals, many of them being human-made (synthetic), amounted to a revolution not only in terms of numbers (more than four million between the mid-1940s and the mid-1960s), but also in their impact on agriculture, on electronics, on industrial processes, and on everyday life.

New insecticides proved to be highly efficient in controlling pests and insect-borne disease. Herbicides have been similarly successful in controlling weeds. Together with the use of fertilizers, these chemicals have brought about huge increases in agricultural productivity as well as other benefits such as reductions in tillage, labor requirements, and soil erosion.

Regrettably, there were costs involved that were not apparent at first. Thus, it became apparent that chemicals that were effective in killing insects and weeds had serious environmental effects. As Rachel Carson explained, in prose which could be readily understood, synthetic pesticides such as DDT (dichloro-diphenyl-trichloroethane) have extraordinary power. Unfortunately, this power is not confined to killing insects; it can have disastrous effects on all forms of life. DDT is highly persistent and can be readily passed from one organism to another through the food chain. Through this process it can become heavily concentrated, with severe and even fatal effects on humans. Such chemicals are (in Carson's memorable phrase) 'elixirs of death'. There are many other proven carcinogens including dioxin, asbestos, and polychlorinated biphenyls (PCBs); more than 500 such chemicals have been prohibited or restricted by EPA.

However, not all harmful toxics can be readily identified, and there is a great deal of uncertainty about the health effects of a large number (though a much greater number are harmless). Though a substance may have serious, even fatal, latent effects, the 'latency period' can be long, and since those affected may also be exposed to other toxics, it is a very complicated matter to isolate the effects of individual substances. It follows that it is also difficult to determine what regulatory controls are appropriate. It took five years of inquiry and debate before Congress was able to decide on an acceptable approach to the control of toxics. The Toxic Substances Control Act was passed in 1976.

The nature of regulation differs between new and existing chemicals. New chemicals gave rise to particularly acrimonious disagreement but, despite opposition from the chemical industry, it was decided that EPA should review all new chemicals before production commences. Manufacturers have to shoulder the burden of proof that the new chemical is safe, and EPA can ban or hold up production until it is satisfied about safety.

By contrast, existing chemicals can continue to be marketed unless EPA invokes a review procedure. It does this where the safety data are deemed to be insufficient and where it is felt that there is an unreasonable safety risk. In such a case, after a lengthy, full rule-making process (which includes public notification, time for public comment, and testing), EPA can require testing. When adequate data are available about a challenged chemical, EPA has very broad powers which it can invoke: these range from stricter labeling requirements to an outright ban. This power was used to ban chlorofluorocarbon (CFC) propellants in aerosols (because of their effect on the ozone layer).

Controls over pesticides have a longer history, though initially this was for the purposes of consumer protection from fraudulent goods. Control for environmental objectives did not arise until the burgeoning of public opinion in the 1960s (the time when Rachel Carson's *Silent Spring* was published).

As with toxic substances, pesticides which are already on the market can be challenged by EPA only by way of a lengthy and involved procedure. New pesticides, however, have to be licensed by EPA before they can be marketed. Licences are given when the manufacturer can show that they pose no unreasonable risk. In determining risk, account is taken of 'the economic, social, and environmental costs and benefits of the use of any pesticide'.

It is inherently difficult to judge the effectiveness of toxic substance policies. The state of scientific knowledge is too inadequate for even a rough judgment to be made. And, of course, to the extent that policies are successful in preventing toxics being introduced to the environment, their effects are not there to be seen.

NUCLEAR WASTE

The reader may be surprised that no mention has been made in this account of nuclear waste. The reason that it appears separately, almost as an appendix to the main discussion, is that it is dealt with quite separately from other wastes. As with a number of other government functions (such as coal-mining control and reclamation), it does not fall within the responsibility of EPA, but of its federal guardian – the Nuclear Regulatory Commission (NRC). This is an independent agency set up in 1974 (taking over the functions of the Atomic Energy Commission), which licenses the operation of nuclear power stations. (It also has the responsibility of licensing the building of new nuclear power stations, but there have been none of these for many years.) All issues relating to the safety and environmental aspects of nuclear power rest with the Commission.

The history of nuclear waste disposal has been a dismal one, even after the shock of Three Mile Island killed the dream of a nuclear age in which energy would be clean, safe, and cheap. It has proved to be none of these, and the regulatory machine has proved incapable of dealing with the increasingly complex and horrendously dangerous problems which have arisen. Many of these problems were quite unexpected: among the list of unanticipated difficulties have been severe operational problems with nuclear power plants (some of which flowed from basic design faults), rapid deterioration of plants, mismanagement, severe safety problems – all accompanied by escalation of both costs and public anxieties. Some of these problems could probably have been avoided (or at least reduced) by better management and planning; but the early days of nuclear power were characterized by a high degree of optimism and a belief that any teething difficulties would be overcome by technological solutions. However, the most troublesome – and unsolved – issues were totally unexpected: above all the question of disposing safely of nuclear waste – which grew increasingly difficult to deal with as public concern (and outright fear) made it impossible to find adequate sites. The original assumption was that spent fuel would simply be reprocessed, and the residue dealt with safely by advanced technology – an assumption which proved to be false. Added problems arose with the temporary storage of waste. For these and other reasons nuclear waste became a huge liability, and bitter interstate battles raged on site selection.

Eventually, Congress was forced to act: the Nuclear Waste Policy Act of 1982 was intended to solve the long battle over sites by introducing a scrupulously fair and open process which, it was hoped, would satisfy everybody. In fact, it satisfied nobody. The initial three sites nominated (Deaf Smith County, Texas; Hanford nuclear military reservation, Washington; and Yucca Mountain, Nevada) were overwhelmed by controversy, and Congress attempted another solution by summarily designating Nevada to be the home of the first site. (The designation was accompanied by a large bribe – $20 million annually.) However, after two years preparatory work involving an expenditure of $500 million, the Department of Energy abandoned the project on the ground of

inadequate technical quality. Difficulties arose with other projects such as the Waste Isolation Pilot Plant near Carlsbad, New Mexico. At the time of writing, no solution to these is on the horizon.

Public fear about the dangers of nuclear waste sites has been a major factor in this sad story. This fear is justified since there are so many uncertainties about making nuclear waste safe; and the unfortunate history to date now bedevils the issue. Even if a solution were to be found, it is likely that the news would be met with disbelief. As if this were not bad enough, the salience of the issue has diverted attention from another emerging problem: that of safely decommissioning nuclear facilities at the end of their useful life. Rosenbaum dismally concludes that 'Waste management and nuclear power decommissioning problems will trouble Americans for centuries and remain a reminder of the technological optimism and mission fixation that inspired Washington's approach to nuclear technology development' (Rosenbaum 1995: 278). Even if this should happily prove to be too pessimistic a judgment, it is clear that the story is likely to continue for some time without a happy ending.

Note: Further Reading and Questions to Discuss for this chapter are located at the end of Chapter 16.

16

THE LIMITS OF ENVIRONMENTAL POLICY

I know of no safe depository of the ultimate powers of society but the people themselves . . . and if we
think them not enlightened enough to exercise their control with a wholesome discretion, the remedy is
not to take it from them, but to inform their discretion.

Thomas Jefferson

INTRODUCTION

In 1988, beaches in New York, New Jersey, and elsewhere on the Atlantic coast had to be closed because
of pollution: among the evidence were hypodermic
needles, syringes, blood bags, and other repulsive
medical waste. Not surprisingly, public alarm was
immediate. The alarm was increased in the localities
affected by the temporary solution of closing the
beaches where the impact on local economies was
sometimes severe. Further political response quickly
followed. Several states passed or debated legislation; EPA established a task force to consider the
problem; and Congress held hearings which led to
the passing of the Medical Waste Tracking Act in the
same year. This was a remarkable demonstration of
rapid governmental responsiveness. Unfortunately,
the action was far from effective since, despite the
apparent obvious evidence, the real culprit of beach
pollution was not medical waste: it was municipal
sewage. Only a small proportion of the beach closings
were due to 'medical-related' waste. The majority were
due to high levels of fecal coliform which was the
result of sewer overflows in periods of heavy rainfall.
The overflowing pollutants were carried by weather
and tide conditions down the coast. The solution
therefore lay, not in the better handling of medical

waste, but in hugely expensive investments in
municipal waste systems (Fiorino 1995: 155).

This incident is of particular interest for the
analyst of environmental policy since it highlights
the importance of three interrelated issues: politics,
ignorance, and public opinion. Reference to these
issues has already been made in earlier chapters, but
they are so important that it is worth examining
them more thoroughly.

The role of politics in environmental policy is
central. One reason is that there are huge areas where
unequivocal solutions to environmental problems
simply do not exist: in the final analysis, the decision
has to be one of judgment – which is another way of
saying that it is political. In a democracy, this means
that the decision is taken openly, with the 'facts'
(such as they are) being freely available and subject
to public discussion. Even when relevant information
is available, questions of interpretation remain. These
involve value judgments which will differ according
to individual and group beliefs and attitudes.

When so much is uncertain, it is important for
the political process to be as free as possible from
undue influence or unjustified restraint. It seems
clear that democratic political systems are more
attuned to environmental needs than are dictator-
ships. The rights of access to information, of free

protest, of electing governments, and all such features of democracy are effective as well as inherently desirable.

Ignorance (or scientific uncertainty, if the term is preferred) is not only widespread in the environmental field: it is not much of an exaggeration to say that it is commonplace. The rate of technological innovation has been so great for so long that the area of uncertainty is now vast. The easy environmental problems are behind us: those that remain are much more difficult to deal with; and they are constantly being joined by new ones opened up by technological advances and by belated discovery of the long-term pollutant effects of earlier innovations. As a result, much policy is based on quicksand rather than on firm scientific ground. This makes it difficult to inform, persuade, or force public opinion; and without supportive (or at least tolerating) public opinion, no policy can work.

However, public opinion is not a slave to scientific fact: indeed, the gradual realization that there is not a clear 'scientific' solution to all problems has increased public distrust of 'official' views, whether these are expressed by scientists, politicians, or any others who purport to have clear answers. But distrust also acts as a safeguard against bogus science or the unwarranted promotion of a particular interpretation. The old adage that knowledge is power now has to be qualified, since unshared knowledge may not be politically acceptable: the very authority of science has been dramatically weakened. (Monuments range from Chernobyl in the old Soviet dictatorship to the nuclear waste sites in the USA.) Thus science, politics, and public opinion intertwine and create a new image of the aligned 'expert' who is recognizably associated with a particular viewpoint – an environmentalist, an economic expansionist, or whatever.

Having summarized in bold terms the essential argument of this chapter, the constituent elements can now be examined in detail. The starting point is the nature of current policies.

TECHNOCRATIC POLICY

Annual expenditure on pollution control now exceeds $140 billion a year – about 2.4 per cent of GNP (Hahn 1994: 319). This may be readily affordable, but it is a very large sum and it could continue to rise to unacceptable levels. As awareness of the range of environmental problems increases, it has become apparent that there will never be sufficient resources to deal with all of them. It is therefore important that policies should be kept under review and consideration given to changes in the pattern of expenditure. That changes might be appropriate is suggested by the fact that the present pattern is not the result of a carefully considered strategy. On the contrary, it largely reflects surges in public opinion – from clean air and water in the 1970s, to toxic pollutants in the food chain and on waste sites in the 1980s, to the current concern for global ecological problems. However justified these peaks in public concern may have been, it is at least questionable whether the resultant array of policies is 'optimal'.

Many argue that too much effort is directed at risks which are small but scary-sounding, while larger, more commonplace ones are ignored (Morgan 1993). Some figures look compelling: if the mathematics (and the underlying assumptions) are correct, the US is spending at a rate of $12 million per potential victim of hazardous waste pollution but only $5,000 per potential victim of indoor radon. This clearly suggests that more lives would be saved by transferring resources from hazardous waste control to radon control. However, public attitudes rate the two dangers quite differently. Indeed, EPA reports have concluded that the Agency follows priorities which are often very different from those which its own experts consider to be the largest environmental risks. Frequently, the public and Congress (reflecting public opinion) focuses on problems that experts consider to be of relatively small importance.

A bill introduced by Senator Moynihan in 1992 (and again in 1993) responded to this by requiring EPA to seek 'ongoing advice from independent experts in ranking relative environmental risk . . . and to use such information in managing available

Plate 37 Industrial Pollution
Courtesy M. Coppin, Viewfinder Colour Photo Library

resources to protect society from the greatest risks to human health, welfare, and ecological resources'. Senator Moynihan's initiative is a good example of the technocratic approach to public policy: what Jonathan Lash, president of the World Resources Institute, has described as 'a nostrum to quell the effects of public ignorance and to prevent the contamination of the domain of experts, with its hard, quantitative, reproducible results, by unscientific values' (Lash 1994: 75).

VALUES AND RISKS

The inadequacy of this technocratic approach is that it marginalizes the crucial issues of value. The idea that there is an objective, scientific, value-free solution to problems is quite false. Science is not like a piece of arithmetic, where the 'answer' can be found by a feat of intelligence. The shortcomings can be seen by examining the difficulties encountered in the attempts to develop 'risk analysis'. At first sight this seems an eminently sensible approach: risks would be scientifically evaluated, and resources then allocated according to the severity of the risk.

The difficulties of this approach are several. In the first place, the scientific information required is lacking and, though efforts to reduce the area of ignorance and uncertainty will be helpful, it is common experience that research findings often raise new questions which demand further research (Unman 1993). Moreover, 'a growing body of experience seems to suggest that, in fact, more research and better technical information actually exacerbate conflict among experts and in the policy process' (Brooks 1988). Herein lies the insurmountable problem facing risk assessment: it involves issues of value – and these cannot be settled by inquiry. Though research can contribute to the debate by making it better informed, no amount of research can avoid ultimate questions of value. And values are not the preserve of the environmental expert: they are essentially a matter for the public.

A few illustrations of the value-laden issues that arise are eloquent. Is the death of a child worse than the death of an adult? Is a slow death by cancer worse than a swift death in a storm-caused flood? Is equity important? Is it relevant to take account of the benefits which accompany a hazard? Are involuntary risks worse than those voluntarily incurred? Does fear of an event increase the seriousness of a risk? (Lash 1994: 79).

The list can easily be lengthened by encompassing other issues: how should a health risk be weighed against employment benefits? what importance should be attached to wildlife, the rivers and forests, landscape, and ecosystems? is beauty relevant?

PERCEPTIONS OF RISK

Differing perceptions of risk underlie many issues in environmental policy. A good example arises with the siting of waste disposal facilities (Mazmanian and Morell 1992: ch. 7). Scientists see the risks in terms of statistical likelihood. Officials are particularly concerned with the problem of alternatives and the relative 'risk' of opposition among the residents of different sites (a factor relevant to environmental equity); they are also prone to consider the electoral implications of different sites. Private developers of waste sites see risk in the normal economic sense of the chances of making a profit.

Residents, however, do not perceive risk in statistical, financial, or policy terms: they see risk in stark terms of frightening danger. Moreover, it is being imposed upon them by human forces which they can counteract. This makes the risk different from natural disasters, which are acceptable because they cannot be predicted or prevented. In moving to an area with a natural risk, that risk is accepted – even if there is a real belief that it will not come to pass. Furthermore, while there is an accompanying benefit to a natural risk (the benefits of the location), there is no such benefit with an imposed environmental risk. Any benefit goes to the developer. (Risks that have accompanying benefits are much more acceptable – as with those jobless communities which welcome nuclear waste for the economic benefit it brings.)

The problems are made more difficult by feelings of

deep distrust. Institutions of government and business that were at one time trusted (or at least accepted) are now subject to what has been termed a 'confidence gap' (Lipset and Schneider 1983). This arises from the inherent complexities of many contemporary issues, the failures to anticipate disasters such as Love Canal and Three Mile Island, and clear evidence of scandals and corruption (such as Watergate and the incredible savings and loan fiasco – not to mention Vietnam). The NIMBY syndrome is not simple parochialism: it has deep and justifiable roots.

The situation is considerably worsened by the huge area of scientific ignorance which surrounds so many environmental issues. If the experts do not really know, how can their judgment be accepted? Increased information and provision for public participation is of little or no assistance in the face of antagonism: indeed, it can serve to confirm suspicions rather than allay them. The battle in Congress over nuclear waste reflects the deadlock accurately.

If this were all that could be said on the matter, the outlook would be bleak indeed. There are, however, some rays of hope, weak though they may be. At one time, it seemed that California had the answer with its YIMBY (yes-in-my-backyard) programme, which was developed (after a very lengthy period of discussion and negotiation) on the basis that each county would provide facilities for dealing with the hazardous waste produced within its boundaries. This both stimulated more strenuous efforts of waste-prevention, and was accepted by all fifty-eight counties as fair. Unfortunately, if fell foul of the state's politics; but it might have a better fate elsewhere (Mazmanian and Morell 1992: 192–203).

Another approach might be by way of a negotiated contract between a community and a site operator: this could be effective if the basic fear is not of hazardous waste itself, but of the way in which it is managed (Elliott 1984).

RISK AND EQUITY

One of these points – that of equity – can be illustrated by some actual cases (Hornstein 1994). Assessments of the risk involved in eating fish ignore the fact that non-whites eat more carcinogenic fish than the population at large, and that non-whites may prepare their food in ways which increase their exposure to contaminants. Similar issues arise with Native Americans who are extremely vulnerable to polluted food (how relevant is it that they are few in number?). Assessment of the health risks from toxic chemicals or hazardous waste facilities are made on an 'aggregate' basis which ignores the much higher risks faced by racial groups who disproportionately work with such chemicals or live near waste facilities.

These are not exceptional cases. Indeed, there is extensive documentation on the widespread extent to which environmental hazards are disproportionately located in minority and low-income areas. A long series of reports have consistent findings: 'Blacks make up the majority of the population in three out of four communities where landfills are located' (US GAO 1983); 'Three out of every five Black and Hispanic Americans lived in communities with uncontrolled toxic waste sites' (United Church 1987); 'Racial minority and low-income populations experience higher than average exposures to selected air pollutants, hazardous waste facilities, contaminated fish, and agricultural pesticides in the workplace' (US EPA 1992). It is also clear that there is discrimination in the enforcement of environmental regulations (Lavelle and Loyle 1992). Much of this discrimination is influenced by the effectiveness of groups of articulate, wealthier groups – an unacceptable aspect of the healthy role to be played by public involvement in environmental policy (Saleem 1994). An Environmental Justice Act has been debated in Congress (Blank 1994).

There are also wider issues. In ranking environmental risks, how should 'environment' be defined? Should it include overcrowding or poverty or lack of access to services? In determining policy, should the emphasis be on preventing pollution – and accepting a major upheaval in socio-economic organization? Or is it better to concentrate on a regulatory system which determines 'acceptable' amounts of pollution, thereby minimizing upset to existing economic processes and structure?

238

Plate 38 'No Littering', Florida
Courtesy Rob Scott, Viewfinder Colour Photo Library

The last point may seem a radical one, though, in fact, some of the most striking environmental improvements have been achieved by prohibition of certain pollutants. Lead was removed from gasoline; DDT and PCBs were banned; phosphate was severely restricted in detergents; and the nuclear test ban treaty led to a major reduction in Strontium 90. Barry Commoner (1994) has argued that this is the route which policy should take. He is prepared to follow his argument through to its logical conclusion: a pollution prevention strategy could be fully implemented 'only by undertaking massive, wholesale transformations of systems of production: energy, transportation, agriculture, and major industries, the petrochemical industry in particular'.

This takes us too far afield for discussion here, though it has to be accepted that there is a severe limit to the amount of environmental improvement that is possible by means of regulating emissions. Waste policy is explicitly moving towards the control of sources, and the mandated electric car is similarly a recognition that emission controls are insufficient. Another straw in the same wind is the verdict of the International Joint Commission on the Great Lakes (1994) that the only safe way of dealing with their pollution is virtually to eliminate the toxics at source.

Such changes cannot be brought about quickly, but they are possible – as is demonstrated by the examples already quoted and from changes in public behavior in relation to smoking, recycling, and garbage sorting.

THE POLLUTION PREVENTION ACT 1990

The force of the arguments in favor of focusing policy on the reduction of pollution at source is now widely accepted, and the 1990 Pollution Prevention Act promotes an integrated approach. EPA has the responsibility of developing a detailed, coordinated strategy to promote source reduction; to promote research to disseminate experience in this field; to advance better methods of data collection and public access to environmental data; and to report biennially on the implementation of the pollution prevention strategy. Manufacturing plants are required to report annually (and publicly) on their releases of toxic chemicals to the environment and on the steps taken for source reduction, recycling, treatment, and disposal.

The strategy is based on the key principle of encouraging initiative in the private sector: voluntary effort rather than slavish adherence to detailed federal codes is considered to be more efficient and cost-effective. EPA has the role of integrating pollution prevention options into strong regulatory and enforcement programmes.

It will take some time for this system to show results, particularly since it involves major changes in approach. Instead of mastering and implementing the detailed rules set by EPA, firms will have the scope for determining the best way of meeting objectives. This means that costs and benefits have to be assessed. It will be interesting to see how far the interests of individual firms harmonize with broader socio-economic interests.

BALANCING COSTS AND BENEFITS

No policy is cost-free and at some point the costs may outweigh the benefits. How is that point to be determined? It might be thought that an obvious goal of environmental policy should be to abolish pollution, or at least reduce it to proportions which have negligible effects on health. Unfortunately, this is simply impossible unless all economic activity is brought to a halt; and even low levels of pollution involve some risk. There is *no* level of risk which is risk-free for everybody. Even if virtually all the population would be unaffected by a low-level pollutant, there will always be some individuals who are so sensitive that any level would affect them; and there may be others who, while not being made ill by a pollutant, may be discomforted by it. (A final point worth remembering is that not all pollution is human-made: volatile organic compounds are emitted from natural sources such as plants and trees: these natural emissions can become a significant contributor to ozone.) It follows that, whether explicitly or by default, a decision has to be taken on the levels of pollution that are to be tolerated. This also means taking a decision on which types of pollution are to receive priority treatment.

Answers to such questions can come from experts or from those who are affected by pollution. As earlier discussion showed, these two groups give markedly different answers, and it was argued that a democratic political system requires decisions to be taken by the public. However, the public needs to have a factual basis on which the favored option can be decided. In the words of Jefferson (quoted at the head of this chapter), discretion should be informed. Is there any objective way in which this can be done? The answer to this has two parts. First, there is the contribution that can be made by economists, who have the techniques for measuring costs and benefits. Second, there is the American decision-making process which ensures that decisions are taken by the operation of pluralist politics.

ECONOMIC ANALYSIS

It is an elementary axiom of economics that expenditure on one good precludes expenditure upon an alternative. Thus, if resources are devoted to cleaning up hazardous waste, the same resources cannot be used for protecting endangered species (or for developing hospital services, or for road building). The cost of any environmental measure can be expressed in terms of forgone alternatives. It follows that an important question for environmental policy is the way in which resources are to be allocated among different specific environmental objectives and, when the objectives have been agreed, how these can most economically be met. There are so many worthwhile environmental objectives, and so many different ways of meeting them, that it is typically difficult to choose among them. Economic analysis provides a useful set of techniques to assist in this complex process.

Despite its utility, there has been little use of economic analysis in environmental policy-making.

Many laws are 'cost-oblivious': they mandate the achievement of a goal (such as 'clean' water or air) with total disregard for the costs involved; and they often require this to be done by a certain date or in a certain way, again regardless of costs. This disdain for economics cannot be explained simply in terms of politics: it often stems from a worrying feeling that there is something inappropriate, unseemly, or even immoral in using economic tools. On this line of thinking, the use of prices debases environmental benefits and puts them on an equivalent basis with market commodities which are bought and sold; and it denies the inherent value of the environment, which is seen as being of a different order from the value of perfume or television sets. Such views are sometimes passionately held, and are supported by reference to an alleged moral imperative of safeguarding the intrinsic values of the wilderness, of nature, and of an uncontaminated environment. On this line of thought, such environmental legacies cannot be valued as if they were market goods. More widespread is the unease (if not repugnance) over the valuation of life in those cost-benefit analyses which compare the costs of different pollution prevention measures in terms of lives saved. Yet assigning a dollar value to life is basically no more difficult for an economist than determining the market value of transplant organs or motor vehicles or hamburgers. This 'commodification' does not sit easily on the public conscience.

These points have weight, and they are increased by an appreciation of the limitations of cost-benefit analysis and the ways in which it can be subverted. One of the major limitations is the paucity of relevant data. There is too large an area of ignorance about the physical, biological and chemical mechanisms which operate in the environment to allow figures to be calculated with confidence; and benefits are elusive, if only because there is a wide variation in the values which different people place on environmental 'goods'. In addition to these problems, opponents of economic analysis tend to be especially worried about the extent to which it can falsify. The classic example is the use of cost-benefit analysis in water resource development projects. It is now widely known that these analyses used methods which understated costs and exaggerated benefits. As a result of these distortions, a number of schemes went ahead that were costly and environmentally damaging. The height of absurdity was reached when serious consideration was given to the building of a dam in the Grand Canyon (Carlin 1973). Clearly, the extremism of which environmentalists are accused is not unique to them.

ECONOMIC INCENTIVES IN ENVIRONMENTAL REGULATION

Public opinion is in favor of regulatory standards because of their apparent fairness: all are required to meet the same target. Polluters may also like them because of the certainty which they give to the market. In fact, the fairness is illusory. Fixed standards impose quite different costs on different firms. Some can meet standards easily, particularly if their machinery is modern. Older firms, by contrast, may need to invest heavily in new plant to meet a standard; and they will understandably seek to negotiate a less-onerous one. More important in terms of effective environmental improvement, firms will tend not to seek anything beyond the regulatory standard even if they can achieve a higher standard at relatively low cost. They will have no incentive to do so, unless they thereby obtain other benefits.

There are considerable advantages to be derived from designing pollution controls in a way which gives firms economic incentives to reduce pollution to the maximum extent. An incentive can make a firm take a totally different approach to its waste. If, for example, a tax is levied on every ton of waste produced, a firm will not be satisfied with calculating the economics of compliance: it will be motivated to review its processes to reduce its waste to the minimum. It has an inducement to calculate the real cost which its waste involves – a cost which otherwise is borne by the environment.

Of course, this is not to suggest that the interests of individual firms and of environmental policy are now uniformly harmonious: most will simply obey the law and follow their self interest. 'It is not the job of companies to decide what values ought to be attached to natural resources and what the priorities of environmental policy ought to be' (Cairncross 1992: 299).

The use of regulatory instruments and an absence of an economic incentive to reduce pollution does not mean that firms simply abide by the dictates of the regulation. Far from it: the incentive is to avoid the costs of compliance. Thus the regulatory agency has to demand detailed records, inspect the record-keeping system, carry out site inspections, and undertake other such control functions. If, however, the resources of the regulatory body are not sufficient to enable this to be done, some firms at least will be tempted to circumvent or even ignore some part of a regulation. Since administrative resources are typically inadequate, this is a significant issue. The laxer the day-to-day controls and the higher the costs of compliance, the greater will this temptation be. Overstretched agencies may be aware that some firms are in default, but they may have some difficulty proving it, or they may have to accept a firm's assurance that it is doing the best it can, or, given the pressure of work, they may simply leave the relevant file in the pending tray. Particularly bad cases may be prosecuted, but this takes even more time and resources, and the court is frequently unpredictable. There is no need to labor the point further. The incentives to adhere to a regulation are weak for the individual firm; the costs of enforcement are high for the agency.

Congress has a penchant for legislating in detail the manner in which environmental policies are to be carried out. Though there have been political explanations for this, particularly in the Reagan years when the Administration was bent on reducing controls over business, detailed congressional provisions have unfortunate long-term effects. They shackle the administering agencies with approaches which are a product of the time; later more-effective approaches, technological innovations, and the lessons to be learned about the weaknesses and strengths of particular types of controls have to be ignored.

THE RANGE OF ECONOMIC INSTRUMENTS

The simplest economic instrument is a tax on pollution, levied at a rate determined, for example, in relation to the damage caused by the pollutants and the costs of cleanup. Such a tax could be levied on lead or carbon content. (Several European countries have such a carbon tax.) The tax provides an immediate incentive to firms to reduce their use of the pollutant – and it is a continuing incentive. The difficulty arises in setting an equitable rate – a problem which also arises with marketable pollution rights.

Economic incentives can be applied to some types of waste with a deposit-refund system. This is essentially the same as the charges on returnable bottles, though rather more complicated. The producer of something which would become a waste after it has been used in a manufacturing process (a solvent for instance) would be required to pay a charge for each unit produced. This would increase its price (thereby introducing an incentive for reduction in its use). A refund of the charge would be payable to anyone who returned the solvent after its use. This system has the advantage of providing a disincentive to illegal tipping. (Clear evidence of its effectiveness with bottles is provided by their disappearance from the countryside after the charge was introduced.) The same system can be applied to motor vehicles.

EMISSIONS TRADING

Several techniques have been established for 'trading' in emissions. One of the earliest was the *offset* technique which was introduced to allow new economic development (and its accompanying new pollution) to take place without increasing the level of pollution in an area. This was achieved by further 'offsetting' reductions in the emissions from existing plants. In effect, this creates a trading market in emissions. It not only enables firms to achieve lower compliance costs, but also results (because of EPA rules) in a net reduction in emissions. An advance on this was the 'bubble' policy. This first arose in the case of large industrial plants which had several smoke stacks notionally grouped in a bubble: controls were operated over the bubble as a whole instead of over each stack. This enabled the company concerned to 'trade' between stacks, thereby achieving a lower cost of compliance. Again, a net reduction in emissions can be obtained. A variation on this technique allows 'emission banking', whereby firms earn credits for achieving reductions in emissions greater than required, and can apply them elsewhere (either to their own emissions or, by sale, to other firms). Techniques such as these encourage technical innovation and more efficient methods of emissions control. It is also suggested that it may encourage the early retirement of older, dirty facilities.

The 1990 Clean Air Act went further with its outright attack on the coal-fired power plants which contribute so significantly to acid rain. (That it was able to do so was a result of a change in the political climate which made it imperative that the new Act should be passed. One result of this was that Midwest politicians found themselves isolated from other parts of the country which now objected to paying for the cost of cleanup in the Ohio Valley – a good example of the territorial politics which underlies environmental policy; Cohen 1992). The Act mandates a major reduction in sulphur dioxide emissions by the electricity generating plants of eighteen states (mainly in the Midwest). This is being effected by an emissions trading program which effectively establishes a market in pollution rights. Each plant is given an emissions permit which sets the limit to its 'allowance'. If this is exceeded, a fine is payable and a reduction made in the following year's allowance. The allowances can be traded, and thus plants that are not able to reduce their emissions by the required amount can buy credits from plants that have made excess reductions. In this way, plants which find it very costly to reduce their emissions can trade with plants that are able to make a reduction more cheaply. The theoretical result is that the maximum reduction in emissions is made at the lowest possible cost. The essential feature of this trade in emissions is that the total is limited; and, if it works successfully, the cap can be reduced.

Emissions trading has been welcomed by economists and policy analysts, who see it as a highly efficient method of pollution control which avoids the problems of a purely regulatory regime. It certainly has political attractions: for conservatives, it is consistent with market principles; for liberals, it promises environmental improvement at an economic cost. It is akin to a user fee 'that ensures that those who benefit from the use of a natural resource pay for those benefits' (Bryner 1993). However, not all would agree. To the layman, there is something distasteful in such a market. Additionally, there may be concern that the market would distribute costs and benefits unequally. As already indicated, there is considerable evidence that minorities bear a disproportionate burden of pollution (for example, in the location of hazardous waste sites).

Economic incentives have an important role to play in environmental policy (though they are not appropriate in all circumstances – in relation to monopolies for example). They have other real advantages. They involve much lower costs for government than do regulatory systems. They require less information – which, as has been repeatedly demonstrated in this discussion, is at a premium. They have an effect on all firms – the small (difficult to regulate) as well as the large (easier to regulate); and they avoid 'negotiation' of deals between regulators and the regulated – as well as suspicion that this is widespread. Above all, they place the costs of pollution where they properly belong: on polluters.

The 'polluter pays principle' is thus a very important one. In practice, individual firms will go beyond what is prompted by price-signals. There are

at least two reasons for this. First, there is the highly efficacious policy instrument of requiring firms to collect and provide information about their activities. Information is a management tool of crucial importance: it can alert firms to facts of which they were totally unaware (see Box 16.1). This is in addition to its public use and the effect which publicity can have on a firm's public standing – which in turn can precipitate action.

Second, the people who run business are subject to the same shifts in opinion and perception as is the general public. Companies have embraced the environmental ethic as clearly (even if not usually as enthusiastically) as dedicated environmentalists. That this is partly a matter of self-interest cannot be denied, but enlightened self-interest works toward social benefit in the same way as does altruism. But it is difficult to believe that the individuals who run the nation's businesses are unaffected by the environmental awareness and concern that is now so widespread. There is an unmistakable concern for the environment at least on the part of many large companies.

THE RIGHT TO KNOW

The 1986 federal Emergency Planning and Community Right-to-Know Act requires certain classes of manufacturing companies to report annually on a wide range of toxic emissions (Hadden 1989). (The data are reported to EPA, maintained in a Toxics Release Inventory, and made available on request.). This elementary idea is intended not only to assist

---

BOX 16.1   INFORMATION AS AN ENVIRONMENTAL INCENTIVE

The British subsidiary of Rhone-Poulenc decided to build a computerized waste-accounting system to keep track of the waste each plant generates and the costs of disposing of it. The data go back to each plant every month. 'The first time I did this', the manager in charge of the system told the newsletter ENDS, 'there was quite a sensation. I was besieged by calls saying "Are you absolutely sure?" It was a revelation. They were jolted from blissful ignorance about their true product costs.'

Source: Cairncross 1992: 291

regulatory bodies and community groups, but also to provide an incentive to firms to keep their emissions low in order to avoid public agitation or disapproval. Though this might be considered wishful thinking, the power of knowledge should not be underestimated. It may even be useful in informing companies (as well as their shareholders) of their environmental behaviour. Portney (1990: 280) recounts the story of the Monsanto Company, whose chairman, on seeing the level of his company's reported emissions, immediately pledged to remove 90 per cent of those emissions over the next five years, even though they were not in violation of any law. Whether such action results from a desire for good public relations, a real concern for environmental quality, or enlightened self-interest does not much matter – though the level of concern of industrialists for the environment may well be higher than they are often given credit for.

This Act also required industry to change the way in which emissions were reported, from parts per million to pounds per year. The changeover revealed 'surprisingly big' numbers – 'not only in terms of lost product and substantial cost savings but also in terms of credibility'. The quotation is from a paper by a spokesman for the Chemical Manufacturers Association (continued in Box 16.2). It is seldom that a technique for reducing pollution proves so effective and cheap.

PUBLIC PARTICIPATION

It is striking that many of the major changes in environmental policy have resulted from public pressure, itself roused by environmental disasters. The nuclear reactor accident at Three Mile Island led to the establishment of the Nuclear Regulatory Commission; Love Canal provided the political base for the 1980 Superfund legislation; the 1984 chemical accident at Bhopal, India, which killed more than three thousand people and injured several hundred thousand, led to the Emergency Planning and Community Right-to-Know Act; the *Exxon Valdez* oil spill in 1989 was followed by the Oil Pollution Act of 1990. However, the role of the public in environmental

policy is much more than that of alerting politicians. As the discussion on waste disposal demonstrates, some problems demand public involvement. Though such involvement may prove inadequate in itself, there may be no possibility of any solution without it.

The discussion on the shortcomings of risk assessment underlined the importance of values and the limited validity – and viability – of technocratic approaches. Putting these various points together adds up to a powerful case for a high degree of public participation. It needs to be stressed that this is not to suggest that the public's role is simply to supply value judgments to experts. Facts and values are inextricably intertwined – and this, in a democracy, is how it should be. The conclusions of a conference on risk assessment (Finkel and Golding 1994: 335) are particularly telling here (see Box 16.3).

Public involvement in environmental policy is an essential feature of 'bottom-up' regulatory systems, as distinct from 'top-down' systems which are run by experts and administrators. To coin a phrase, the issues are too important to be left to experts. However, it is a mistake to think solely in terms of active participation in the actual administration of pollution control. More typically, the involvement is through the normal channels of pluralistic politics. Usually this involves the state, the local governments, and citizen groups. Citizen groups often enter the scene by way of protest, either in a NIMBY way (campaigning against a waste disposal site, for instance) or against the desecration of some natural feature (objecting to the flooding of a valley or to the continued pollution of a river). In the classic case, they start by vociferous opposition, graduate through learning about the problems involved, and eventually finish as powerful participants in a workable solution to the environmental problem which first motivated them.

If the necessary political framework exists, such a process can develop into a wider environmental movement. This framework might actually be brought into existence by the coming together of like-minded citizen groups, as with state environmental coordinating councils. Donald Snow, in his

---

BOX 16.2   EXPLAINING TO THE PUBLIC

It was clear that explaining those numbers to the public would be difficult. No matter how comfortable a company or plant had become with the safety of a parts-per-million concentration at the fence, the emissions of hundreds of thousands, or even millions, of pounds per year was tough to explain – even if it was only a fraction of a per cent of the company's throughput. This revelation caused most plant managers – often for the first time – to put themselves in their neighbours' place and consider what other people might be thinking about the perceived safety of a plant's operations. Real confrontation of issues began. Risk communication experts were retained and plant managers learned about the importance of listening to their neighbour's concerns . . . and about how to understand and deal with an outraged public.

Source: Cairncross (1992)

BOX 16.3   FACTS AND VALUES

The conflict between 'expert' rankings and those of the public-at-large is less than one of 'facts versus values' than one of 'values versus values'. We tend to talk about 'non-quantifiable factors' as if they were accessories to risk estimates that might marginally change these estimates . . . But what if the variation contributed by values is equal to or greater than the variation among risk estimates? The 'fact' that indoor radon may cause 100 times the death toll of Superfund sites moves from foreground to background if citizens view preventing each injury from the latter cause as thousands of times more important than the former.

Source: Finkel and Golding 1994: 336

---

account of *Inside the Environmental Movement* (1992: 29), has noted:

> The existence of a long-lived and successful 'coordinating council' with a paid staff and a grass-roots board is a good indicator of whether a given state has a rich enough corps of environmental activists. In most states where they occur, these coordinating councils often arose from the need to enhance the representation of many local conservation organizations before state legislatures and administrations.

Such groups operate as a political link between local groups and the state government. The benefit is reciprocal. Not only do the local groups gain from access to the machinery of state environmental policy, but the state gains a concerned active partner which can inform and advise it of the perspectives of localities and bring new questioning and ideas before it. Since public support can be an important ingredient of state environmental policy, this is no mean input.

The states are at the cutting edge of the implementation of environmental policy. Discussions of 'federal policy' can easily underplay the importance of the states. Though federal powers and funding are obviously important, they are only part of the total public policy operation. Environmental policy is largely carried out by the states; and this is inevitable since there is a definite limit to what can be done by the federal government on its own. Even with environmental issues which clearly fall within federal responsibility (such as transboundary pollution), the states play an important role. Indeed, the general complaint is not that the federal government takes the lead, but that it gives insufficient resources to the states to enable them to carry out their responsibilities.

States, of course, have constitutional powers; but there is more to federal–state relations than these.

A major power of the states lies in their knowledge of their diverse territories. Washington can rarely operate effectively (at least for any length of time) at the state level. It lacks the effective power, knowledge, and 'feel' that are needed for state administration. Moreover, the federal government is a many-headed monster that has the greatest difficulty in coordinating its own organs: there is no way in which it can find its way around the fifty state labyrinths, let alone go beyond these to the localities and communities where environmental policy actually operates.

THE PROGRESS OF
ENVIRONMENTAL POLICIES

It seems that everything in the field of environmental policy is beleaguered by problems of lack of knowledge, scientific uncertainty, and scarcity of data. And so it is when one tries to assess what progress has been made. It is virtually impossible to determine the effect that particular policies have had on health, amenity, and standards of living, though it is generally held that the overall effect has been a very positive one. Even the extent to which environmental quality has improved is problematic. The relevant data are sparse and scattered. The frustrated attempt by the US General Accounting Office to assess water quality (Box 16.4) is illustrative.

Nevertheless, it needs no sophisticated analysis to show that progress has been both positive and very uneven. Air quality has shown the greatest improvement, though little progress has been made with the persistent problem of ozone, and there are still 100 million people living in areas which fail to meet national standards. However, even where air quality has not improved, it is likely that it would have been very much worse had it not been for the extensive pollution controls that have been introduced over the last quarter of a century. Firm judgments are difficult because of data problems, due mainly to the lack of adequate monitoring. (It should be noted that some improvement is due to the changing industrial scene – with major decline in heavy polluting industries such as steel.)

Changes in water quality have been very varied. Much publicity has attended the cleansing of a number of rivers to which the fish have returned, but such improvements are restricted to a few areas. More generally, there has been a deterioration in water quality, largely due to the lack of success (and even effort) in tackling non-point pollution from agricultural and urban areas. On the other hand, there has been an improvement in the quality of drinking water supply and sewage disposal, though this has been far from uniform across the country, and increasing contamination is caused by the growing number of new pollutants. An overall judgment is rendered even more difficult by the crude nature of the monitoring system. It does seem, however, that such progress as has been made is frequently offset by an increase in the number of pollutants created by technological development, and even by better understanding of the processes of contamination.

Hazardous waste policies have been faced with great technical, legal, and political problems, which would remain even if greater resources were made available. They have had the result of reducing the number of sites for disposal, which may have been beneficial; but, in the longer run, the greater benefit

---

BOX 16.4   THE INFORMATION GAP

GAO was not able to draw definitive, generalizable conclusions . . . because evaluating changes in the nation's rivers and streams is inherently difficult, the empirical data produced by the studies sparse, and the methodological problems reduce the usefulness of the findings. Therefore, little conclusive information is available to the Congress to use in policy debates on the nation's water quality.

Source: Rosenbaum 1995: 52

will come from the trend away from land disposal to alternative methods of disposal and waste reduction. Superfund has proved extremely costly and cumbersome; and the problems it was designed to meet have proved vastly larger than initially anticipated. Indeed, even more than with other pollution problems, little is known about the nature and size of the problem, except that it is enormous and will absorb large cleanup resources for many years to come. EPA has identified more than 27,000 abandoned sites, of which 1,275 are on the National Priority List; but only 149 were cleaned up between 1980 and 1993 (Rosenbaum 1995: 59). The average cost has been about $25 million per site. Progress has been modest, to say the least.

Any assessment of environmental policies is clouded by the shortage of data and a lack of adequate measures of environmental quality. It seems clear, however, that current approaches are failing to achieve anticipated benefits. There are many reasons for this, several of which stem from the sheer complexity of the problems and the limits of our knowledge. Better policies, better administration, greater resources, more research – all these would help; but there is no quick fix.

FURTHER READING (CHAPTERS 15 AND 16)

It is not easy to keep up to date with advances (and reverses) in environmental affairs: events can move quickly, and there is an enormous outpouring of relevant reports. It is important to use the latest editions of standard texts. At the time of writing, the best is Rosenbaum (1995) *Environmental Politics and Policy* 2nd edn. Also to be recommended is Smith (1995) *The Environmental Policy Paradox*, 2nd edn.

An informative collection of essays is Vig and Kraft (1990) *Environmental Policy in the 1990s*. For a wide discussion of environmentalism, see Gottlieb (1993) *Forcing the Spring: The Transformation of the American Environmental Movement*.

There are many important post-war writings on the environment. Lewis Mumford's books of the period include *The City in History* (1961); and (1962) *Technics and Civilization*. René Dubos wrote a column in the *American Scholar*, as well as *So Human an Animal* (1968); see also Piel and Segerberg (1990) *The World of René Dubos*. Also important are Aldo Leopold (1949) *Sand County Almanac* (published posthumously); Paul Ehrlich (1968) *The Population Bomb*; and Rachel Carson (1962) *Silent Spring*.

A history of environmental policy is given in Lacey (1990) *Government and Environmental Politics: Essays on Historical Development since World War II*. A revealing book of readings is Nash (1990) *American Environmentalism: Readings in Conservation History*. For insights into the problems of translating scientific research into policy, see Unman (1993) *Keeping Pace with Science and Engineering: Case Studies in Environmental Regulation*. Assessment of risk plays a major role in environmental policy. An excellent review is provided by a range of essays in Finkel and Golding (1994) *Worst Things First? The Debate over Risk-based National Environmental Priorities*. A fascinating, detailed case study is given in Cole (1993) *Element of Risk: The Politics of Radon*.

There is a rich literature on the subject of economic incentives. The classic statement is Dales (1968) *Pollution, Property and Prices*. A well-known work is Kneese and Schultze (1975) *Pollution, Prices, and Public Policy*. Highly recommended as thoughtful, informative and readable is Cairncross (1992, paperback edition 1993) *Costing the Earth*. Also useful is Kelman (1981) *What Price Incentives?*. See also the articles in *Columbia Journal of Environmental Law* 13 (1988).

A particularly useful book on environmental law is Findlay and Farber (1992) *Environmental Law in a Nutshell*. This small book provides a succinct and accessible summary. An excellent, detailed, but clear, account of the EIS process is given in Bass and Herson (1993) *Mastering NEPA: A Step-by-Step Approach*. For a good example of a sophisticated state environmental protection act, see Remy *et al.* (1994) *Guide to the California Environmental Quality Act*.

Supplementary references for particular policy areas are:

Waste

An early popular and highly readable account is Packard (1960) *The Waste Makers*. An accessible, digestible account of the Love Canal incident and its legislative influences is given in the chapter on 'Passing Superfund' in Landy *et al.* (1994) *The Environmental Protection Agency: Asking the Wrong Questions*. A detailed account of the Love Canal controversy is given in Levine (1982) *Love Canal: Science, Politics and People*. On packaging, see Rousakis and Weintraub (1994) 'Packaging, environmentally protective municipal solid waste management, and limits to the economic premise'. A recent study of public involvement in decisions on hazardous waste siting is Rabe (1994) *Beyond NIMBY: Hazardous Waste Siting in Canada and the United States*.

Nuclear waste

Lenssen (1991) *Nuclear Waste: The Problem that Won't Go Away*; US GAO (1993a) *Department of Energy: Cleaning up Inactive Facilities Will Be Difficult*, and (1993b) *Much Work Remains to Accelerate Facility Cleanups*.

Clean air

Bryner (1993) *Blue Skies, Green Politics: The Clean Air Act of 1990*.

Water

Adler (1993) *The Clean Water Act: 20 Years Later*.

QUESTIONS TO DISCUSS

1 Discuss the ways in which environmental concerns have developed since the 1960s.

2 Why is it necessary to have environmental controls over federal agencies?

3 Describe the environmental review process. What are its limitations?

4 Explain the structure of clean air controls.

5 What are the problems of water pollution, and how can they be remedied?

6 'With waste, treatment is better than disposal; and reduction is better than treatment.' Discuss.

7 What is 'Superfund'? Does it work well?

8 Compare regulatory controls and economic incentives. Do we need both?

9 Discuss the nature of risk. Why is it relevant to environmental policy?

10 Discuss the significance of vehicle emissions in environmental pollution.

11 What is the case for a greater degree of public participation in environmental policy?

CONCLUSION

17

CONCLUSION:
SOME FINAL QUESTIONS

The search for scientific bases for confronting problems of social policy is bound to fail, because of the nature of these problems. Policy problems cannot be definitively described. Moreover, in a pluralistic society there is nothing like the indisputable public good; there is no objective definition of equity; policies that respond to social problems cannot be meaningfully correct or false . . . There are no 'solutions' in the sense of definitive and objective answers.

Rittel and Webber 1973: 155

DETERMINING THE QUESTIONS

It would be satisfying for both the reader and the author if this final chapter could set out solutions to the problems raised in this book. It should be clear, however, from the emphasis laid on the nature of these problems, that they admit of no easy solution. In fact, many of them are not 'solvable' in any real sense of the term. One could go further and suggest that the term 'problem' is unhelpful in that it suggests that a solution is possible. *Webster's Dictionary* gives helpful alternatives: 'difficult to deal with' and 'a source of perplexity, distress, or vexation'.

There is an important distinction between a difficulty and a problem: a difficulty is a problem only if it is thought that something can be done about it. In the scientific world, 'difficulties' can be changed into problems because of advances in knowledge; and expectations of such advances stimulate optimistic expectations and further research. There is little to mirror this in the social world: with many issues, the more that social issues are studied, the more complex they seem to become. Indeed, part of the difficulty facing the policy process is that public interest and support (mirrored in Congress) wanes in the face of

lack of progress. As has been pointed out by Downs (1972), issues rise and fall in public interest. Concern arises with the publicity given to particular events (environmental disasters, an apparent rise in crime, race riots, and so on), and a demand quickly develops for strong action to be taken to deal with the 'problem'. But, as the complexities and the resource implications become apparent, public attention wanes – often as other 'problems' claim center stage. An issue may even be transformed into its opposite, as when environmental concerns are overwhelmed by worries about unemployment.

An added complication has come about as a result of the growth of pluralist politics, the demotion of the professional and the expert in public opinion, the increased mistrust of government, and the demand for greater public participation. At one time, problems were for experts to solve. There was justification for this optimism in the nineteenth century: the successes of science and the growth of the professions provided the necessary instruments to deal with the problems of the developing industrial society. The problems were – or became – capable of solution. (See the discussion in Chapter 3 on the rise of the professions in the nineteenth century and the widespread belief in both progress and efficiency: this was the era in which planning was born.) Unprecedented progress was made in engineering, sanitation, control of infectious disease, and the building of roads, bridges, schools, and hospitals. Some issues, of course, such as poverty and working-class crime, proved unamenable to the forces of progress; but these could be ascribed to moral inadequacy, or catered for by charity, or dealt with by penal measures. Hopefully, the tides of progress would eventually raise up the deviants to the new standards of cleanliness, probity, and efficiency.

Overdrawn though this analysis may be, it contrasts sharply with the contemporary scene. In retrospect, the problems that were solved in those simpler times were easy. They were 'definable, understandable and consensual' (Rittel and Webber 1973: 156). There is now far greater concern for equity, for debating what the important questions are, for accommodating – and even facilitating – differences of opinion and attitude, and for mediation. Questions can no longer be left to experts, since their formulation is of prior importance and answers involve 'unscientific' value judgments.

Rittel and Webber (1973), in a paper which makes a major contribution to an understanding of these issues, have termed the older, easy problems 'tame' ones, in contrast to the 'wicked' problems we face today. These are 'wicked' in the sense that they are incredibly complicated, multi-faceted, and elusive; to quote *Webster* again, 'a source of perplexity, distress, or vexation'. They are so difficult to deal with that one is tempted to suspect that evil forces are at work to prevent solution. Indeed, 'social problems are never solved. At best they are only re-solved – over and over again'. (See Box 17.1.)

All this is very depressing, but it does point to the direction in which the problems can be approached. Since there is now a major concern with the legitimacy and equity of policies, means have to be found to provide for these in the policy-making process. Encouragingly, this takes us into some familiar territory, such as public participation, mediation, and intergovernmental relations. Moreover, maintaining this book's focus on land use, urban development, and environmental issues helps to make the discussion manageable. It needs to be stressed, however, that the intention is not to provide a compelling, comprehensive approach even to these selected areas. Rather is it intended to provide some thoughts that emerge from the previous analyses which, it is submitted, are worth considering for the improvement

---

BOX 17.1   SOLUTIONS DEFINE PROBLEMS

What would be necessary in identifying the nature of the poverty problem? Does poverty mean low income? Yes, in part. But what are the determinants of low income? Is it deficiency of the national and regional economies, or is it deficiencies of cognitive and occupational skills within the labor force? If the latter, the problem statement and the problem 'solution' must encompass the educational process. But, then, where within the educational system does the real problem lie? What then might it mean to 'improve the educational system'?

Or does the poverty problem reside in deficient physical and mental health? If so, we must add those etiologies to our information package, and search inside the health services for a plausible cause. Does it include cultural deprivation? spatial distribution? problems of ego identity? deficient political and social skills? – and so on. If we can formulate the problem by tracing it to some sorts of sources – such that we can say 'Aha! That's the locus of the difficulty' – then we have thereby also formulated a solution. To find the problem is thus the same thing as finding the solution: the problem can't be defined until the solution has been found.

Source: Rittel and Weber 1973: 161

of the planning process. Added confidence is given to this claim by the fact that the thoughts are not new: some (such as public involvement) have been shown to have some modest successes, though others (such as regional government) have proved to be more problematic in practice than they are convincing in theory.

PROPERTY RIGHTS – 'WISE USE'

For the most part, Americans are not prepared to support a land use planning system that deprives them of control over development in their local area. There is a long tradition of belief in the sanctity of property rights. There is also a strong concern that those who are affected by government policy should be able to influence that policy; and where the effect is to impose an unreasonable burden on individuals, there should be relief from that hardship either by way of exception or by payment of compensation. The mainstream of land use planning is thus essentially local, with sensitivity to local opinion and the avoidance of hardship to individuals.

Zoning meets these narrow concerns neatly, and it specifically provides a safety valve with a system of variances in cases of hardship (discussed in Chapter 6). It is important to remember that the support for zoning stemmed from its essential role in the protection of property values. The broader conception of comprehensive planning as a framework for zoning signally failed to obtain similar support.

However, even the restricted conception of zoning has given rise to controversy over its operation. Protecting property rights also involves reducing those rights. Zoning at a low density protects property values of owners but at the same time precludes them from maximizing potential development values. Much conflict has occurred between owners who want to preserve the status quo and those who anticipate large increases in value due to changes in the local land market.

Such controversy has burgeoned with the growth of regulatory controls over environmental quality. These are now very wide in scope: in addition to more sophisticated zoning controls, they include restrictions aimed at preserving historic buildings, wildlife, coastlines, and wetlands. They pertain to waste disposal, the protection of groundwater, and air quality. They regulate commuting, parking, pollution, and aesthetics. Every new 'protection' can involve owners in loss of value (immediate or potential). Moreover, uncertainties abound: these include the extent to which loss is acknowledged by regulatory agencies; the degree to which relief from hardship is permitted; the amount of compensation which may be payable; and the indeterminate cost and outcome of litigation. Such unknowns have created a widespread anxiety, and have added to the long-standing opposition of those who are strongly opposed in principle to government regulation. As a result, a new 'property rights' movement has developed claiming guaranteed compensation for diminution in property values.

Between 1991 and 1995, forty-four states considered, and nine passed, legislation providing for such compensation – typically for cases where a regulation reduces potential value by half (Tibbetts 1995: 5). The support for such (tendentiously self-styled 'wise use') legislation comes from a wide range of opinion fueled by the fear of the uncertainties of the current situation as well as by the publicity attendant on a number of court cases. At the time of writing, however, most legislative proposals (including those before Congress) had been defeated. The middle ground points to the scope – and the need – for procedural remedies (akin to the zoning variance) that would deal with cases of real hardship, and for a range of legal and administrative reforms. Other issues in the debate range from the greater certainty which would be afforded by an extension of comprehensive plans, to the need to balance any system of compensation for 'wipeouts' with the 'logical corollary' of a system for capturing 'windfalls'. (The latter argument is the long-established one that if owners are to be compensated for serious losses in value due to government action, they should logically be required to pay for major increases in the value of their land caused by government action. In fact, neither policy is easy to devise and operate fairly.)

BEYOND LOCALISM

Though much of land use planning is essentially local in scope and intent, it is sometimes possible to marry local interests with wider concerns. Indeed, occasionally a local community can organize itself to accommodate a wider public good with profit to itself. Such is the case, for instance, with communities which accept hazardous waste disposal sites in return for community gains. In one case these amounted to '$4.2 million in benefits that would include a new town park, trust funds for the library and the fire department, road improvements, and a scholarship fund for local youths headed to college'. Additionally, the town would receive around $1.5 million more in taxes and fees, roughly double its current tax revenues (*New York Times*, 28 June 1991).

This is by no means a unique example. The problem of finding sites for nuclear waste has increased since the federal government mandated every state to store its waste within its own boundaries, a move necessitated by the closure to out-of-state waste of the only existing disposal sites, in South Carolina, Nevada, and Washington. Different states are following different tactics, and no doubt other bargain offers will be made. There is even a publication, the *Radioactive Exchange*, which is keeping track of efforts.

This may be a telling illustration of how a market works: increased prices bring about additional supply. But it remains to be seen whether it circumvents the difficulties of efficiency and equity that attend zoning, and indeed whether it works at all, other than in exceptional circumstances. And is there anyone asking whether the bargained site is desirable on wider social, economic, or environmental grounds? Surely this is more than a private matter between 'neighbors'?

A market mechanism operating in a different way is the neighborhood buyout, or 'assemblages' as they are known in the real estate world (Haar and Wolf 1989: 1094). There are a number of variations; in one, a group of home owners, realizing that commercial development is creeping up on them, get together and negotiate with the developer for the sale of the whole area. When this works, all home owners receive higher prices than they could have otherwise expected, while at the same time the deal also makes negotiations simpler for the developer, who may well end up paying more, but some time is saved and, of course, 'time is money'. Presumably, on completion of this market transaction the zoning designation is obligingly changed.

Examples of this nature do not take us very far, but they do illustrate a point: in deciding local futures, residents are concerned with the balance of costs and benefits. Much opposition stems from a view that a development proposal involves severe costs and few or no benefits. Alternatively, the costs which the local community are being asked to bear are unfair. Is there a mechanism which could translate this idea of local fairness into wider social advantage? A good test case is the siting of hazardous waste sites.

LOCAL ACCEPTANCE OF UNWANTED LAND USES

There are several reasons why the siting of hazardous waste facilities is problematic (discussed in Chapter 15), but there is no question that a crucially important one is public attitudes. These attitudes may be based on fear of the implications for health or property values (which may be quite rational), and it can be strengthened by a perception that other people's waste is being unfairly imposed on them. Two possible responses to this are to compensate affected localities to offset their costs; another is to attempt to distribute sites in a fair way.

Compensation for community disbenefits is a traditional economic response. In the exceptional case noted above, it is noteworthy that the scale of the compensation was very large. An experiment in Massachusetts foundered on this very issue. As part of state-mandated negotiations between facility developers and local communities, it was found that virtually no reasonable amount of compensation was sufficient to persuade local residents to accept sites. On the other hand, there is some evidence to suggest that offsetting compensation at an acceptable level might be negotiable as part of a wider package that includes a reduction in the risks (Hadden 1989: 52). Since mitigation of risks is, in any case, a constructive approach, the idea seems worthy of further study.

Fair shares for all is a morally attractive policy: who can object to a scheme in which all bear the unwanted costs of hazardous waste? Mention was made in Chapter 16 of the California YIMBY (yes-in-my-backyard) program which was to locate facilities in such a way as to ensure that each county would deal with the waste produced within its boundaries. Though accepted by all the counties, the scheme became a casualty of state politics; but it might have better luck elsewhere (Mazmanian and Morell 1992: 192–203).

These are small crumbs of comfort. Are people so blind, ignorant, and selfish that they will ignore the public necessity of finding adequate facilities for disposing of waste? A fundamental problem is the widespread distrust of governments, experts, business – indeed, everyone in power. Though community and environmental groups may employ their own experts to assist in their campaigns of opposition, this only serves (outside the courtroom) to exacerbate the difficulties. As Harvey Brooks (1984: 48) has put it, the conflict becomes among 'noncommunicating publics that each rely on different sources and talk to different experts. Thus many public policy discussions become dialogues of the deaf'. (See Box 17.2.)

These basic issues of democracy have long been debated: they acquire a new urgency in the complex, interdependent, global society of today. Since neither governments nor experts are apparently to be trusted, perhaps one way forward lies with a transformation of the environmental movement into one with a positive concern for seeking acceptable and workable solutions? But, sadly, the path may be too easily diverted into yet another area of expertise.

UNWANTED NEIGHBORS

If environmental issues arouse strong passions, unwanted social neighbors create fury. Three examples are discussed here: two which offer hope (day-care facilities and group homes), and one which seems almost impossible to resolve (low-income housing). The provision of day care has given rise to neighborhood opposition on account of traffic, disturbance, and the general effect on the character of the neighborhood. The large increase in the number of working mothers, and the changed attitudes to the propriety of this, have reduced this opposition. However, a number of states have found it necessary to pass legislation to prevent municipalities from zoning against child-care facilities in residential areas.

With group homes it seems clear that either an earnest and positive community consultation approach or the legal sanctions of the Fair Housing Act (or both) can be effective, though a neighborhood battle may be involved. Local fears can be allayed by frank and persuasive explanations of what is involved,

---

BOX 17.2   PUBLIC FEARS AND MISTRUST

It is understandable that local residents and facility sponsors would fail to arrive at common understandings and agreements on siting because they have focused on different aspects of the issue – technical capabilities of a facility versus the potential harm to an individual member of the community, for example – and talked past one another. At the root of the problem is a lack of shared understanding, perceived mutual interests, and trust. Still excluded from most decisions that lead to initiating proposed new facilities, people today can and do just say 'no'. The resulting gridlock inhibits commencement of any serious improvement in toxics policy.

Source: Mazmanian and Morell 1992: 242

Plate 39 Neighborhood Change in Newark, New Jersey: 1980
Courtesy of Camilo Vergara

Plate 40 Neighborhood Change in Newark, New Jersey: 1985
Courtesy of Camilo Vergara

Plate 41 Neighborhood Change in Newark, New Jersey: 1986
Courtesy of Camilo Vergara

Plate 42 Neighborhood Change in Newark, New Jersey: 1994
Courtesy of Camilo Vergara

who will live in the home, and how it will be controlled. Of course, this is easier in the case of homes for disabled children than it is for one for ex-prisoners. However, even with groups such as the latter, where there may be real neighborhood fear, good groundwork can allay apprehensions.

Low-income housing is a much more difficult issue. As with group homes, a major fear is a fall in property values, but the apprehensions go further: community character, unwelcome neighbors, as well as the usual litany of tax and overcrowding impacts (ranging from schools to the fire service). Whether or not these fears are justified, there is no doubt about the underlying opposition to people of a different color or class.

Given the strength of the prejudice against low-income housing and the underlying racial discrimination, there is no easy remedy. The sheer force of this opposition has to be appreciated; there are innumerable stories which demonstrate it. Historically, an important one dates from the 1960s when the federal government was under pressure from many sources (including the Justice Department, the Equal Employment Opportunities Commission, and many officials in Washington, as well as numerous groups such as Paul Davidoff's Suburban Action Institute) to 'open up the suburbs' to low-income groups. A scheme of rent supplements was introduced to enable low-income households to obtain access to federally subsidized housing built with FHA mortgages by bodies such as non-profit organizations. This was a neat way of bypassing local opposition since no formal participation by local government was necessary. In the words of one leading lobbyist, this would 'help penetrate the wall of exclusion erected by many suburban communities against the introduction of housing for low and moderate income families' (Keith 1973: 161).

In the Nixon administration, the program was coupled with *Operation Breakthrough*. This was a program aimed at facilitating the mass production of factory-built housing by large corporations. The 'breakthrough' was to be a major reduction in the cost of housing, from which large numbers would benefit. It was this program which federal officials stressed

at the local level - studiously avoiding any mention of race, though developers would be required to employ fair-marketing practices to maximize housing opportunities for minorities. The major emphasis, however, was on the benefits of *Operation Breakthrough* as well as generous federal grants for infrastructure and other worthy local causes. With these attractions, it was hoped that local governments would waive local zoning and building codes. It was not that easy. Though the initial reaction seemed promising, opposition mounted as it became clear that the program would involve subsidized housing for blacks. It reached fever pitch in the Detroit suburb of Warren where George Romney, the enthusiastic Secretary of HUD, met at first hand the fury of local people threatened by what they saw as federal enforcement of integration. The experience was referred to as 'the political education of George Romney', as a result of which he quickly changed his stance and argued that there was a danger of 'setting things back as a result of pushing too hard too fast' (Danielson 1976: 225).

One of the fears of white suburbs is that they may be overwhelmed by minorities. To allay this fear, while promoting a small degree of integration, various programs have been evolved. Several of these are based on the Gautreaux model. The name originates from a class action suit by Dorothy Gautreaux and other public-housing residents and applicants which alleged deliberate discrimination by the Chicago Housing Authority in siting public-housing developments and assigning tenants. (Virtually all public housing in Chicago was located in areas where more than a half of the residents were black.) After many years of legislation, the federal courts approved a plan to scatter public housing projects, tenants and applicants throughout the metropolitan area.

This effectively breached the barriers defended by suburban communities. The schemes include modest housing developments planned (in scale and design) to blend into the local area. There are also 'mobility' programs which provide Section 8 certificates for housing in areas where no more than 30 per cent of the population is black. Counseling is available to assist families to adjust to their new environment.

Between 1976 and 1994, such programs have

placed around 5,600 low-income black families, of which over a half have located in predominantly white neighborhoods. Research undertaken by the Center for Urban Research and Policy Studies at Northwestern University, has shown that these schemes can be successful: contrary to expectations, most black families are not isolated; racial harassment may be experienced by some at first, but this quickly diminishes; movers have better employment experience; and, after initial difficulty in adjusting to the higher standards, children do well in school. The most important lesson drawn from the research is that the program works because its effect on the racial composition of the neighborhood is negligible. Other areas are following the Gautreaux model (Joseph 1993).

Two other mobility programs were initiated by HUD: the Areawide Housing Opportunity Plan and the Regional Housing Mobility Program, both of which were designed to promote the voluntary cooperation of regional bodies and suburban governments in desegregating federally assisted housing. These were small in scale, and were abandoned by the Reagan administration. However, another program was authorized by Congress on a demonstration basis in 1991. This *Moving to Opportunity* program has the objective of assisting families to move from high to low poverty areas, where employment opportunities are much better. Such programs offer the prospect of modest progress in a field where progress has been rare, though their impact on the problems of the inner city seems unlikely to be significant.

THE ROLE FOR THE STATES

States have their own housing programs which range from construction to mortgage lending and overseeing local zoning provisions for affordable housing. Typically they involve concerns which are wider (or resources which are greater) than those of local government. New Jersey has a Fair Housing Act which established a Council on Affordable Housing: this body assesses regional housing needs and municipal 'fair shares'. Oregon's state planning agency reviews

local plans to ensure adequacy of planning provision for affordable housing. Illinois has a citizens' advisory body which reviews the effects of local zoning and building controls on the provision of affordable housing. Connecticut encourages regional agreements on low- and moderate-income housing (facilitated by financial incentives for infrastructure). It also has a system of appeals against local zoning decisions on housing. Several other states have a similar system, using either the courts or a state agency. Virginia requires local governments to assess the need for affordable housing as part of the comprehensive planning process.

Some of these mechanisms may be more impressive on paper than they are in fact, but they do point to an important role which state governments can play if they are prepared to accept the political responsibility – and if they can persuade their local governments to cooperate.

The inherent shortcomings of local government emerge from debates on a number of public services: housing, economic development, environmental planning, and transportation planning. The rationale for a larger authority was stated in magisterial terms by Madison over two centuries ago (see Box 17.3). Larger communities tend to embrace a wider range of interests, needs, and abilities, a greater concern for the common good, and a broader conception of the role and purpose of public policy. In short, they are less selfish. Of course, exclusionary attitudes do not wither away; but they do not flourish as they do when isolated from other influences. Instead, they are kept in rein by 'the sheer number whose concurrence is necessary'.

Unfortunately, this can be too simple a view of the matter, and increased size by itself is not necessarily enough. A large constituency may be nothing more than an assembly of small bodies, each of which works the political process to achieve its particular aims. Where this is so, mutual accommodation of interests forces out broader thinking. A large constituency may also be so homogeneous that its character is essentially narrow-minded. Such is the chemistry of democracy; but generally, the larger the authority, the more likely it is to represent a range of

---

BOX 17.3   MADISON ON GOVERNMENTS AND REPRESENTATION

The smaller the society, the fewer will be the distinct parties and interests composing it; the fewer probably will be the distinct parties and interests, the more infrequently will a majority be found of the same party; and the smaller the number of individuals composing a majority, and the smaller the compass within which they are placed, the more easily will they concert and execute their plans of oppression. Extend the sphere and you will take in a greater variety of parties and interests; you make it less probable that a majority of the whole will have a common motive to invade the rights of other citizens; or if such a common motive exists, it will be more difficult for all who feel it to discover their own strength and to act in unison with each other. Besides other impediments, it may be remarked that, where there is a consciousness of unjust or dishonorable purposes, communication is always checked by distrust in proportion to the number whose concurrence is necessary.

Source: James Madison, *The Federalist*, 1788

---

interests, and thus more predisposed to seek to widen public policy goals.

Given the insularity of local governments, there is a need for a compelling framework of social responsibility within which they should operate. This has two advantages. First, it provides policy objectives which have to be followed. These move to the forefront of political attention: they can develop, to use Donald Schon's (1971: 123) phrase, into 'ideas in good currency', and they become a spur to action (in much the same way as environmental protection is currently espoused by all). Second, it gives local representatives a defense against those who would have them work to a narrow interest: 'we have no alternative', the feint-hearted can plead, 'we must do what we are required to'.

Thus the rationale for policy can be a matter of high principle or of base political calculation: it makes no difference. What matters is that a public authority is operating in the broad public interest, not to the advantage of a powerful minority; and this means that it is including all needs in its civic policy.

Measures taken by states to rein in the narrow-minded policies of localities include a ban on commercial rent control, regularization of local impact fees, and overriding local prohibitions on manufactured housing and child-care facilities. (The federal government has acted in similar ways, of course, as with racial and other forms of discrimination.)

So far as land use planning is concerned, such an approach paves the way for devising a system in which zoning becomes the servant, rather than the master, of local land use planning. Here the local plans approved by a state or regional authority can assume great significance, as is the case in Oregon. Oregon provides a set of goals and procedures that have been agreed to be desirable for the future of the people of the state. These provide a policy basis for the preparation of local plans, which local governments are required to prepare and implement (see Chapter 11).

Goals, objectives, declarations and suchlike are, of course, the stuff of political discourse, and they can be totally devoid of meaning. There are many plans like this: they adorn the libraries of planning schools, and sometimes they are useful to a planning historian – as long as his or her concern is with plans and not with planning. Planning is a process, and like all processes must have an engine to drive it. To prepare plans and to articulate goals is one thing: to give them substance is another. Thus, without in any way belittling the symbolic value of declarations, it is also necessary to devise mechanisms to prompt, attract, bribe, and if necessary force, local governments to formulate and implement local plans which accord with the state goals.

The precise way in which this can be done will vary from state to state. Oregon has an effective system for consultation, agreeing plans, and enforcing and

---

BOX 17.4   SHORTCOMINGS OF LOCAL LAND USE CONTROL

1 The absence of a comprehensive planning framework (inevitable with 'non-comprehensive jurisdictional entities');
2 the predominance of municipal self-interest and the lack of a mechanism to allocate undesirable but socially necessary land uses to optimal sites;
3 the inherent inability of local governments to address larger environmental questions;
4 the essentially negative character of local controls. Little of a positive nature (e.g. affirmative action) can be achieved.

Source: Delogu 1984

---

reviewing plans by means of a Land Conservation and Development Commission and the Land Use Board of Appeals (Buchsbaum and Smith 1993: 53). This is one model, but it may not work elsewhere: each state has to evolve ways of devising, debating, and agreeing goals, and of implementing them.

PUBLIC PARTICIPATION AND PUBLIC CONFIDENCE

A recurring theme in this book has been that the effectiveness of policies depends upon public support. This support can be enthusiastic (though often not for long), or it can be simply passive, but in either case a degree of public understanding is necessary. As discussed in the previous chapter, this has become increasingly important as the limits (and limitations) of technical knowledge have become more widely appreciated. The trust in value-free scientific solutions is now on a par with that which is placed in political judgments: neither can be accepted at face value. At the same time, the limited authority of planners has been further reduced by a greater appreciation of the unknowability of the future. Since the future cannot be predicted, the role of planning is the modest one of attempting to accommodate change efficiently, to maximize benefits, and to minimize unwanted side-effects. The role of the planner thus becomes facilitative: to assist in the public debate on planning policies; to point up the costs and benefits of alternative courses of action; and to articulate the uncertainties which have to be faced.

This is no easy task, and there are moral dimensions which make it even more difficult. Problems of equity, of non-discrimination, of respect for differences of color and creed, of toleration for non-traditional life-styles: these and a host of other matters which are important in human relationships cannot be ignored – at least not in the long term.

The long term is seldom at the heart of debate: the short term is too urgent. Yet the cumulative effect of policies and prejudices can have disastrous effects. Poorer areas become increasingly deprived as their innate abilities are overwhelmed by the sheer enormity of their problems. The social divisions of American society are increasing as those who can escape to suburban locations continue to do so. Walled communities with private policing provide a physical haven of security which may prove to be unsustainable. The social balkanization of metropolitan areas guarantees only an uncertain truce between the advantaged and the disadvantaged.

Certainly, the problems of some inner cities are viewed by suburbanites as overwhelming. The frightening description of some of these as 'no-go areas' underlines both the severity of their problems and the perceived hopelessness of public action. But much of the descriptive language obscures the truth. The *magnitude* of the problems is greatly exaggerated by its *concentration*. As Henry Cisneros (1995) has pointed out, though Hartford, Connecticut, is one of America's most distressed cities, its problems are not overwhelming when viewed on a metropolitan scale (see Box 17.5). Cisneros presents an optimistic view, but there are many who may find it difficult to accept

---

BOX 17.5   HARTFORD'S METROPOLITAN POVERTY PROBLEM IS SOLUBLE

Hartford, the capital of Connecticut, has become one of America's most distressed cities. Between 1950 and 1990, the city's population dropped 21 per cent to 139,000. In 1989 city residents' average income was 53 per cent of suburbanites' income, and over 27 per cent of city residents were poor. Crime rates have soared, and school failure rates were so high that last year the city brought in a private management company to run the city's school system. Hartford's social agony seems unsolvable.

Yet viewed from a regional perspective, problems in the 1-million Hartford metropolitan area are not so unsurmountable: out of every 100 residents only 3 are poor and white, 2 are poor and Hispanic, and less than 2 are poor and African American. Poor whites are scattered throughout the metropolitan areas: only 12 per cent live in Hartford, and only 13 per cent live in poor neighborhoods. By contrast, 76 per cent of poor Hispanics and 80 per cent of poor African Americans live in city neighborhoods, and nearly 9 out of 10 poor minorities live in neighborhoods of concentrated poverty.

The problem is not the region's overall level of poverty – only seven out of every hundred area residents are poor – but the high concentration of minority poor in inner-city areas. Viewed in that light, greater Hartford is capable of absorbing poor minorities into the region's prosperous, middle-class society just as it already integrates poor whites into that society.

Source: Cisneros 1995: 10

---

this. For example, in his book *Understanding Urban Unrest*, Gale (1996) argues that effective policies for the urban poor require fundamental changes in the economic structuring of the country and a new federal agenda. Like many commentators, he sees few signs of this coming to pass.

IN CONCLUSION

This title of this chapter was chosen purposely. It has been no part of the objective of the chapter (or of the book) to set out a program of reform. Others have embarked on this difficult task with careful thought and enthusiasm. (Recent notable examples have included Calthorpe 1993, Cisneros 1993, Downs 1994, Rusk 1993; and many more will follow.) The present intention has been to highlight a number of questions which arise in designing and tackling policies of reform. Several matters stand out in importance: freedom of information; public debate and involvement in the planning process; establishing appropriate mechanisms for dealing with different types of issue; seeking methods of persuasion and

mediation where there is reluctance to abide by policies for the common good (and, where such methods are inadequate, designing regulatory instruments).

The overriding objective, however, has been to demonstrate the huge difficulties involved in dealing with urbanization, transportation, land use planning, discrimination, and those complexes of issues which are labeled 'urban' or simply 'social'. It is because so much effort is put into trying to solve huge insoluble matters that insufficient attention is given to the more restricted issues on which progress can be made.

Fortunately, there are some who will find this approach far too clinical and unsatisfying. They will strive passionately for reform with a disdain for the difficulties which this book has identified. Such activists have a crucial role to play in the political process.

Plate 43 Pittsburgh Renaissance
Courtesy Alex MacLean/Landslides

FURTHER READING

On the nature of policy 'problems' and the difficulties that these pose for the policy process, see Rittel and Webber (1973) 'Dilemmas in a general theory of planning'; Kingdon (1984) *Agendas, Alternatives and Public Policies*; and Wildavsky (1987) *Speaking Truth to Power*. Downs' 1972 paper 'Up and down with ecology – the *issue–attention* cycle', is an interesting analysis of the fickleness of public concern.

Group homes are dealt with in Jaffe and Smith (1986) *Siting Group Homes for Developmentally Disabled Persons*. On child-care facilities, see Cibulskis and Ritzdorf (1989) *Zoning for Child Care*.

An excellent review of the issues involved in current debates on the taking issue and compensation for affected property owners is provided by Yandle (1995) *The Property Rights Rebellion: Land Use Movements in the 1990s*, and by Strong *et al.* (1996) 'Property rights and takings'. See also Tibbetts (1995) 'Everybody's taking the Fifth'. A collection of papers on the legal issues involved is Callies (1996) *Takings: Land-development Conditions and Regulatory Takings after Dollan and Lucas*.

A major study of class and racial prejudice is Danielson (1976) *The Politics of Exclusion* which, though not up to date, provides an excellent perspective of the 1960s and early 1970s. The details have changed since then, but the basic themes are the same. Just how little has changed can be judged from Keating (1994) *The Suburban Racial Dilemma: Housing and Neighborhoods*.

The chapter contains a quotation from Cisneros (1995) *Regionalism: The New Geography of Opportunity*, but more easily obtainable is the same author's edited collection of essays: *Interwoven Destinies* (1993). There is a vast literature on the problems of the inner city, segregation and discrimination. A useful overview is given in Downs (1994) *New Visions for Metropolitan America*. See also Gale (1996) *Understanding Urban Unrest*, and Darby and Anderson (1996) *Reducing Poverty in America*. A collection of papers on various aspects of urban policy is to be found in a special issue of *Urban Affairs*, May 1995.

QUESTIONS TO DISCUSS

1 'Since the problems are so difficult to define, wouldn't it be better simply to get on with doing something constructive, rather than escaping into clever arguments?' Discuss.

2 Why does public interest in important problems wax and wane?

3 Do you think that it is a good idea to compensate residents who have obnoxious hazardous waste facilities located close to them?

4 'Forcing local governments to allow child-care facilities is acceptable; but it is not right that locally elected representatives should be made to accept low-income housing.' Discuss.

5 What roles can states play in urban planning?

6 How far do you think that resolution of urban problems must involve the reduction of discrimination?

MAIN CASES

REFERENCES

Abbott, C. (1991) 'Urban design in Portland, Oregon, as policy and process, 1960–1989', *Planning Perspectives* 6: 1–18

Abbott, C., Howe, D. and Adler, S. (eds) (1994) *Planning the Oregon Way: A Twenty-Year Evaluation*, Corvallis, OR: Oregon State University Press

Abbott, W. W., Moe, M. E., and Hanson, M. (1993) *Public Needs and Private Dollars: A Guide to Dedications and Development Fees*, Point Arena, CA: Solano Press (*Supplement* 1995)

Ackerman, B. A. and Hassler, W. T. (1981) *Clean Coal, Dirty Air, or How the Clean Air Act Became a Multibillion Dollar Bail-out for High-Sulphur Coal Producers and What Should Be Done About It*, New Haven, CT: Yale University Press

Adler, R. (1993) *The Clean Water Act: 20 Years Later*, Washington, DC: Island Press

Adler, R. (1994) 'The Clean Air Act: has it worked?', *EPA Journal* 20 (1/2): 10–14

Advisory Commission on Intergovernmental Relations (1964) *Impact of Federal Urban Development Programs on Local Government Organization and Planning*, Washington, DC: US Government Printing Office

Advisory Committee on Regulatory Barriers to Affordable Housing (1991) *Not in My Backyard*, Washington, DC: US Government Printing Office

Advisory Council on Historic Preservation (annual) *Report to the President and Congress*, Washington, DC: ACHP

Alexander, E. R. (1981) 'If planning isn't everything, maybe it's something', *Town Planning Review* 52:131–42

Alexander, E. R. (1992) *Approaches to Planning: Introducing Current Planning Theories, Concepts and Issues* (second edition), Philadelphia: Gordon & Breach

Allen, G., Alley, D., and Hicks, E. (1994) *Development, Marketing, and Operation of Manufactured Home Communities*, Washington, DC: Urban Land Institute

Altshuler, A. A. (1965) *The City Planning Process: A Political Analysis*, Ithaca, NY: Cornell University Press (There is an extract from this book in Stein 1995)

American Farmland Trust (1988) *Protecting Farmland through Purchase of Development Rights: The Farmers' Perspective*, Washington, DC: American Farmland Trust

American Planning Association (1994) *Planning and Community Equity*, Chicago: Planners Press

Ames, D. L., Callahan, M. H., Herman, B. L., and Siders, R. J. (1989) *Delaware Comprehensive Historic Preservation Plan*, Newark, DE: Center for Historic Architecture and Engineering, University of Delaware

Anderson, L. T. (1995) *Guidelines for Preparing Urban Plans*, Chicago: Planners Press

Audirac, I., Shermyen, A. H., and Smith, M. T. (1990) 'Ideal urban form and visions of the good life: Florida's growth management dilemma', *Journal of the American Planning Association* 56: 470–82

Ausubel, J. H. and Sladovich, H. E. (eds) (1989) *Technology and Environment*, Washington, DC: National Academy Press

Babcock, R. F. (1966) *The Zoning Game: Municipal Practices and Policies*, Madison, WI: University of Wisconsin Press

Babcock, R. F. and Larsen, W. U. (1990) *Special Districts: The Ultimate in Neighborhood Zoning*, Cambridge, MA: Lincoln Institute of Land Policy

Babcock, R. F. and Siemon, C. L. (1985) *The Zoning Game Revisited*, Boston, MA: Oelgeschlager, Gunn & Hain

Bair, F. H. (1984) *The Zoning Board Manual*, Chicago: Planners Press

Banach, M. and Canavan, D. (1987) 'Montgomery County agricultural preservation program', in Brower and Carol (1987)

Banfield, E. C. (1959) 'Ends and means in planning', *International Social Science Journal* 11; reprinted in Faludi (1973)

Banfield, E. C. (1961) 'The political implications of metropolitan growth', *Daedalus* 90: 61–78

Banfield, E. C. (1975) 'Corruption as a feature of governmental organization', *Journal of Law and Economics* 18: 587–615

Barnekov, T., Boyle, R., and Rich, D. (1989) *Privatism and Urban Policy in Britain and the United States*, New York: Oxford University Press

Barnett, J. (1982) *An Introduction to Urban Design*, New York: Harper & Row

Barrett, S. and Hill, M. J. (1993) unpublished research report quoted in Ham, C. and Hill, M. *The Policy Process in the Modern Capitalist State*, New York: Simon & Schuster

Bartik, T. J. (1991) *Who Benefits from State and Local Economic Development Policies?*, Kalamazoo, MI: W. E. Upjohn Institute for Employment Research

Barton, S. E. and Silverman, C. J. (eds) (1994) *Common Interest Communities: Private Governments and the Public Interest*, Berkeley, CA: Institute of Governmental Studies Press

Bass, R. E. and Herson, A. I. (1993) *Mastering NEPA: A Step-by-Step Approach*, Point Arena, CA: Solano Press

Bassett, E. M. (1940) *Zoning: The Laws, Administration, and Court Decisions during the First Twenty Years* (revised edition), New York: Russell Sage Foundation

Bay Vision 2020 Commission (1991) *Bay Vision 2020: The Commission Report*, San Francisco: The Commission

Beatley, T. (1988) 'Ethical issues in the use of impact fees to finance community growth', in Nelson (1988)

Beatley, T. (1994) *Ethical Land Use: Principles of Policy and Planning*, Baltimore, MD: Johns Hopkins University Press

Berger, L. (1991) 'Inclusionary zoning as takings: the legacy of the *Mount Laurel* cases', *Nebraska Law Review* 70: 186–228

Birch, E. L. and Roby, D. (1984) 'The planner and the preservationist: an uneasy alliance', *Journal of the American Planning Association* 50: 194–207

Bishir, C. W. (1989) 'Yuppies, Bubbas, and the politics of culture', in Carter and Herman (1989)

Blair, J. P. (1995) *Local Economic Development: Analysis and Practice*, Thousand Oaks, CA: Sage

Blake, P. (1964) *God's Own Junkyard: The Planned Deterioration of America's Landscape*, New York: Holt, Rinehart & Winston

Blakely, Edward, J. (1994) *Planning Local Economic Development: Theory and Practice* (second edition), Thousand Oaks, CA: Sage

Blank, L. D. (1994) 'Seeking solutions to environmental inequity: the Environmental Justice Act', *Environmental Law* 24: 1109–36

Blumenthal, S. and Siler, B. (1990) *Tax Incentives for Rehabilitating Historic Buildings: Fiscal Year 1989 Analysis*, Washington, DC: National Trust for Historic Preservation

Boorstin, D. (1958) *The Americans: The Colonial Experience*, New York: Random House

Boorstin, D. (1965) *The Americans: The National Experience*, New York: Random House

Boorstin, D. (1973) *The Americans: The Democratic Experience*, New York: Random House

Bosselman, F. (1973) 'Can the town of Ramapo pass a law to bind the rights of the whole world?', *Florida State University Law Review* 1: 234–65

Bosselman, F. and Callies, D. (1972) *The Quiet Revolution in Land Use Control*, Washington, DC: Council on Environmental Quality

Bosselman, F., Callies, D., and Banta, J. (1973) *The Taking Issue: An Analysis of the Constitutional Limits of Land Use Control*, Washington, DC: Council on Environmental Quality

Boston, City of (1988) *Design Guidelines for Neighborhood Housing*, Boston, CA: City of Boston Public Facilities Department

Bourne, L. S. (1981) *The Geography of Housing*, London: Edward Arnold

Boyer, M. C. (1983) *Dreaming the Rational City: The Myth of American City Planning*, Cambridge, MA: MIT Press

Brogan, H. (1986) *The Pelican History of the United States of America*, New York: Viking Penguin

Brooks, A. V. N. (1989) 'The office file box – emanations from the battlefield' [on the *Euclid* case], in Haar and Kayden (1989a)

Brooks, H. (1984) 'The resolution of technically intensive public policy disputes', *Science, Technology and Human Values* 9 (1)

Brooks, H. (1988) Foreword to Graham, J. D., Green, L., and Roberts, M. J. *In Search of Safety: Chemicals and Cancer Risk*, Cambridge, MA: Harvard University Press

Brower, D. J. and Carol, D. S. (eds) (1987) *Managing Land Use Conflicts: Case Studies in Special Area Management*, Durham, NC: Duke University Press

Bryner, G. C. (1993) *Blue Skies, Green Politics: The Clean Air Act of 1990*, Washington, DC: Congressional Quarterly Press

Buchsbaum, P. A. and Smith, L. J. (eds) (1993) *State and Regional Comprehensive Planning: Implementing New Methods for Growth Management*, Chicago: American Bar Association

Bucknall, B. (1988) *Of Deals and Distrust: The Perplexing Perils of Municipal Zoning*, Toronto: Canadian Bar Association. unpublished paper

Burchell, R. W. and Sternlieb, G. (1978) *Planning Theory in the 1980s*, New Brunswick, NJ: Center for Urban Policy Research, Rutgers University Press

Burchell, R. W., Listokin, D., and Pashman, A. (1994) *Regional Housing Opportunities for Lower-income Households: An Analysis of Affordable Housing and Regional Mobility Strategies*, New Brunswick, NJ: Center for Urban Policy Research, Rutgers University Press

Burrows, T. (ed.) (1989) *A Survey of Zoning Definitions*, Planning Advisory Service Report 421, Chicago: American Planning Association

Cairncross. F. (1992) *Costing the Earth: The Challenge for Governments, The Opportunities for Business*, Boston, MA: Harvard Business School Press

Callies, D. L. (1980) 'The quiet revolution revisited', *Journal of the American Planning Association* 46: 135–44

Callies, D. L. (1984) *Regulating Paradise: Land Use Controls in Hawaii*, Honolulu: University of Hawaii Press

Callies, D. L. (1994) *Preserving Paradise: Why Regulation Won't Work*, Honolulu: University of Hawaii Press

Callies, D. L. (ed.) (1996) *Takings: Land-development Conditions and Regulatory Takings after Dolan and Lucas*, Chicago: American Bar Association

Callies, D. L. and Grant, M. (1991) 'Paying for growth and planning gain: an Anglo-American comparison of development conditions, impact fees, and development agreements', *Urban Lawyer* 23: 221 (also reprinted in Freilich and Bushek 1995)

Callies, D. L., Freilich, R. H., and Roberts, T. E. (1994) *Land Use: Cases and Materials* (second edition), St Paul, MN: West

Calthorpe, P. (1993) *The Next American Metropolis: Ecology, Community, and the American Dream*, Princeton, NJ: Princeton Architectural Press

Campbell, S. and Fainstein, S. (eds) (1996) *Readings in Planning Theory*, Cambridge, MA: Blackwell (See also companion volume on urban theory edited by Fainstein and Campbell 1996)

Carlin, A. (1973) 'The Grand Canyon controversy; or how reclamation justifies the unjustifiable', in Enthoven, A. and Freeman, A. M. *Pollution, Resources, and the Environment*, New York: W. W. Norton

Carson, R. (1962) *Silent Spring*, New York: Houghton Mifflin (Penguin Edition 1965, reprinted 1991) (There is an extract from this in Stein 1995)

Carter, T. and Herman, B. L. (1989) *Perspectives in Vernacular Architecture III*, Columbia, MI: University of Missouri Press

Caves, R. W. (1992) *Land Use Planning: The Ballot Box Revolution*, Newbury Park, CA: Sage

Cervero, R. (1989) *America's Suburban Centers: The Land Use–Transportation Link*, Boston, MA: Unwin Hyman

Cervero, R. (1994) 'Transit-based housing in California: evidence on ridership impacts', *Transport Policy* 1: 174–83

Champion, A. G. (ed.) (1989) *Counterurbanization: The Changing Pace and Nature of Population Deconcentration*, London: Edward Arnold

Checkoway, B. (ed.) (1994) 'Paul Davidoff and advocacy planning in retrospect', *Journal of the American Planning Association* 60: 139–61

Cheney, C. H. (1920) 'Zoning in practice', *National Municipal Review* 9: 31–43

Chinitz, B. (1990) 'Growth management: good for the town, bad for the nation?', *Journal of the American Planning Association* 56: 3–8

Cibulskis, A. and Ritzdorf, M. (1989) *Zoning for Child Care*, Planning Advisory Service Report 422, Chicago: American Planning Association

Cisneros, H. G. (ed.) (1993) *Interwoven Destinies: Cities and the Nation*, New York: Norton

Cisneros, H. G. (1995) *Regionalism: The New Geography of Opportunity*, Washington, DC: US Department of Housing and Urban Development

Clavel, P., Forester, J., and Goldsmith, W. (eds) (1980) *Urban and Regional Planning in an Age of Austerity*, Oxford: Pergamon

Cohen, R. E. (1992) *Washington at Work: Back Rooms and Clean Air*, New York: Macmillan

Cole, L. A. (1993) *Element of Risk: The Politics of Radon*, New York: Oxford University Press

Commoner, B. (1971) *The Closing Circle: Nature, Man and Technology*, New York: Alfred A. Knopf

Commoner, B. (1990) *Making Peace with the Planet*, New York: Pantheon

Commoner, B. (1994) 'Pollution prevention: putting comparative risk assessment in its place', in Finkel and Golding (1994)

Costonis, J. J. (1989) *Icons and Aliens: Law, Aesthetics, and Environmental Change*, Urbana, IL: University of Illinois Press

Council of State Community Development Agencies (1994) *Making Housing Affordable: Breaking Down Regulatory Barriers. A Self-assessment Guide for States*, Washington, DC: US Department of Housing and Urban Development

Cowden, R. (1995) 'Power to the zones: HUD offers a new twist on an old standby', *Planning* 61 (February): 8–10

Dales, J. H. (1968) *Pollution, Property and Prices: An Essay in Policy-making and Economics*, Toronto: University of Toronto Press

Daniels, T. L. (1991) 'The purchase of development rights: preserving agricultural land and open space', *Journal of the American Planning Association* 57: 421–31

Danielson, M. N. (1976) *The Politics of Exclusion*, New York: Columbia University Press

Darby, M. R. and Anderson, J.E. (eds) (1996) *Reducing Poverty in America*, Thousand Oaks, CA: Sage

Davidoff, P. (1965) 'Advocacy and pluralism in planning', *Journal of the American Institute of Planners* 31: 331–8; reprinted in Faludi (1973), Stein (1995), and Campbell and Fainstein (1996)

Davidoff, P. (1975) 'Working toward redistributive justice', *Journal of the American Planning Association* 41: 317–18

Dear, M. J. and Wolch, J. R. (1987) *Landscapes of Despair: From Deinstitutionalization to Homelessness*, Princeton, NJ: Princeton University Press

DeGrove, J. M. (1984) *Land Growth and Politics*, Chicago: Planners Press

DeGrove, J. M. and Miness, D. A. (1992) *The New Frontier for Land Policy: Planning and Growth Management in the States*, Cambridge, MA: Lincoln Institute of Land Policy

Delogu, O. E. (1984) 'Local land use control: an idea whose time is passed', *Maine Law Review* 36: 261–310

Derthick, M. (1972) *New Towns In-town: Why a Federal Program Failed*, Washington, DC: Urban Institute

Downs, A. (1962) 'The law of peak-hour expressway convergence', *Traffic Quarterly* 16: 393–409

Downs, A. (1967) *Inside Bureaucracy*, Boston, MA: Little, Brown

Downs, A. (1970) *Urban Problems and Prospects*, Chicago: Markham

Downs, A. (1972) 'Up and down with ecology – the issue–attention cycle', *Public Interest* 28 (Summer): 38–50

Downs, A. (1992) *Stuck in Traffic: Coping with Peak-hour Traffic Congestion*, Washington, DC: Brookings Institution, and Cambridge, MA: Lincoln Institute of Land Policy

Downs, A. (1994) *New Visions for Metropolitan America*, Washington, DC: Brookings Institution, and Cambridge MA: Lincoln Institute of Land Policy

Dubos, R. (1968) *So Human an Animal*, New York: Charles Scribner's Sons

Duerksen, C. J. (ed.) (1983) *A Handbook on Historic Preservation Law*, Washington, DC: Conservation Foundation and National Center for Preservation Law

Duerksen, C. J. (ed.) (1986) *Aesthetics and Land Use Controls: Beyond Ecology and Economics*, Planning Advisory Service Report 399, Chicago: American Planning Association

Dwight, P. (ed.) (1992) *Landmark Yellow Pages: Where to Find All the Names, Addresses, Facts, and Figures You Need*, Washington, DC: Preservation Press

Ehrlich, P. (1968) *The Population Bomb*, New York: Ballantine

Eisinger, P. (1988) *The Rise of the Entrepreneurial State: State and Local Economic Development Policy in the United States*, Racine, WI: University of Wisconsin Press

Elazar, D. J. (1984) *American Federalism: A View from the States*, New York: Harper & Row

Ellickson, R. C. (1977) 'Suburban growth controls: an economic and legal analysis', *Yale Law Journal* 86: 385–511

Elliott, M. L. P. (1984) 'Improving community acceptance of hazardous waste facilities through alternative systems of mitigating and managing risk', *Hazardous Waste* 1: 397–410

Erie, S. P. (1988) *Rainbow's End: Irish-Americans and the Dilemmas of Urban Machine Politics, 1840–1985*, Berkeley, CA: University of California Press

Etzioni, A. (1967) 'Mixed-scanning: a "third" approach to decision-making', *Public Administration Review*, December 1967 (reprinted in Faludi 1973)

Fainstein, S. and Campbell, S. (eds) (1996) *Readings in Urban Theory*, Cambridge, MA: Blackwell (see also the companion volume on planning theory, edited by Campbell and Fainstein 1996)

Fainstein, S. S., Fainstein, N. I., Smith, M. P., Hill, R. C., and Judd, D. R. (1986) *Restructuring the City: The Political Economy of Urban Development* (second edition), New York: Longman

Faludi, A. (1973) *A Reader in Planning Theory*, Oxford: Pergamon

Farley, R. (ed.) (1995) *State of the Union: America in the 1990s*, vol. 2: *Social Trends*, New York: Russell Sage Foundation

Feagin, J. R. (1989) 'Arenas of conflict: zoning and land use reform in critical economic perspective', in Haar and Kayden (1989a)

Ferguson, E. (1990) 'Transportation demand management', *Journal of the American Planning Association* 56: 442–56

Findlay, R. W. and Farber, D. A. (1992) *Environmental Law in a Nutshell* (third edition), St Paul, MN: West

Finkel, A. M. and Golding, D. (eds) (1994) *Worst Things First? The Debate over Risk-based National Environmental Priorities*, Washington, DC: Resources for the Future

Fiorino, D. J. (1995) *Making Environmental Policy*, Berkeley, CA: University of California Press

Fischel, W. A. (1982) 'The urbanization of agricultural land: a review of the National Agricultural Lands Study', *Land Economics* 58: 236–59

Fischel, W. A. (1990) *Do Growth Controls Matter? A Review of Empirical Evidence*, Cambridge, MA: Lincoln Institute of Land Policy

Fischer, M. L. (1985) 'California's coastal program: larger-than-local interests built into local plans', *Journal of the American Planning Association* 51: 312–21

Fishman, R. (1987) *Bourgeois Utopias: The Rise and Fall of Suburbia*, New York: Basic Books

Fitch, J. M. (1990) *Historic Preservation: Curatorial Management of the Built World*, Charlottesville, VA: University Press of Virginia

Flack, T. A. (1986) '*Euclid* v. *Ambler*: a retrospective', *Journal of the American Planning Association* 52: 326–37

Floyd, C. F. (1979a) 'Billboard control under the Highway Beautification Act', *Real Estate Appraiser and Analyst* July–August: 19–26

Floyd, C. F. (1979b) 'Billboard control under the Highway Beautification Act – a failure of land use controls', *Journal of the American Institute of Planners* 45: 115–26

Floyd, C. F. (1979c) *Highway Beautification: The Environmental Movement's Greatest Failure*, Boulder, CO: Westview

Freeman, A. M. (1990) 'Water pollution policy', in Portney (1990)

Freilich, R. H. and Bushek, D. W. (1995) *Exactions, Impact Fees and Dedications: Shaping Land-use Development and Funding Infrastructure*, Chicago: American Bar Association

Frey, W. H. (1989) 'United States: counterurbanization and metropolis depopulation', in Champion (1989)

Frey, W. H. (1994a) 'The new urban revival in the United States', in R. Paddison, J. Money, and B. Lever, *International Perspectives in Urban Studies 2*, London: Jessica Kingsley

Frey, W. H. (1994b) 'Minority suburbanization and continued "white flight" in US metropolitan areas: assessing findings from the 1990 Census', *Research in Community Sociology* 4: 15–42

Frey, W. H. (1995) 'The new geography of population shifts: trends toward balkanization', in Farley (1995)

Frey, W. H. and Fielding, E. L. (1994) *New Dynamics of Urban–Suburban Change: Immigration, Restructuring and Racial Separation*, Ann Arbor, MI: Population Studies Center, University of Michigan

Frey, W. H. and Speare, A. (1992) 'The revival of metropolitan growth in the US: an assessment of findings from the 1990 census', *Population and Development Review* 18: 129–46

Frieden, B. J. (1979) *The Environmental Protection Hustle*, Cambridge, MA: MIT Press

Frieden, B. J. and Kaplan, M. (1977) *The Politics of Neglect: Urban Aid from Model Cities to Revenue Sharing* (second edition), Cambridge, MA: MIT Press

Frieden, B. J. and Sagalyn, L. B. (1989) *Downtown Inc: How America Rebuilds Cities*, Cambridge, MA: MIT Press

Friedmann, J. (1987) *Planning in the Public Domain: From Knowledge to Action*, Princeton, NJ: Princeton University Press (There is an extract from this book in Stein 1995)

Fulton, W. (1990) 'The trouble with slow-growth politics: it wins elections, but the subdivisions keep going up', *Governing* 3 (7): 27–33

Fulton, W. (1991) *Guide to California Planning*, Point Arena, CA: Solano Press

Gale, D. E. (1996) *Understanding Urban Unrest: From Reverend King to Rodney King*, Thousand Oaks, CA: Sage

Galster, G. (ed.) (1996) *Reality and Research: Social Science and US Urban Policy since 1960*, Washington, DC: Urban Institute Press

Garner, J. F. and Callies, D. L. (1972) 'Planning law in England and Wales and in the United States', *Anglo-American Law Review* 1: 292–334

Garreau, J. (1991) *Edge City: Life on the New Frontier*, New York: Doubleday

Gelfand, M. I. (1975) *A Nation of Cities: The Federal Government and Urban America, 1933–1965* New York: Oxford University Press

Gerckens, L. C. (1988) 'Historical development of American city planning', in So and Getzels (1988)

Getzels, J. and Jaffe, M. (1988) *Zoning Bonuses in Central Cities*, Planning Advisory Service, Report 410, Chicago: American Planning Association,

Gilbert, J. L. (1995) 'Selling the city without selling out: new legislation on development incentives emphasize account-ability', *Urban Lawyer* 27: 427–93

Glaab, C. N. and Brown, A. T. (1983) *A History of Urban America* (third edition), New York: Macmillan

Godschalk, D. R., Brower, D. J., McBennett, L. D., Vestal, B. A., and Herr, D. C. (1979) *Constitutional Issues of Growth Management* (revised edition), Chicago: Planners Press

Goldberger, P. (1981) *The Skyscraper*, New York: Alfred A. Knopf

Goldman, S. and Jahnige, T. P. (1985) *The Federal Courts as a Political System*, New York: Harper & Row

Gordon, R. J. and Gordon, L. (1990) 'Neighborhood responses to stigmatized urban facilities', *Journal of Urban Affairs* 12: 437–47

Gottlieb, R. (1993) *Forcing the Spring: The Transformation of the American Environmental Movement*, Washington, DC: Island Press

Green, R. E. (ed.) (1991) *Enterprise Zones: New Directions in Economic Development*, Newbury Park, CA: Sage

Grigsby, W. G. (1963) *Housing Markets and Public Policy*, Philadelphia: University of Pennsylvania Press

Grigsby, W. G., Baratz, M., Galster, G., and Maclennan, D. (1987) 'The dynamics of neighborhood change and decline', *Progress in Planning* 28 (1): 1–76

Haar, C. M. and Kayden, J. S. (1989a) *Zoning and the American Dream: Promises Still to Keep*, Chicago: Planners Press

Haar, C. M. and Kayden, J. S. (1989b) *Landmark Justice: The Influence of William J. Brennan on America's Communities*, Washington, DC: Preservation Press

Haar, C. M. and Wolf, M. A. (1989) *Land Use Planning: A Case Book on the Use, Misuse, and Re-use of Urban Land* (fourth edition), Boston, MA: Little, Brown

Habe, R. (1989) 'Public design control in American communities', *Town Planning Review* 60: 195–219

Hadden, S. G. (1989) *A Citizen's Right to Know: Risk Communication and Public Policy*, Boulder, CO: Westview

Hahn, R. W. (1994) 'United States environmental policy: past, present and future', *Natural Resources Journal* 34: 305–48

Haider, D. H. (1974) *When Governments Come to Washington: Governors, Mayors, and Intergovernmental Lobbying*, New York: Free Press

Hall, P. (1988) *Cities of Tomorrow: An Intellectual History of Urban Planning and Design in the Twentieth Century*, New York: Blackwell

Hall, P. (1992) *Urban and Regional Planning* (third edition), New York: Routledge, Chapman & Hall

Hambleton, R. and Thomas, H. (1995) *Urban Policy Evaluation: Challenge and Change*, London: Paul Chapman

Handy, S. L. and Mokhtarian, P. L. (1995) 'Planning for tele-commuting: measurement and policy issues', *Journal of the American Planning Association* 61: 99–111

Harrigan, J. J. (1993) *Political Change in the Metropolis* (fifth edition), New York: HarperCollins

Hartman, C. (1994) 'On poverty and racism, we have had little to say', *Journal of the American Planning Association* 60: 158–9

Healey, R. G. and Rosenberg, J. S. (1979) *Land Use and the States*, Baltimore, MD: Johns Hopkins University Press

Hedman, R. and Jaszewski, A. (1984) *Fundamentals of Urban Design*, Chicago: Planners Press

Hendler, S. (ed) (1995) *Planning Ethics: A Reader in Planning Theory, Practice, and Education*, Piscataway, NJ: Center for Urban Policy Research, Rutgers University

Hetzel, O. (1994) 'Some historical lessons for implementing the Clinton Administration's empowerment zones and enterprise communities program: experiences from the model cities program', *Urban Lawyer* 26: 63–81

Heyman, I. M. and Gilhool, T. K. (1964) 'The constitutionality of imposing increased community costs on new subdivision residents through subdivision exactions', *Yale Law Journal* 73: 1119–57

Hiemstra, H. and Bushwick, N. (eds) (1989) *Plowing the Urban Fringe: An Assessment of Alternative Approaches to Farmland Preservation*, Miami, FL: Florida Atlantic University and Florida International University Joint Center for Environmental and Urban Problems

Hill, R. C. (1986) 'Crisis in the Motor City: the politics of economic development in Detroit', in Fainstein *et al.* (1986)

Hiscock, K. (1995) 'Groundwater pollution and protection', in O'Riordan (1995)

Hofstadter, R. (1948) *The American Political Tradition*, New York: Vintage

Hornstein, D. T. (1994) 'Paradigms, process, and politics: risk and regulatory design', in Finkel and Golding (1994)

Hosmer, C. F. (1965) *Presence of the Past: A History of the Preservation Movement in the United States before Williamsburg*, New York: G. P. Putnam's Sons

Hosmer, C. F. (1981) *Preservation Comes of Age: From Williamsburg to the National Trust, 1926–1949*, Charlottesville, VA: University Press of Virginia

Hough, D. E. and Kratz, C. G. (1983) 'Can "good" architecture meet the market test?' *Journal of Urban Economics* 14: 40–54

Howe, E. (1994) *Acting on Ethics in City Planning*, New Brunswick, NJ: Center for Urban Policy Research, Rutgers University

Hubbard, T. K. and Hubbard, H. V. (1929) *Our Cities of Today and Tomorrow: A Survey of Planning and Zoning Progress in the United States*, Cambridge, MA: Harvard University Press

Ingram, H. M. and Mann, D. E. (1984) 'Preserving the Clean Water Act: the appearance of environmental victory', in

Vig, N. J. and International City Management Association *Transportation Planning under ISTEA*, Washington, DC: ICMA

International Joint Commission (on the Great Lakes) (1993) *A Strategy for Virtual Elimination of Persistent Toxic Substances*, Windsor, Ontario: IJC

International Joint Commission (on the Great Lakes) (1994) *Seventh Biennial Report on Great Lakes Water Quality*, Washington, DC: IJC

Jackson, K. T. (1985) *Crabgrass Frontier: The Suburbanization of the United States*, New York: Oxford University Press

Jaffe, M. and Smith, T. P. (1986) *Siting Group Homes for Developmentally Disabled Persons*, Planning Advisory Service Report 397, Chicago: American Planning Association

Jefferson, T (1892–9) *The Writings of Thomas Jefferson*, (ed. P. L. Ford), New York: G. Putnam's Sons

Joseph, L. B. (ed.) (1993) *Affordable Housing and Public Policy: Strategies for Metropolitan Chicago*, Chicago: Center for Urban Research and Policy Studies, Northwestern University

Journal of the American Planning Association (1992) Issue largely devoted to growth management, 58, 4 (autumn)

Judd, D. R. (1988) *The Politics of American Cities: Private Power and Public Policy* (third edition), Glenview, IL: Scott, Foresman

Judd, D. R. and Swanstrom, T. (1994) *City Politics: Private Power and Public Policy*, New York: HarperCollins

Kain, J. (1994) 'Impacts of congestion pricing on transit and carpool demand and supply', in National Research Council, Transportation Research Board, *Moving Urban America*, Washington, DC: National Academy Press

Kaplan, H., Gans, S. P., and Kahn, H. M. (1970) *The Model Cities Program: The Planning Process in Atlanta, Seattle, and Dayton*, New York: Praeger

Kayden, J. S. (1990) 'Zoning for dollars: new rules for an old game? Comments on the *Municipal Art Society* and *Nollan* cases', in Lassar (1990)

Keating, W. D. (1994) *The Suburban Racial Dilemma: Housing and Neighborhoods*, Philadelphia: Temple University Press

Keith, N. S. (1973) *Politics and the Housing Crisis since 1930*, New York: Universe Books

Kelly, B. M. (1993) *Expanding the American Dream: Building and Rebuilding Levittown*, Albany, NY: State University of New York Press

Kelman, S. (1981) *What Price Incentives? Economics and the Environment* Boston, MA: Auburn House

Kent, T. J. (1964) *The Urban General Plan*, Chicago: American Planning Association reprint (1990)

Kettl, D. F. (1981) 'Regulating the cities', *Publius* 11: 111–25

Kingdon, J. W. (1984) *Agendas, Alternatives, and Public Policies*, New York: HarperCollins

Knaap, G. and Nelson, A.C. (1992) *The Regulated Landscape: Lessons on State Land Use Planning from Oregon*, Cambridge, MA: Lincoln Institute of Land Policy

Knack, R. E. (1995) 'BART's village vision', *Planning* 61 (1): 18–21

Kneese, A. and Schultze, C.L. (1975) *Pollution, Prices, and Public Policy*, Washington, DC: Brookings Institution

Koenig, J. (1990) 'Down to the wire in Florida: concurrency is the byword in the nation's most elaborate statewide growth management scheme', *Planning* 56 (10): 4–11

Kolis, A. B. (1979) 'Architectural expression: police power and the first amendment', *Urban Law Annual* 16: 272–304

Krumholz, N. and Clavel, P. (1994) *Reinventing Cities: Equity Planners Tell Their Stories*, Philadelphia: Temple University Press

Krumholz, N. and Forester, J. (1990) *Making Equity Planning Work: Leadership in the Public Sector*, Philadelphia: Temple University Press (There is an extract from this in Stein 1995)

Lacey, M. J. (ed.) (1990) *Government and Environmental Politics: Essays in Historical Development since World War II*, Lanham, MD: University Press of America

Landis, J. D. (1992) 'Do growth controls work? A new assessment, *Journal of the American Planning Association* 58: 489–508

Landy, M. K., Roberts, M. J., and Thomas, S. R. (1994) *The Environmental Protection Agency: Asking the Wrong Questions – From Nixon to Clinton* (expanded edition), New York: Oxford University Press

Lapping, M. B. and Leutwiler, N. R. (1987) 'Agriculture in conflict: right-to-farm laws and the peri-urban milieu for farming', in Lockeretz (1987)

Lash, J. (1994) 'Integrating science, values, and democracy through comparative risk assessment', in Finkel and Golding (1994)

Lassar, T. J. (1989) *Carrots and Sticks: New Zoning Downtown*, Washington, DC: Urban Land Institute

Lassar, T. J. (1990) *City Deal Making*, Washington, DC: Urban Land Institute

Lavelle, M. and Loyle, M. (1992) 'Unequal protection – the racial divide in environmental law', *National Law Journal*, September 21

Lenssen, N. (1991) *Nuclear Waste: The Problem that Won't Go Away*, Washington, DC: Worldwatch Foundation

Leopold, A. (1949) *Sand County Almanac*, New York: Oxford University Press

Levine, A. (1982) *Love Canal: Science, Politics and People*, Lexington, MA: Lexington Books

Levy, F., Meltsner, A., and Wildavsky, A. (1973) *Urban Outcomes*, Berkeley, CA: University of California Press

Lindblom, C.E. (1959) 'The science of "muddling through"', *Public Administration Review* (Spring); reprinted in Faludi (1973) and Campbell and Fainstein (1996)

Lipset, S. and Schneider, W. (1983) *The Confidence Gap: Business, Labor, and Government in the Public Mind*, New York: Free Press

Lockeretz, W. (ed.) (1987) *Sustaining Agriculture Near Cities*, Ankenny, IA: Soil and Water Conservation Society

Lowry, I. S. and Ferguson, B. W. (1992) *Development Regulation and Housing Affordability*, Washington, DC: Urban Land Institute

Lubove, R. (1962) *The Progressives and the Slums: Tenement House Reform in New York City 1890–1917*, Pittsburgh, PA: University of Pittsburgh Press

MacKenzie, J. J., Dower, R. C., and Chen, D. T. (1992) *The Going Rate: What It Really Costs to Drive*, Washington, DC: World Resources Institute

Mallach, A. (1984) *Inclusionary Housing Programs: Policies and Practices*, New Brunswick, NJ: Center for Urban Policy Research, Rutgers University Press

Mandelbaum, S. J., Mazza, L., and Burchell, R. W. (1996) *Explorations in Planning Theory*, New Brunswick, NJ: Center for Urban Policy Research, Rutgers University Press

Mandelker, D. R. (1962) *Green Belts and Urban Growth*, Madison, WI: University of Wisconsin Press

Mandelker, D. R. (1993) *Land Use Law* (third edition), Charlottesville, VA: Michie

Mandelker, D. R. and Cunningham, R. A. (1990) *Planning and Control of Land Development: Cases and Materials* (third edition), Charlottesville, VA: Michie

Markusen, A. R., Hall, P., Campbell, S., and Deitrick, S. (1991) *The Rise of the Gunbelt: The Military Remapping of Industrial America*, New York: Oxford University Press

Marsh, B. C. (1909) *An Introduction to City Planning: Democracy's Challenge to the American City*, New York: Marsh

Mazmanian, D. and Morell, D. (1992) *Beyond Superfailure: America's Toxics Policy for the 1990s*, Boulder, CO: Westview

McCormack, A. (1946) 'A law clerk's recollections' [on the *Euclid* case], *Columbia Law Review* 46: 710–18

Mckenzie, E. (1994) *Privatopia: Homeowner Associations and the Rise of Residential Private Government*, New Haven, CT: Yale University Press

Merriam, D., Brower, D. J., and Tegeler, P. D. (1985) *Inclusionary Zoning Moves Downtown*, Chicago: Planners Press

Merrill, T. W. (1986) 'The economics of public use', *Cornell Law Review* 72: 61–116.

Meshenberg, M. J. (1976) *The Language of Zoning: A Glossary of Words and Phrases*, Planning Advisory Service Report 322, Chicago: American Planning Association

Meyerson, M. (1956) 'Building the middle-range bridge for comprehensive planning', *Journal of the American Institute of Planners* 22: 58–64

Meyerson, M. and Banfield, E. C. (1955) *Politics, Planning and the Public Interest: The Case of Public Housing in Chicago*, New York: Free Press

Miles, M. E., Haney, R. L., and Berens, G. (1996) *Real Estate Development: Principles and Process* (second edition), Washington, DC: Urban Land Institute

Miller, Z. L. and Melvin, P. M. (1987) *The Urbanization of Modern America* (second edition), New York: Harcourt Brace Jovanovich

Mills, E. S. and Hamilton, B. W. (1994) *Urban Economics* (fifth edition), New York: HarperCollins

Mitchell, W. J. (1995) *City of Bits: Space, Place, and the Infobahn*, Cambridge, MA: Massachusetts Institute of Technology

Mogulof, M. B. (1971) 'Regional planning, clearance and evaluation: a look at the A-95 process', *Journal of the American Planning Association* 37: 418–22

Mollenkopf, J. H. (1983) *The Contested City*, Princeton, NJ: Princeton University Press

Moore, C. G. and Siskin, C. (1985) *PUDs in Practice*, Washington, DC: Urban Land Institute

Moore, T. and Thorsnes, P. (1994) *The Transportation/Land Use Connection*, Planning Advisory Service Report 448/449, Chicago: American Planning Association

Morgan, M. G. (1993) 'Risk analysis and management', *Scientific American* 169: 2–41

Mumford, L. (1961) *The City in History: Its Origins, Its Transformations, and Its Prospects*, New York: Harcourt Brace (Penguin edition 1991)

Mumford, L. (1962) *Technics and Civilization*, New York: Harcourt, Brace & World

Muse, D. E., Driscoll, R. L., and Green, R. L. (1995) 'A municipal landfill Superfund response cost and contribution action – the city's response', *Urban Lawyer* 27: 129–61

Nash, R. F. (1990) *American Environmentalism: Readings in Conservation History* (third edition), New York: McGraw-Hill

National Commission on the Environment (1993) *Choosing a Sustainable Future: The Report of the National Commission on the Environment*, Washington, DC: Island Press

National Congress for Community Economic Development (1995) *Tying It All Together: The Comprehensive Achievements of Community-based Development Organizations*, Washington, DC: NCCED

National Research Council, Transportation Research Board (1993) *Moving Urban America*, Washington, DC: National Academy Press

National Research Council, Transportation Research Board (1994) *Curbing Gridlock: Peak-period Fees to Relieve Traffic Congestion*, 2 vols, Washington, DC: National Academy Press

Nelson, A. C. (1988) *Development Impact Fees: Policy Rationale, Practice, Theory and Issues*, Chicago: Planners Press

Nelson, A. C. (1992) 'Preserving prime farmland in the face of urbanization: lessons from Oregon', *Journal of the American Planning Association* 58: 467–88

Nelson, A. C. (1994) 'Development impact fees: the next generation', *Urban Lawyer* 26: 541–62

Nelson, R. R. (1977) *The Moon and the Ghetto*, New York: W. W. Norton

Nicholas, H. G. (1986) *The Nature of American Politics*, Oxford: Oxford University Press

Nicholas, J. C., Nelson, A. C., and Juergensmeyer, J. C. (1991) *A Practitioner's Guide to Development Impact Fees*, Chicago: Planners Press

Oliver, G. (1992) '1000 Friends are watching: checking out the record of Oregon's pace-setting public interest group', *Planning* 58 (11): 9–13

O'Riordan, T. (ed.) (1995) *Environmental Science for Environmental Management*, Harlow, Essex: Longman

Packard, V. (1960) *The Waste Makers*, New York: David McKay

Palen, J. J. (1995) *The Suburbs*, New York: McGraw-Hill

Paul, E. F. (1987) *Property Rights and Eminent Domain*, New Brunswick, NJ: Transaction Books

Peiser, R. (1990) 'Who plans America? Planners or developers?' *Journal of the American Planning Association* 56: 496–503

Philbrick, F. S. (1938) 'Changing concepts of property in law', *University of Pennsylvania Law Review* 86: 691–732

Pickering, K. T. and Owen, L. A. (1994) *An Introduction to Global Environmental Issues*, New York: Routledge

Piel, G. and Segerberg, O. (eds) (1990) *The World of René Dubos*, New York: Holt

Pinchot, G. (1910) *The Fight for Conservation*, New York: Doubleday

Poole, S. E. (1987) 'Architectural appearance review regulations and the first amendment: the good, the bad, and the consensus ugly', *Urban Lawyer* 19: 287–344

Porter, D. R. (1986) *Growth Management: Keeping on Target?* Washington, DC: Urban Land Institute

Porter, D. R. (ed.) (1992) *State and Regional Initiatives for Managing Development: Policy Issues and Practical Concerns*, Washington, DC: Urban Land Institute

Porter, D. R. (ed.) (1995) *Housing for Seniors: Developing Successful Projects*, Washington, DC: Urban Land Institute

Porter, M. (1995) 'The competitive advantage of the inner city', *Harvard Business Review*, May–June: 55–71

Portney, P. R. (ed.) (1990) *Public Policies for Environmental Protection*, Washington, DC: Resources for the Future

President's Commission on a National Agenda for the Eighties (1980a) *A National Agenda for the Eighties*, Washington, DC: US Government Printing Office

President's Commission on a National Agenda for the Eighties (1980b) *Urban America*, Washington, DC: US Government Printing Office

Pressman, J. L. and Wildavsky, A.B. (1984) *Implementation: How Great Expectations in Washington Are Dashed in Oakland* (third edition), Berkeley, CA: University of California Press

Rabe, B. G. (1994) *Beyond NIMBY: Hazardous Waste Siting in Canada and the United States*, Washington, DC: Brookings Institution

Rapp, D. (1988) *How the US Got into Agriculture: and Why It Can't Get Out*, Washington, DC: Congressional Quarterly

Remy, M. H., Thomas, T. A., and Moose, J. G. (1994) *Guide to the California Environmental Quality Act*, Point Arena, CA: Solano Press

Reps, J. W. (1965) *The Making of Urban America: A History of City Planning in the United States*, Princeton, NJ: Princeton University Press

Rittel, H. W. J. and Weber, M. M. (1973) 'Dilemmas in a general theory of planning', *Policy Sciences* 4: 155–69

Robertson, D. B. and Judd, D. R. (1989) *The Development of American Public Policy: The Structure of Policy Restraint*, Glenview IL: Scott Foresman

Robinson, C. M. (1903) *Modern Civic Art, or the City Made Beautiful*, New York: George Putnam's Sons

Robinson, C. M. (1907) *The Improvement of Towns and Cities*, New York: George Putnam's Sons

Rohse, M. (1987) *Land Use Planning in Oregon: A No-nonsense Handbook in Plain English*, Corvallis, OR: Oregon State University Press

Rosenbaum, W. A. (1995) *Environmental Politics and Policy* (third edition), Washington, DC: Congressional Quarterly

Rosenberg, G. N. (1991) *The Hollow Hope: Can Courts Bring About Social Change?*, Chicago: University of Chicago Press

Ross, B. H., Levine, M. A., and Stedman, M. S. (1991) *Urban Politics: Power in Metropolitan America* (fourth edition), Itasca, IL: Peacock

Rousakis, J. and Weintraub, B. A. (1994) 'Packaging, environmentally protective municipal solid waste management, and limits to the economic premise', *Ecology Law Quarterly* 21: 947–1005

Rusk, D. (1993) *Cities Without Suburbs*, Washington, DC: Woodrow Wilson Center

Saleem, O. (1994) 'Overcoming environmental discrimination: the need for a disparate impact test and improved notice requirements on facility siting decisions', *Columbia Journal of Environmental Law* 19: 211–47

Savas, E. S. (1983) 'A positive urban policy for the future', *Urban Affairs Quarterly* 18: 447–53

Scheer, B. C. and Preiser, W. (1994) *Design Review: Challenging Urban Aesthetic Control*, New York: Chapman & Hall

Schiffman, I. (1990) *Alternative Techniques for Managing Growth*, Berkeley, CA: Institute of Governmental Studies

Schnidman, F., Smiley, M., and Woodbury, E. G. (1990) *Retention of Land for Agriculture: Policy, Practice and Potential in New England*, Cambridge, MA: Lincoln Institute of Land Policy

Schon, D. (1971) *Beyond the Stable State*, New York: W. W. Norton

Schwab, J. (1994) 'Environmental LULUs: is there an equitable solution?', in American Planning Association (1994)

Scott, M. (1969) *American City Planning since 1890*, Berkeley, CA: University of California Press

Shabecoff, P. (1993) *A Fierce Green Fire: The American Environmental Movement*, New York: Hill & Wang

Smith, Z. A. (1995) *The Environmental Policy Paradox* (second edition), Englewood Cliffs, NJ: Prentice-Hall

Snow, D. (ed.) (1992) *Inside the Environmental Movement*, Washington, DC: Island Press

Snyder, T. and Stegman, M. A. (1986) *Paying for Growth: Using Development Fees to Finance Infrastructure*, Washington, DC: Urban Land Institute

So, F. S. and Getzels, J. (eds) (1988) *The Practice of Local Government Planning* (second edition), Washington, DC: International City Management Association

Squires, G. D. (ed.) (1989) *Unequal Partnerships: The Political Economy of Urban Redevelopment in Postwar America*, New Brunswick, NJ: Rutgers University Press

Squires, J. F. (1992) 'Growth management redux: Vermont's Act 250 and Act 200', in Porter (1992)

Stanfield, R. L. (1991) 'Strains in the family', *National Journal* 23, 39: 2316–33

Stanilov, K., Pailthorp, M., Carlson, D., and Pivo, G. (1993) *A Literature Review of Community Impacts and Costs of Urban Sprawl*, Washington, DC: National Trust for Historic Preservation

Stegman, M. A. and Holden, J. D. (1987) *Nonfederal Housing Programs: How States and Localities are Responding to Federal Cutbacks in Low-income Housing*, Washington, DC: Urban Land Institute

Stein, J. M. (ed.) (1993) *Growth Management: The Planning Challenge of the 1990s*, Newbury Park, CA: Sage

Stein, J. M. (ed.) (1995) *Classic Readings in Urban Planning*, New York: McGraw-Hill

Stein, J. M. (ed.) (1996) *Classical Readings in Real Estate and Development*, Washington, DC: Urban Land Institute

Steinman, L. D. (1988) 'The impact of zoning on group homes for the mentally disabled: a national survey', in Gordon, N. J. *1988 Zoning and Planning Law Handbook*, New York: Clark Boardman

Stipe, R. (1987) *The American Mosaic: Preserving a National Heritage*, Washington, DC: US Committee, International Council on Monuments and Sites

Strong, A. L., Mandelker, D. R., and Kelly, E. K. (1996) 'Property rights and takings', *Journal of the American Planning Association* 62: 5–16

Suchman, D. R. (1995) *Manufactured Housing: An Affordable Alternative*, Washington, DC: Urban Land Institute

Suchman, D. R., Scott, D., and Giles, S. L. (1990) *Public–Private Housing Partnerships*, Washington, DC: Urban Land Institute

Teaford, J. C. (1990) *The Rough Road to Renaissance: Urban Revitalization in America*, Baltimore, MD: Johns Hopkins University Press

Teaford, J. C. (1993) *The Twentieth-century American City* (second edition), Baltimore, MD: Johns Hopkins University Press

Tibbetts, R. (1995) 'Everybody's taking the Fifth', *Planning* 61 (1): 4–9

Tocqueville, A. de (1848) *Democracy in America*; ed. J. P. Mayer, London: Fontana Press, 1969

Toll, S. I. (1969) *Zoned American*, New York: Grossman

Tomioka, S. and Tomioka, E. M. (1984) *Planned Unit Developments: Design and Regional Impact*, New York: John Wiley

Torma, C. (1987) 'Assessing the work to date', in *Proceedings of the Workshop on Historic Mining Resources*, Pierre, SD: South Dakota State Historical Preservation Society

Toronto City (1988) *Section 36 Guidelines: Further Report on Guidelines for Bonusing Pursuant to Section 36 of the Planning Act*, Toronto: City of Toronto Planning and Development Department

Tribe, L. H. (1985) *God Save this Honorable Court: How the Choice of Supreme Court Justices Shapes Our History*, New York: New American Library

Tschinkel, V. J. (1989) 'The rise and fall of environmental expertise', in Ausubel and Sladovich (1989)

United Church of Christ, Commission on Racial Justice (1987) *Toxic Wastes and Race in the United States*, New York: United Church of Christ

US Bureau of Census (1994) *Statistical Abstract of the United States*, Washington, DC: US Government Printing Office

US Conference of Mayors (1966) *With Heritage So Rich: A Report of a Special Committee on Historic Preservation*, New York: Random House

US Congressional Budget Office (1994) *Cleaning up the Department of Energy's Nuclear Weapons Complex*, Washington, DC: CBO

US Department of Agriculture (1981) *National Agricultural Lands Study*, Washington, DC: US Government Printing Office

US Department of Housing and Urban Development (1994) *Impact Fees and the Role of the State: Guidance for Drafting Legislation*, Washington, DC: HUD

US Department of Housing and Urban Development (1995) *Empowerment: A New Covenant with America's Communities. President Clinton's National Urban Policy Report*, Washington, DC: HUD

US Department of Transportation (1984) *Report on Highway Beautification Program*, Washington, DC: DOT

US Department of Transportation (1993) *Transport Implications of Telecommuting*, Washington, DC: Department of DOT

US Environmental Protection Agency (1987) *Unfinished Business*, Washington, DC: EPA

US Environmental Protection Agency (1992) *Environmental Equity – Reducing Risk for All Communities*, Washington, DC: EPA

US Environmental Protection Agency (1993) *Technical and Economic Capacity of States and Public Water Systems to Implement Drinking Water Regulations: Report to Congress*, Washington, DC: EPA

US Environmental Protection Agency (1994a) *Quality of Our Nation's Rivers 1992*, Washington, DC: EPA

US Environmental Protection Agency (1994b) *President Clinton's Clean Water Initiative*, Washington, DC: EPA

US Environmental Protection Agency (1994c) *Clean Water: A Memorial Day Perspective*, Washington, DC: EPA

US Environmental Protection Agency (1994d) *Safe Drinking Water Act Reauthorization Review*, Washington, DC: EPA

US General Accounting Office (1983) *Siting of Hazardous Waste Landfills and their Correlation with Racial and Economic Status of Surrounding Communities*, Washington, DC: GAO

US General Accounting Office (1986) *The Nation's Water: Key Unanswered Questions about the Quality of Rivers and Streams*, Washington, DC: GAO

US General Accounting Office (1993a) *Department of Energy: Cleaning-up Inactive Facilities Will Be Difficult*, Washington, DC: GAO

US General Accounting Office (1993b) *Much Work Remains to Accelerate Facility Cleanups*, Washington, DC: GAO

US Office of Technology Assessment (1989) *Facing America's Trash: What Next for Municipal Solid Waste?*, Washington, DC: OTA

Unman, M. F. (ed.) (1993) *Keeping Pace with Science and Engineering: Case Studies in Environmental Regulation* (National Academy of Engineering), Washington, DC: National Academy Press

Upton, D. and Vlach, J. M. (eds) (1986) *Common Places: Readings in American Vernacular Architecture*, Athens, GA: University of Georgia Press

Urban Institute (1995) *Federal Funds, Local Choices: An Evaluation of the Community Development Block Grant Program*, Washington, DC: US Department of Housing and Urban Development

Urban Land Institute (1991) *The Case for Multifamily Housing*, Washington, DC: Urban Land Institute

Urban Mobility Corporation (1995) *Newletter*, Washington, DC: UMC

Varady, D. and Raffel, J. (1995) *Selling Cities: Attracting Homebuyers through Schools and Housing Programs*, Albany, NY: State University of New York Press

Vig, N. J. and Kraft, M. E. (1990) *Environmental Policy in the 1990s*, Washington, DC: Congressional Quarterly Press

Wachs, M. (ed.) (1985) *Ethics in Planning*, New Brunswick, NJ: Center for Urban Policy Research

Wachs, M. (1990) 'Regulating traffic by controlling land use: the South California experience', *Transportation* 16: 249–50

Wallis, A. D. (1991) *Wheel Estate: The Rise and Decline of Mobile Homes*, New York: Oxford University Press

Waltman, J. L. and Holland, K. M. (eds) (1988) *The Political Role of Law Courts in Modern Democracies*, New York: St Martin's Press

Warner, S. B. (1968) *The Private City: Philadelphia in Three Periods of Its Growth*, Philadelphia: University of Pennsylvania Press

Warner, S. B. (1972) *The Urban Wilderness: A History of the American City*, New York: Harper & Row (reprinted Berkeley, CA: University of California Press, 1995)

Warner, S. B. (1978) *Streetcar Suburbs: The Process of Growth in Boston 1870–1900* (second edition), Cambridge, MA: Harvard University Press

Weaver, C. L. and Babcock, R. F. (1979) *City Zoning: The Once and Future Frontier*, Chicago: Planners Press

Weiss, M. A. (1980) 'The origins and legacy of urban renewal', in Clavel *et al.* (1980)

Weiss, M. A. (1987) *The Rise of the Community Builders: The American Real Estate Industry and Urban Land Planning*, New York: Columbia University Press

Wells, C. (ed.) (1986) *Perspectives in Vernacular Architecture II*, Annapolis, MD: Vernacular Architecture Forum

White, M. S. (1992) *Affordable Housing: Proactive and Reactive Planning Strategies*, Planning Advisory Service Report 441, Chicago: American Planning Association

Whyte, W. H. (1988) *City: Discovering the Center*, New York: Doubleday

Wickersham, J. H. (1994) 'The Quiet Revolution continues: the emerging new model for state growth management statutes', *Harvard Environmental Law Review* 18: 489–548

Wildavsky, A. (1987) *Speaking Truth to Power: The Art and Craft of Policy Analysis* (second edition), New Brunswick, NJ: Transaction

Williams, N. (1990) *American Land Planning Law*, Deerfield, IL: Callaghan

Williams, S. F. (1977) 'Subjectivity, expression, and privacy: problems of aesthetic regulation', *Minnesota Law Review* 62: 1–58

Wolman, H. (1988) 'Local economic development policy: what explains the divergence between policy analysis and political behavior?', *Journal of Urban Affairs* 10: 19–28

Wright, R. R. and Gitelman, M. (1982) *Land Use: Cases and Materials* (third edition), St Paul, MN: West

Wright, R. R. and Wright, S. W. (1985) *Land Use in a Nutshell* (second edition), St Paul, MN: West

Wurtzebach, C. H. and Miles, M. E. (1991) *Modern Real Estate* (fourth edition), New York: John Wiley

Wylie, J. (1989) *Poletown: Community Betrayed*, Urbana, IL: University of Illinois Press

Yandle, B. (ed.) (1995) *The Property Rights Rebellion: Land Use Movements in the 1990s*, Lanham, MD: Rowman & Littlefield

Yinger, J. (1995) *Closed Doors, Opportunities Lost: The Continuing Costs of Housing Discrimination*, New York: Russell Sage Foundation

INDEX